Analytical Biochemistry

Third Edition

Analytical Biochemistry

Third Edition

David J. Holme and Hazel Peck

An imprint of **Pearson Education**

Harlow, England · London · New York · Reading, Massachusetts · San Francisco
Toronto · Don Mills, Ontario · Sydney · Tokyo · Singapore · Hong Kong · Seoul
Taipei · Cape Town · Madrid · Mexico City · Amsterdam · Munich · Paris · Milan

Pearson Education Limited
Edinburgh Gate
Harlow
Essex CM20 2JE
England

and Associated Companies throughout the world

Visit us on the World Wide Web at:
www.pearsoned.co.uk

© Addison Wesley Longman Limited 1998

First published   1983
Second edition   1993
This edition   1998

ISBN 978-0-582-29438-7

British Library Cataloguing-in-Publication Data

A catalogue record for this book is available from the British Library

Library of Congress Cataloging-in-Publication Data
A catalog record for this book is available from the Library of Congress

Typeset by 55 in 10/12 Times
Printed and bound by CPI Group (UK) Ltd, Croydon, CR0 4YY

Transferred to Digital Print on Demand 2011

Contents

Preface

TO THE THIRD EDITION

The technology associated with biochemical analysis is rapidly changing and new laboratory instruments are constantly being introduced. However, with a few exceptions, the innovations are not based on new principles of analysis, but offer analytical benefits often through a 'mix and match' approach. For example, modern HPLC instruments may use columns that combine various features to effect separation and offer a range of detector options; the boundaries between electrophoresis and chromatography become blurred in such techniques as capillary chromatography; and mass spectrometry, previously only associated with GLC, is now linked to a much wider variety of chromatographic techniques. These 'state of the art' instruments are normally microprocessor controlled, offer some degree of automation and are attractively designed for ease of use.

Against this background it is perhaps tempting for analysts to underestimate the importance of understanding the principles of the techniques they are using. Unless this is the case they will be unlikely to be able to select, optimize and develop new methods, troubleshoot existing ones and be confident in the quality of their results. With increasing importance being attributed to quality assurance and laboratory accreditation, in addition to the fact that employers require their staff to work efficiently, an appreciation of fundamental principles of analysis is vital. We therefore make no apologies for again concentrating on these in this third edition.

We have deleted some sections that contained detailed accounts of techniques that are rarely encountered in modern laboratories, while retaining reference to the important classical methods that do provide the basis of current methodology. New material has been added to bring some topics up to date and these include increased coverage of laboratory quality, safety and accreditation, use of kits, mass spectrometry, and capillary electrophoresis. Many other changes have been made, not least of which is a completely new layout of the typescript with boxed areas for emphasis. We hope this will aid understanding and make the book more 'user friendly'. Two types of self test question are also included, which are designed to be simple indicators of an understanding of the basic concepts of the section and not a comprehensive test of knowledge of the topics. We have decided not to include photographs of particular instruments, as they are often not particularly informative and the designs change so rapidly. We would like to thank Dr Susan Laird and Dr Robert Smith for revising their chapters on nucleic acids and immunological methods respectively. We are also indebted to the many colleagues who have shared their knowledge and expertise with us over the years and whose advice has been invaluable.

Analytical biochemistry is an extensive subject and both the actual content and the balance of coverage in such a book as this is open to debate. With this in mind, our aim in each edition has been to give a clear account of the principles of the subject that will aid the understanding of a wide range of scientists who are either studying for a qualification or who are working in a laboratory, or perhaps both. The reading lists at the end of each chapter suggest additional texts for readers who require more details of specific topics.

David J. Holme
Hazel Peck

April 1997

Preface

TO THE SECOND EDITION
Since the publication of the first edition in 1983, several specialist books which cover a range of specific techniques in detail have been published. However, the ability to select an appropriate technique for a particular analytical problem still remains fundamental and the first edition of this book evidently proved useful in this respect. Thus the principal objective for this second edition remains unchanged.

Much of the information has been updated for the second edition to reflect substantial changes in the subject. The edition of a chapter on nucleic acids was considered essential and complements the original chapters on the chemical nature and methods of analysis of other important biological molecules. We are indebted to Dr Susan Laird for compiling this chapter and also to Mr Robert Smith for the major update on immunoassays in the immunological methods chapter.

We have maintained the same balance of information in the new chapter and therefore details of specific applications of techniques are not discussed, for example, DNA fingerprinting. Where appropriate, we have included titles of books which have an emphasis on applications in the further reading list at the end of each chapter. These lists are not intended to be fully comprehensive, nor are the chapters referenced as we consider this to be inappropriate for the level of potential readership.

We have received many pleasing reports of the usefulness of the first edition in a range of analytical laboratories, in areas such as pharmaceuticals, biotechnology, agrochemicals, clinical biochemistry, molecular biology, etc. Our own experience and comments from colleagues in other universities have reinforced our initial purpose of writing a book for students on a range of courses that include the analytical aspects of biochemistry. We are therefore delighted that this softback edition is now available which will encourage wider access for student use.

Hazel Peck
David Holme
Sheffield Hallam University

July 1992

Preface

TO THE FIRST EDITION

The initial stimulus for writing this book arose out of difficulties experienced in recommending a single suitable textbook for students on courses in which the analytical aspects of biochemistry were a major component. Although there are many books on analytical chemistry in general and clinical chemistry in particular, many omit the biochemical aspects of analysis such as enzymology and immunology while others do not cover the basic science of the subject. The objective was to bring together in one book those topics which we consider to be essential to the subject of analytical biochemistry.

In the introductory section to each chapter, there is a brief explanation of the scientific basis of the topic and this is followed by a discussion of the analytical methods which are relevant. While it is not intended that it should be a book of 'recipes', technical details for many of the methods described are given. This will help those readers with no practical experience to appreciate the steps involved in the analysis while at the same time giving sufficient detail for the method to be developed in practice. It is intended that the book will provide enough information to enable a student to select a technique or series of techniques which would be appropriate for a particular analytical problem and to be able to develop a valid and reliable analytical method.

The topics covered in this book fall into three main groups. Analytical techniques such as spectroscopy, chromatography, etc. are particularly important in analytical biochemistry as well as in analytical chemistry generally. The principles of each technique are explained and the scope and applications are discussed. There are chapters on enzymes, antibodies and radioisotopes, substances which it may be necessary to detect and measure but which also can be very useful in a variety of analytical methods. Here again, the basic theory is explained before discussing their applications in analytical biochemistry. Finally, there are four chapters which explain the chemical nature and methods of analysis of the major groups of biologically important compounds, namely, carbohydrates, amino acids, proteins and lipids. While it is appreciated that the range of compounds in this final section could be considerably extended it has been deliberately restricted to those groups which we consider to be of particular biochemical importance.

At the end of each chapter, several books are listed for further reading on the subject but it is suggested that the following books would be suitable for further reading on the topic of biochemistry of amino acids, carbohydrates, proteins and lipids.

J.W. Suttie, *Introduction to biochemistry*. Holt, Rinehart and Winston, New York, USA.

H.R. Mahler and E.H. Cordes, *Biological chemistry*. Harper and Row, New York, USA.

A. White, P. Handler and E.L. Smith, *Principles of biochemistry*. McGraw-Hill Book Co., New York, USA.

We would like to thank Dr Rodney Pollitt for reading the draft text and for his invaluable comments. In addition, we would like to thank those colleagues who have helped in various ways and Mrs P. Holme for typing the manuscript.

David J. Holme
Hazel Peck
Sheffield City Polytechnic

February 1982

About the boxes and self test questions

The book contains margin notes and two types of boxes, which are designed to enable the reader to identify certain types of information easily.

The margin boxes highlight important points in the text with a short statement or definition to give some background to the topic or they refer to other sections in the book which give additional information on the topic.

The procedure boxes give technical details of some procedure either to illustrate a technique or to provide technical details for readers who wish to use it.

The self test questions are at the end of most sections in a box. The four questions are designed to test the reader's understanding of the basic principles of the topic without going into the details of the subject. There are two types of questions:

1. Multiple choice question – any number or none of the alternatives may be correct.
2. Relationship analysis – consists of two statements joined by the word BECAUSE. Each statement should be considered separately and identified as being either TRUE or FALSE. If both statements are true, then the whole sentence should be considered to decide whether, overall, it is correct (YES) or not (NO), i.e. whether the second statement provides a correct explanation for the first statement. It should be appreciated that one short statement will not provide a complete explanation but the overall sentence can still be true.

1 General principles of analytical biochemistry

<table>
<tr><td>Key topics</td><td>
• The selection of a valid method of analysis
• The quality of data
• The production of results
</td></tr>
</table>

Analytical biochemistry involves the use of laboratory methods to determine the composition of biological samples and it has applications in many widely differing areas of biological science. The information gained from an analysis is usually presented as a laboratory report, which may simply say what substances are present (a qualitative report) or may specify the precise amount of a substance in the sample (a quantitative report).

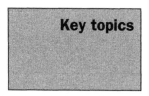

➤ A qualitative method enables only the presence of the substance to be detected.

A **qualitative report** will often indicate whether a particular substance or group of substances is present without commenting on the complete composition of the sample. In many cases the report will also specify the individual members of that group of substances. It might, for instance, name only the different carbohydrates present although the sample contained other substances, e.g. lipids and proteins. It is possible, when using some qualitative methods, to compare the amount of substance in the sample with the amount in a reference sample and to report the presence of either increased or decreased quantities. Such a report is said to be **semi-quantitative**. Chromatographic and electrophoretic methods often give results which can be interpreted in this way.

➤ A quantitative method enables the amount of substance present in a sample to be determined.

A **quantitative report** will state the amount of a particular substance present in the sample and it is important that the units of measurement are meaningful and appropriate in order to prevent subsequent misunderstandings. When reporting quantitative results it is desirable to indicate their reliability, a feature which can often be assessed statistically. In practice it may not be necessary to present this information with each report but it should be readily available for reference.

1.1   The selection of a valid method of analysis

In order to be able to choose a suitable analytical method it is essential to know something about the chemical and physical properties of the test substance (Table 1.1). Because the relationship between the property and the amount of substance is not always a simple one, some methods are only suitable for the detection of the substance (qualitative) while others may be quantitative. For any method it is important to appreciate the nature of the relationship between

Table 1.1   Physical basis of analytical methods

Physical properties that can be measured with some degree of precision	Examples of properties used in the quantitation of		
	Protein	Lead	Oxygen
Extensive			
Mass	+	+	
Volume			+
Mechanical			
Specific gravity	+		
Viscosity	+		
Surface tension	+		
Spectral			
Absorption	+	+	
Emission			
Fluorescence			
Turbidity	+		
Rotation			
Electrical			
Conductivity			
Current/voltage			+
Half-cell potential			+
Nuclear			
Radioactivity			

Proteins are the major components by bulk in many biological samples and hence the weighing of a dried sample should give an estimate of the amount of protein present. Similarly, solutions that contain protein show values for specific gravity and surface tension which are in some way related to protein content. Measurements of the turbidity resulting from the precipitation of protein and the absorption of radiation at specific wavelengths have all been used quantitatively.

The lead content of biological samples is usually very small, rendering gravimetric methods impracticable, and methods have often relied upon the formation of coloured complexes with a variety of dyes. More recently, the development of absorption spectroscopy using vaporized samples has provided a sensitive quantitative method. Oxygen measurements using specific electrodes offer a level of sensitivity which is unobtainable using volumetric gas analysis.

the measurement obtained and the amount of substance in the sample.

Most analytical methods involve several preparative steps before the final measurement can be made and it is possible to produce a flow diagram representing a generalized method of analysis (Table 1.2). Not all the steps may be necessary in any particular method and it may be possible to combine two or more by careful choice of instrumentation. It is important when selecting a particular method to consider not only its analytical validity but also the cost of the analysis in terms of the instrumentation and reagents required and the time taken.

Table 1.2 Generalized method of analysis

The major manipulative steps in a generalized method of analysis
Purification of the test substance
↓
Development of a physical characteristic by the formation of a derivative
↓
Detection of an inherent or induced physical characteristic
↓
Signal amplification
↓
Signal measurement
↓
Computation
↓
Presentation of result

1.1.1 Instrumental methods

The most convenient methods are those that permit simultaneous identification and quantitation of the test substance. Unfortunately these are relatively few in number but probably the best examples are in the area of atomic emission and absorption spectroscopy, where the wavelength of the radiation may be used to identify the element and the intensity of the radiation used for its quantitation.

➤ Atomic spectroscopy – see Sections 2.2 and 2.5.

If a compound does not show an easily detectable characteristic it may be possible to modify it chemically to produce a compound which can be measured more easily. In the early part of this century, this approach to analysis led to the development of many complex reagents designed to react specifically with particular test substances. Generally these reagents resulted in the formation of a colour which could be measured using visual comparators. Most of these reagents have been superseded by improved instrumental methods but some very reliable ones still remain in use. They were often named after the workers associated with their development, e.g. Folin and Ciocalteu's reagent, originally described in 1920 for the detection of phenolic compounds.

Interference occurs when other substances, as well as the test compound, are also detected, resulting in erroneously increased values. Occasionally interference effects can result in suppression of the test reaction. For any method it is important to be aware of substances that may cause interference and to know if any are likely to be present in the sample.

> Interference occurs when other substances as well as the test substance are detected by the method.

If interference is a major problem the sample must be partially purified before analysis. This breaks the analysis into preparatory and quantitative stages. In order to reduce the technical difficulties resulting from such two-stage methods much work has gone into the development of analytical techniques such as gas and liquid chromatography in which separation and quantitation are effected sequentially.

> Chromatography – see Section 3.2.

1.1.2 Physiological methods

While it may be possible to devise quantitative methods of analysis for many biochemical compounds, the only practical method of measurement for others is through their physiological effects. A **bioassay** involves the measurement of a response of an organism or a target organ to the test compound and may be conducted *in vivo* using live animals or *in vitro* using isolated organ or tissue preparations. Many bioassays are quantitative but those that give only a positive or negative result are said to be quantal in nature.

> Bioassays measure the response of living cells to external factors.

A satisfactory bioassay demands that the response of the animal to the substance can be measured in some fairly precise manner but it must be remembered that different animals respond in different ways to the same stimulus. Bioassays must therefore be designed to take account of such variations and replicate measurements using different animals must be made. In all assays it is important that the external factors that may influence the response are standardized as much as possible. The age and weight of an animal may affect its response as may also the environmental conditions, route of injection and many other factors.

In the absence of absolute chemical identification it is often necessary to establish that different samples contain the same physiologically active substance. This may be achieved by comparing the dose–response relationship for both samples. This involves measuring the response to varying amounts of each sample and demonstrating that the slope of the resulting relationship is the same in both cases. In such graphical or statistical methods it may be necessary to use the logarithm of the amount in order to produce a straight line rather than a curve. It is often necessary to use such a technique to confirm the validity of using synthetic or purified preparations as standards in quantitative assays.

► Tissue culture – see Section 8.4.1.

The use of cells from specific tissues grown in cultures, rather than freshly isolated, provides a technique of bioassay which reduces the need for the use of animals with all the implications of costly resources and ethical conflict. Alternatively there is a wide range of cell lines available of different tissue origin (Table 1.3).

Table 1.3 Bioassays using cell lines

Hormone	System used	Parameter measured
Prolactin	Rat lymphoma cells	Cell growth
Interleukin 1 (IL1)	Human myeloma cells	Cell growth
Transforming growth factor β (TGFβ)	Erythroleukaemic cell line	Inhibition of interleukin 5 (IL5) stimulated growth

The measurement of the catalytic activity of an enzyme is also a bioassay despite the fact that chemical methods may be used to measure the amount of substrate of product. Although the use of radioimmunoassays may enable the determination of the molar concentration of an enzyme, the problem of the relationship between molar concentration and physiological effects still remains.

1.1.3 Assay kits

In recent years many methods for a wide variety of analytes have been developed by reagent or instrument manufacturers and are marketed as 'kits'. These range from relatively simple colorimetric assays, which only require the addition of the chemical solutions provided to the test sample, to more sophisticated procedures involving complex reagents such as labelled antibodies or nucleic acids. The kits include all the necessary standards and assay components. They may be designed to be used in a manual procedure or, more commonly, on a particular automated instrument. Full assay protocols are given, together with details of the composition of all the reagents, any associated hazard data and specified storage conditions.

► Assay kits – see Section 7.4.

The increased availability of kits has greatly reduced the necessity for individual laboratories to develop their own methods. It is a requirement that all kits are validated before they can be sold and that details of the expected analytical performance are included with the product. Nevertheless, each laboratory is responsible for its own results and staff should ensure that the manufacturer's instructions are followed and that they are satisfied with all aspects of any kit that they use. This may necessitate checks on the quoted analytical performance being made.

Self test questions

Section 1.1
1. Which of the following measurements are said to be spectral?
    (a) Colour.
    (b) Voltage changes.
    (c) Elasticity.
    (d) Phosphorescence.

2. Which of the following could be said to be a quantitative result?
    (a) The test is positive.
    (b) The sample contains more than 5 g of glucose.
    (c) The sample contains 0.3 g of glucose.
    (d) The sample contains both glucose and lactose.
3. Weighing a sample is a qualitative method
    BECAUSE
    weighing a substance will not give the identity of the substance.
4. Many hormones lend themselves to bioassays
    BECAUSE
    bioassays involve measuring the effect of the test substance on living cells.

1.2   The quality of data

All data, particularly numerical, are subject to error for a variety of reasons but because decisions will be made on the basis of analytical data, it is important that this error is quantified in some way.

1.2.1   Variability in analytical data

The results of replicate analyses of the same sample will usually show some variation about a mean value and if only one measurement is made, it will be an approximation of the true value.

Random error
Variation between replicate measurements may be due to a variety of causes, the most predictable being random error which occurs as a cumulative result of a series of simple, indeterminate variations. These are often due to instrument design and use, e.g. the frictional effects on a balance, variable volumes delivered by auto-pipettes owing to wear, and operator decision when reading fluctuating signals. Such error gives rise to results which, unless their mean value approaches zero, will show a normal distribution about the mean. Although random error cannot be avoided, it can be reduced by careful technique and the use of good quality instruments.

> ➤ A set of replicate measurements is said to show a normal or Gaussian distribution if it shows a symmetrical distribution about the mean value.

The plotting of a histogram is a convenient way of representing the variation in such a set of replicate measurements. All the values obtained are initially divided into a convenient number of uniform groups, the range of each group being known as the **class interval**. In Figure 1.1 there are 11 such groups and the class interval is 1.0 unit. The number of measurements falling within a particular class interval is known as the **frequency** (f) and is plotted as a rectangle in which the base represents the particular class interval and the height represents the frequency of measurements falling within that interval. The class interval with the greatest frequency is known as the **modal class** and the measurement occurring with the greatest frequency is known as the **mode**. The average of all measurements is known as the **mean** and, in theory, to

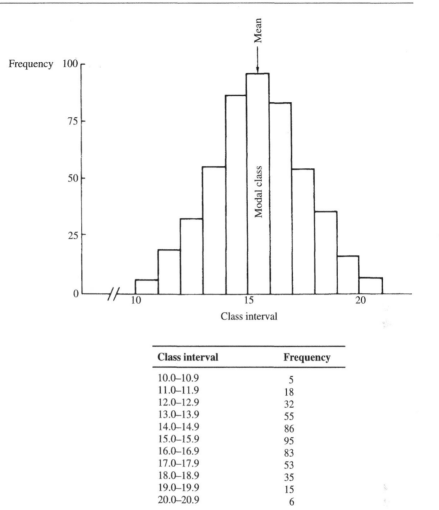

Figure 1.1 **Normal distribution of replicate measurements about a mean.**

Class interval	Frequency
10.0–10.9	5
11.0–11.9	18
12.0–12.9	32
13.0–13.9	55
14.0–14.9	86
15.0–15.9	95
16.0–16.9	83
17.0–17.9	53
18.0–18.9	35
19.0–19.9	15
20.0–20.9	6

determine this value (μ), many replicates are required. In practice, when the number of replicates is limited, the calculated mean ($\bar{x}$) is an acceptable approximation of the true value.

The most acceptable way of expressing the variation that occurs between replicate measurements is by calculating the **standard deviation** (s) of the data:

> ➤ Standard deviation is a statistical parameter expressing the scatter of replicate readings around their mean value.

$$s = \sqrt{\frac{\Sigma(\bar{x} - x)^2}{n}}$$

where x is an individual measurement and n is the number of individual measurements. An alternative formula which is more convenient for use with calculators is:

$$s = \sqrt{\frac{\Sigma x^2 - \frac{(\Sigma x)^2}{n}}{n - 1}}$$

The calculation of standard deviation requires a large number of replicates. For any number of replicates less than 30 the value for s is only an approximate value and the function $(n - 1)$ is used in the equation rather than (n). This function $(n - 1)$ is also known as the degrees of freedom (ϕ) associated with the mean and is important when tests of significance are used.

Knowledge of the standard deviation permits a precise statement to be made regarding the distribution of the replicate measurements about the mean value. Table 1.4 lists the relationship between a standard deviation and the pro-

Table 1.4   Normal distribution about a mean

Defined limits about the mean in terms of the standard deviation(s)	Percentage of total measurements lying within the defined limits
± 0.5	38.30
± 1.0	68.27
± 1.5	86.64
± 2.0	95.45
± 2.5	98.76
± 3.0	99.73
± 3.5	99.96
± 4.0	99.99

portion of measurements lying within defined ranges. Two convenient limits often used are $\pm 1s$ and $\pm 2s$ of the mean value. Out of 100 replicate measurements, for instance, approximately 95 will fall within the range of $\pm 2s$ of the mean value. This allows the probability of a single measurement lying within specified limits of the mean value to be predicted. For example, the probability of a single measurement lying within a range of approximately $\pm 2s$ of the mean value would be 0.95 (95%). If a limited number of replicates were done instead of a single measurement, a greater degree of confidence could be placed in the resulting mean value. This confidence can be expressed as the standard error of the mean (SEM), in which the standard deviation is reduced by a factor of the square root of the number of replicates taken:

➤ Standard error of the mean (SEM) is a statistical assessment of the increased confidence in a result when it is the mean of a limited number of replicates rather than a single measurement.

$$\mathrm{SEM} = \frac{s}{\sqrt{n}}$$

There is therefore a considerable advantage in making a limited number of replicate analyses rather than a single analysis but, in practice, it is necessary to balance the improved confidence that can be placed in the data against the increased time and effort involved.

Systematic error
Systematic errors are peculiar to each particular method or system. They are constant in character and although they can be controlled to some extent, they cannot be assessed statistically. A major effect of the introduction of systematic error into an analytical method may be to shift the position of the mean of a set of readings relative to the original mean. It may not obviously affect

> A method shows bias when it consistently gives results which are either higher (positive bias) or lower (negative bias) than the true value.

the distribution of readings about the new mean and so the data would show similar values for the standard deviation. Such a method is said to show **bias** towards either the positive (an increase in the mean) or the negative (a decrease in the mean) depending upon the direction of displacement.

Instrumental factors

Instability in instruments contributes to random error described earlier but sometimes features of their design or the fatigue or failure of components may result in readings being consistently lower or greater than they should be. This may sometimes be seen as a gradual drift over a period of time. Such variations are said to be systematic in origin and will result in biased results. It is essential in order to minimize the danger of such systematic error that great care is exercised in the choice and use of analytical instruments.

Errors of method

The chemical or biological basis of an analytical method may not permit a simple, direct relationship between the reading and the concentration of the analyte and failure to appreciate the limitations and constraints of a method can lead to significant systematic error. Carbohydrates may be quantified, for instance, using a method based on their reducing properties but results will tend to be higher than they should be if non-carbohydrate reducing substances are also present in the samples.

It is also possible for a perfectly valid analytical method to become less valid when used under different conditions. For example, the potentiometric measurement of pH is temperature-dependent and the use of reference and test solutions at different temperatures without any compensation will result in values being consistently higher or lower than they should be.

1.2.2 The assessment of analytical methods

Analytical methods should be precise, accurate, sensitive and specific but, because of the reasons outlined earlier, all methods fail to meet these criteria fully. It is important to assess every method for these qualities and there must be consistency in the definition and use of these words.

Precision

The precision, or reproducibility, of a method is the extent to which a number of replicate measurements of a sample agree with one another and is affected by the random error of the method. It is measured as imprecision, which is expressed numerically in terms of the standard deviation of a large number of replicate determinations (i.e. greater than 30), although for simplicity in the calculation shown in Procedure 1.1 only a limited number of replicates are used. The value quoted for s is a measure of the scatter of replicate measurements about their mean value and must always be quoted relative to that mean value.

The significance of any value quoted for standard deviation is not immediately apparent without reference to the mean value to which it relates.

> ➤ Coefficient of variation (V) expresses the standard deviation of a set of replicate readings as a percentage of their mean value.

The **coefficient of variation** or **relative standard deviation** expresses the standard deviation as a percentage of the mean value and provides a value which gives an easier appreciation of the precision:

$$\text{Coefficient of variation } (V) = \frac{s}{\text{mean}} \times 100\%$$

It is difficult to appreciate what the statement that a standard deviation of 1.43 gl^{-1} implies but quoting it as a coefficient of variation of 10.3% signifies that the majority of the replicate results (68%) are scattered within a range of $\pm 10.3\%$ of the mean value. Compared with quoting the values for

Procedure 1.1: Assessment of precision

Thirty replicate analyses of the protein content of a sample gave the following results. Calculate the standard deviation of the results.

Protein content in grams per litre

10.6	11.2	11.7	12.3	12.4	12.7
12.8	12.8	13.2	13.2	13.2	13.4
13.5	13.7	13.7	13.8	13.9	14.0
14.1	14.2	14.4	14.6	14.6	14.8
15.3	15.3	15.9	16.1	16.3	16.6

Calculation

$$s = \sqrt{\frac{\Sigma x^2 - \frac{(\Sigma x)^2}{n}}{n-1}}$$

Mean value ($\bar{x}$)	$= 13.81 \ gl^{-1}$
Sum of x (Σx)	$= 414.3$
Sum of x^2 (Σx^2)	$= 5782.89$
$\frac{(\Sigma x)^2}{n}$	$= 5721.483$
Standard deviation (s)	$= 1.43 \ gl^{-1}$
Coefficient of variation	$= \dfrac{1.43}{13.81} \times 100\% = 10.3\%$

Conclusion

The value calculated for the standard deviation indicates that 68% of the replicate readings lie within the range 12.38 gl^{-1} to 15.24 gl^{-1} (i.e. 13.81 $\pm$ 1.43 gl^{-1}) and 95% lie within the range 13.81 $\pm$ 2.86 gl^{-1}.

The coefficient of variation value expresses the same relationship as a percentage and indicates that 68% of the replicate readings lie within the range of the mean value $\pm 10.3\%$ of the mean value.

standard deviation, the implication of coefficient of variation values for two analytical methods, for instance, of 5% and 10% are immediately obvious.

While the value for coefficient of variation is a general statement about the imprecision of a method, only the value for standard deviation can be used in any statistical comparison of two methods. The use of coefficient of variation assumes a constant relationship between standard deviation and the mean value and this is not always true (Table 1.5).

Table 1.5 An example of standard deviation and coefficient of variation for different mean values

Mean ($\bar{x}$) (gl^{-1})	Experimental	
	s (gl^{-1})	V (%)
50	4.0	8
100	5.0	5
150	7.5	5
200	12.0	6

The use of coefficient of variation (V) reveals that maximum precision in the example given is achieved at the mid-range concentration values, a fact that would not be so obvious if only the standard deviation(s) were quoted.

It may be possible to demonstrate a high degree of precision in a set of replicate analyses done at the same time and in such a situation the **within batch** imprecision would be said to be good. However, comparison of replicate samples analysed on different days or in different batches may show greater variation and the **between batch** imprecision would be said to be poor. In practice this may more closely reflect the validity of the analytical data than would the within batch imprecision.

Procedure 1.2: Statistical comparison of the relative precision of two methods using the variance ratio or 'F' test

Method
A sample is analysed in replicate using the two methods of analysis and the variance of each set of data calculated and compared.

Theory
The basic assumption, or null hypothesis, made is that there is no significant difference between the variance (s^2) of the two sets of data and hence in the relative precision of the methods. Hence, if such a hypothesis is true the ratio of the two values for variance will be unity or almost unity.

$$F_{\phi}^{\phi} = \frac{s_1^2}{s_2^2} = \frac{\text{largest variance estimate}}{\text{smallest variance estimate}}$$

However, because the values for s_1 and s_2 are calculated from a limited number of replicates and as a result are only an approximate value, the value calculated for F will vary from unity even if the null hypothesis is true.

Critical values for F are available for different degrees of freedom and if the test value for F exceeds the critical value for F with the same degrees of freedom, then the null hypothesis can be rejected.

Calculation

Analytical method 1
10.0, 10.2, 10.4, 10.5, 10.6

$$\overline{x} = 10.34$$
$$\Sigma x = 51.7$$
$$\Sigma x^2 = 534.91$$
$$s^2 = 0.058$$
$$s = 0.24$$

degrees of freedom
$$(n - 1) = 4$$

Analytical method 2
9.2, 10.5, 10.8, 11.6, 12.1

$$\overline{x} = 10.84$$
$$\Sigma x = 54.2$$
$$\Sigma x^2 = 592.5$$
$$s^2 = 1.243$$
$$s = 1.11$$

$$(n - 1) = 4$$

$$F^{\phi}_{\phi} = \frac{1.243}{0.058} = 21.4$$

Conclusion

The critical value for F from the table below at $\phi_1 = 4$ and $\phi_2 = 4$ is 6.39 at 95% confidence level and because the calculated value for F (21.4) exceeds the critical value for F, it can be concluded with 95% confidence that a significant difference does exist between the precision of the two methods. This information would be very much more reliable if many more replicate values had been used.

Critical values of 'F' at the 95% confidence level (an abbreviated table)

Degrees of freedom of the denominator	Degrees of freedom of the numerator					
	1	2	3	4	5	10
1	161	200	216	225	230	242
2	18.5	19.0	19.2	19.2	19.3	19.4
3	10.1	9.55	9.28	9.21	9.01	8.79
4	7.71	6.94	6.59	6.39	6.26	5.96
5	6.6!	5.79	5.41	5.19	5.05	4.75
10	4.96	4.10	3.71	3.48	3.33	2.98

➤ A test of significance is designed to assess whether the difference between two results is significant or can be accounted for merely by random variations.

A comparison of the imprecision of two methods may assist in the choice of one for routine use. Statistical comparison of values for the standard deviation using the 'F' test (Procedure 1.2) may be used to compare not only different methods but also the results from different analysts or laboratories. Some caution has to be exercised in the interpretation of statistical data and particularly in such tests of significance. Although some statistical tests are outlined in this book, anyone intending to use them is strongly recommended to read an appropriate text on the subject.

From a knowledge of the imprecision of a method it is possible to assess the number of significant figures to quote in any numerical result. There is always the temptation to imply a high degree of precision by quoting numerical data to too many decimal places. On the other hand, the error due to 'rounding off' decimals must not be allowed to impair the precision which is inherent in the method. A zero at the end of a series of decimals is often omitted but it may be significant. It is a convenient rule of thumb not to report any data to significant figures less than a quarter of the standard deviation, e.g.

$$s \quad = \quad 1.43 \text{ gl}^{-1}$$
$$\frac{s}{4} \quad = \quad 0.35 \text{ gl}^{-1}$$

Report data to the nearest 0.5 gl^{-1}.

Accuracy

Accuracy is the closeness of the mean of a set of replicate analyses to the true value of the sample. In practice most methods fail to achieve complete accuracy and the inaccuracy of any method should be determined. It is often only possible to assess the accuracy of one method relative to another which, for one reason or another, is assumed to give a true mean value. This can be done by comparing the means of replicate analyses by the two methods using the 't' test. An example of such a comparison is given in Procedure 1.3 with the comment that only a limited number of replicates are used solely to simplify the calculation.

Some authors use the word 'trueness' instead of 'accuracy' to describe the closeness of the mean of many replicate analyses to the true value. This allows the word 'accuracy' to carry a more general meaning which relates to the accuracy or difference of a single result from the true value, as a conse-

Procedure 1.3: Statistical comparison of the relative accuracy of two methods using the 't' test

Method

A sample is analysed in replicate using the two methods of analysis and the mean value of each set of data calculated and compared.

Theory

The basic assumption, or null hypothesis, made is that there is no significant difference between the mean value of the two sets of data. This is assessed as the number of times the difference between the two means is greater than the standard error of the difference ('t' value). Critical values for 't' have been calculated for different degrees of freedom and if the test value for 't' exceeds the critical value with the same degrees of freedom, the null hypothesis can be rejected. When comparing the means of replicate determinations it is desirable that the number of replicates is the same in each case when the degrees of freedom for the test (ϕ) are $(n - 1) + (n - 1)$.

$$t_\phi = \frac{\bar{x}_1 - \bar{x}_2}{\dfrac{\sqrt{\Sigma(x_1 - \bar{x}_1)^2 + \Sigma(x_2 - \bar{x}_2)^2}}{n(n - 1)}}$$

Calculation
Using the data quoted in Procedure 1.2

$$t_8 = \frac{10.84 - 10.34}{\sqrt{\dfrac{0.232 + 4.972}{5 \times 4}}} = \frac{0.5}{\sqrt{0.2602}} = 0.98$$

Conclusion
The critical value for 't' from the table below with eight degrees of freedom and at the 95% confidence level is 2.31. Because the calculated value for 't' does not exceed the critical value for 't', it can be concluded with 95% confidence that no significant difference exists between the mean values for the two sets of data and the two methods have similar degrees of accuracy. As with Procedure 1.2, this information would be very much more reliable if more replicate values had been available.

Critical values of 't' at the 95% confidence level (an abbreviated table)

Degrees of freedom	Value for 't'	Degrees of freedom	Value for 't'
1	12.71	11	2.20
2	4.30	12	2.18
3	3.18	13	2.16
4	2.78	14	2.14
5	2.57	15	2.13
6	2.45	16	2.12
7	2.36	17	2.11
8	2.31	18	2.10
9	2.26	19	2.09
10	2.23	20	2.09

quence of both the imprecision (random error) and any bias (systematic error) of a method. Throughout this book the word is used as initially defined above.

The method under study is usually compared with an accepted reference method or, if one is not available, with a method which relies on an entirely different principle. In the latter case the two methods are unlikely to show exactly the same degree of bias and if they give very similar results it can be assumed that neither shows any significant bias.

Procedure 1.4: Statistical comparison of the relative accuracy of two methods using the paired 't' test

Method

A series of different samples are analysed using each of the two methods and the difference between each pair of results is calculated and compared.

Theory

The null hypothesis to be tested states that there is no significant difference between the pairs of results. Using a paired 't' test, the standard deviation for the difference (d) between the pairs (s_d) is calculated and the value for 't' derived from the formula:

$$t_\phi = \frac{\bar{d}}{\sqrt{\dfrac{s_d^{\,2}}{n}}}$$

Calculation

Sample	Sample		Difference (d)
	Method 1	Method 2	
1	5.4	6.2	+0.8
2	10.8	11.6	+0.8
3	25.4	23.8	−1.6
4	37.3	36.7	−0.6
5	46.2	48.2	+2.0
			$\Sigma d = 1.4$

$$\Sigma d^2 = 8.2$$
$$\bar{d} = 0.28$$
$$s_d^{\,2} = 1.952$$

$$t_4 = \frac{0.28}{\sqrt{\dfrac{1.952}{5}}} = 0.448$$

Conclusion

The critical value for 't' with four degrees of freedom is 2.78 and as the calculated value for 't' does not exceed this value it can be concluded with 95% confidence that there is no significant difference between the mean values for the two sets of data and the two methods show similar degrees of accuracy.

The formula for the 't' test described in Procedure 1.3 compares the mean of replicate analyses of only one sample but it may be preferable to compare the accuracy over the analytical range of the method. To do this a paired 't' test may be used in which samples with different concentrations are analysed using both methods and the difference between each pair of results is compared. A simplified example is given in Procedure 1.4.

One criticism of such a paired 't' test is that for a wide range of concentrations the difference between the pairs is accorded equal significance regardless of the size of the numerical value. The difference between values of 8 and 10, for instance, is more significant than, although numerically equal to, the difference between 98 and 100.

An additional approach to handling paired data is to assess the degree of correlation between the pairs. The data can be presented as a graph in which one axis is used for the results obtained by one method and the other axis for the results of the same samples obtained by the other method. If each sample analysed gave an identical result by both methods then a characteristic graph would result (Figure 1.2(a)). The closeness of the fit between all the points and

> ➤ Correlation coefficient is a measure of the closeness of fit between a series of points and a straight line through the points.

the straight line can be assessed by the correlation coefficient (r). Perfect correlation as illustrated in Figure 1.2(a) gives a correlation coefficient of 1. A coefficient of 0 indicates no correlation between the data; values between 0 and 1 indicate varying degrees of correlation. In general a correlation coefficient greater than 0.9 indicates fair to good correlation and together with an acceptable result for the paired 't' test would provide strong evidence for a common degree of accuracy between the two methods. The method of calculating the correlation coefficient is illustrated in Procedure 1.5.

Procedure 1.5: Comparison of the relative accuracy between two methods using the correlation coefficient

Method

A series of different samples, preferably with concentration values over the working range of the methods, is each analysed using both methods.

The correlation coefficient (r) is calculated using the following equation:

$$r = \frac{\Sigma xy - n\bar{x}\bar{y}}{\sqrt{(\Sigma x^2 - n\bar{x}^2)(\Sigma y^2 - n\bar{y}^2)}}$$

where n is the number of samples, x is the individual values by method 1, y is the individual values by method 2 and $\bar{x}$ and $\bar{y}$ are appropriate mean values.

Calculation

Using the data from Procedure 1.4

Σx	=	125.1	Σy =	126.5
$\bar{x}$	=	25.02	$\bar{y}$ =	25.30
Σx^2	=	4316.69	Σy^2 =	4409.57
Σxy	=	4359.03		

$$r = \frac{4359.03 - (5)(25.02)(25.3)}{\sqrt{[4316.69 - (5)(626.00)][4409.57 - (5)(640.09)]}}$$

$$= \frac{1194}{\sqrt{(1186.69)(1209.12)}}$$

$$= \frac{1194}{1197.85}$$

$$r = 0.997$$

Conclusion

A correlation coefficient of 0.997 indicates a good agreement between the data, and taken together with the positive result from the 't' test supports the view that the two methods show no significant difference in their accuracy.

It is possible that two methods that differ significantly in their accuracy may give a good correlation of data but would fail by the 't' test. Plotting the data as a regression plot would show a divergence from the pattern of Figure 1.2(a). Many complex variations might occur but two fairly simple ones would show characteristic features. If one method gave results that were consistently high or low by a fixed amount (owing to a lack of specificity for instance), a graph similar to Figure 1.2(b) would result, in which the intercept was not zero but corresponded to the fixed error involved. Similarly if one method showed

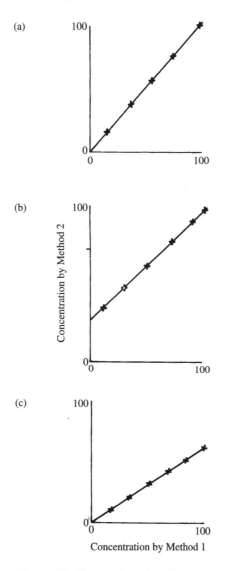

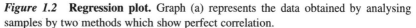
Concentration by Method 1

Figure 1.2 **Regression plot.** Graph (a) represents the data obtained by analysing samples by two methods which show perfect correlation.
Method 2 in graph (b) shows a constant positive bias compared with Method 1.
In graph (c), Method 2 shows a negative bias which is proportional to the concentration of the analyte.

a systematic error which was proportional to the sample concentration in some way, the slope would be different (Figure 1.2(c)). These values can be calculated using the method of linear regression analysis (Procedure 1.6). Using this statistical method, the equation for the straight line is determined and the val-

Procedure 1.6: Determination of the equation for a straight line using the method of linear regression analysis

Method

When used for comparing two analytical methods, a series of different samples are each analysed using both methods. The method can be used generally for any two sets of data which show a linear relationship to each other.

Theory

A straight line relationship can be described by the equation

$$y = ax + b$$

where x and y are the variables and a is the slope and b the intercept of the line. The best straight line through a series of points is that line for which the sum of the squares of the deviations of the points from that line are a minimum. Hence the name for the method of 'least squares'.

The values for a and b can be calculated using the following equations:

$$a = \frac{\Sigma xy - \dfrac{(\Sigma x\, \Sigma y)}{n}}{\Sigma x^2 - \dfrac{(\Sigma x)^2}{n}}$$

$$b = \bar{y} - a\bar{x}$$

Calculation

Using the data from Procedure 1.4

$$a = \frac{4359.03 - \dfrac{(125.1)(126.5)}{5}}{4316.69 - \dfrac{(15\,650.01)}{5}}$$

$$= \frac{4359.03 - 3165.03}{4316.69 - 3130.002}$$

$$= \frac{1194.00}{1186.688}$$

$$= 1.006$$

$$b = 25.3 - (1.006)(25.2)$$

$$= 0.13$$

Conclusion

The values for the slope and the intercept enable a line to be drawn on a graph. The arbitrary values for x can be selected and the corresponding values for y calculated from the equation

$$y = 1.006x + 0.13$$

and the values used to draw the line.

In comparing the two analytical methods, the additional evidence of a correlation coefficient of 0.997 and the paired 't' test result (Procedure 1.4) confirm the observation that a slope of 1.006 and an intercept of 0.13 do not significantly differ from the theoretical values of 1.0 and 0 respectively.

ues for the slope and intercept calculated. If these differ from 1.0 and zero respectively, the graph differs from the characteristic one of Figure 1.2(a) and the two methods differ in their accuracy.

Sensitivity

The sensitivity of a method is defined as its ability to detect small amounts of the test substance. Some confusion may arise from the ways in which sensitivity is measured. It can be assessed by quoting the smallest amount of substance that can be detected; for example, the smallest reading after zero that can consistently be detected and measured. The slope of the calibration graph is a conventional way of expressing sensitivity and is particularly useful when comparing two methods. It is essential that for such a comparison, the units of both axes are the same for each method. While there may be a significant difference between the mean values of replicate determinations of two samples with slightly different concentrations, there may not be a significant difference between single or even duplicate analyses of these two samples. In such cases the lack of precision is more significant than the sensitivity of the method, which cannot be better than the precision if only single or duplicate analyses are undertaken.

Specificity

Specificity is the ability to detect only the test substance. Lack of specificity will result in false positive results if the method is qualitative and positive bias in quantitative results. The nature of any interfering substance for particular methods will be discussed in the appropriate sections but it is important to appreciate that specificity is often linked to sensitivity. It is possible to reduce the sensitivity of a method with the result that interference effects become less significant and the method is specific although less sensitive to the test substance. In such a situation false positives (interfering substances) will not occur, but false negatives (undetected low concentrations) may and it is necessary to decide whether maximum sensitivity or specificity is required.

1.2.3 Quality assurance in analytical biochemistry

In order to produce reliable results, all analytical methods should be carefully designed and their precision and accuracy must be determined. The stability of samples should be investigated and their subsequent handling controlled in an appropriate manner. The attitude of the staff involved is of vital importance: they must be motivated to produce valid data and to take a pride in the quality of the final product. All of these factors together with the scheme for monitoring performance will be scrutinized if a laboratory seeks formal accreditation.

Quality control

Quality control refers to an internal scheme that will give a warning when unforeseen factors cause a reduction in the analytical performance of the method. This allows an immediate decision to be made on whether the test results are acceptable or must be rejected. This is usually done by the analysis of a control sample with each batch of tests.

Control samples

A control sample is a sample for which the concentrations of the test analyte is known and which is treated in an identical manner to the test samples. It should ideally be of a similar overall composition to the test samples in order to show similar physical and analytical features. For instance, if serum samples are being analysed for their glucose content, the control sample should also be serum with a known concentration of glucose. A control sample will be one of many aliquots of a larger sample, stored under suitable conditions and for which the between batch mean and standard deviation of many replicates have been determined. It may be prepared within the laboratory or purchased from an external supplier. Although values are often stated for commercially available control samples, it is essential that the mean and standard deviation are determined from replicate analyses within each particular laboratory.

Control samples should be analysed along with the test samples but the analyst should not know which are the control samples. Knowing the mean value for the control sample and the precision expected from the method, it is possible to forecast limits within which a single control result should normally fall. The basis of a quality control programme is the assumption that if a single control result falls within these defined limits, the method is under control and the test results produced at the same time are valid. If the control value falls outside the defined limits it is likely that the test results are in error and must be rejected.

Control charts

It is often helpful to record the results of control samples in a visible manner not only because of the greater impact of a visual display but also for the relative ease with which it is possible to forecast trends. A variety of styles of quality control charts have been suggested but the most commonly used are those known as Levey–Jennings or Shewart charts, which indicate the scatter of the individual control results about the designated mean value (Procedure 1.7).

Incorporated in the chart are control limits set at $\pm 2s$ and $\pm 3s$, which approximate to the 95% and 99% confidence limits respectively. If a control result falls outside the 95% limit there is only a maximum probability of 0.05 (5%) that the result lies in a normal distribution about the accepted mean. The

Procedure 1.7: Construction of a quality control chart (Levey–Jennings or Shewart)

Data

A control sample was analysed 50 times and gave a mean value of 98 mg l^{-1} with a standard deviation of 5 mg l^{-1}. The same control sample was analysed on twelve successive days in batches of test samples and gave the following results.

Day	Result (mg l^{-1})	Day	Result (mg l^{-1})
1	101	7	106
2	95	8	109 (repeat 97)
3	100	9	95
4	94	10	99
5	103	11	94
6	101	12	114

Method

A quality control chart was plotted on day 1 with a mean line set at the mean value (98 mg l^{-1}) and warning limits set at ± 2 SD (88 mg l^{-1} and 108 mg l^{-1}) and action limits set at ± 3 SD (83 mg l^{-1} and 113 mg l^{-1}).

Each day the control value was plotted on the chart.

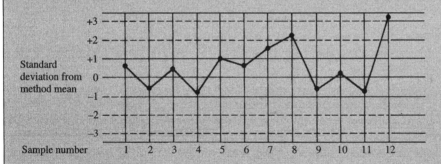

Conclusion

The control sample results up to and including day 7 all lie within the warning limits and are satisfactory and the test results can be reported. On day 8 the control result lies outside the warning limit. The test results are withheld and the control sample immediately retested. This gives a value well within the limits and the test results are reported. The control values fall within the acceptable range for the following three days. On day 12 the control value lies outside the action limits indicating that a gross error was likely. The test results are discarded, fresh standards are prepared and the control sample re-analysed. Assuming that this solves the problem, the test samples are also re-analysed and the new results reported.

most obvious implication is that the result could be wrong. Similar deductions, with an increased degree of confidence, can be made about the 99% control limits. The precise position of the control limits can be set according to the demands of the situation and in some instances it may be necessary to set more stringent control levels ± 1s and ± 2s.

> ➤ Warning limits are the maximum and minimum values within which a single control sample result is normally expected to lie.

> ➤ Action limits are specified maximum and minimum values outside which a single control sample result is extremely unlikely to lie without there being a serious error in the analysis.

The narrower limits are usually known as the **warning limits**. Failure to meet these limits implies that the method must be investigated and any known weakness, such as unstable reagents, temperature control, etc., should be rectified. However, results obtained at the same time as the control result can still be accepted. Probably the first step in a case like this is to repeat the control analysis. If the original result was a valid random point about the mean then the repeat result should be nearer to the mean value. If the repeat analysis shows no improvement or the original control result lay outside the wider control limits (known as **action limits**) then it must be assumed that all the results are wrong. The method must be investigated, the fault rectified and the analysis of samples and controls repeated.

The chart can give additional information about any change in the accuracy of the method. It would normally be expected that a series of control results would show scatter about the mean value. If the points showed a tendency to lie to one side of the mean but still within the accepted range, this would be an indication that the method was showing a bias in one particular direction.

An alternative visual method of presenting quality control results is known as the **Cusum plot** (cumulative sum). This is particularly useful when control samples are not available (possibly owing to their lack of stability) and also when large numbers of analyses which give a constant mean value are undertaken. The mean value for each large batch of analyses will vary in a normal distribution about the 'mean of means' calculated from a large number of batches over a period of time. The difference between the mean of a batch and the 'mean of means' should therefore be zero and the cumulative sum of these differences should also be zero. A graph is plotted of the cumulative sum of the differences against the date of the analysis or the batch number (Procedure 1.8). For quality control samples the cumulative sum of the individual result compared to the previously calculated mean value is plotted.

In many cases a Cusum plot will not show the expected horizontal line but rather a line with a small but constant slope owing to the value attributed to the 'mean of means' being incorrect. The plot is still acceptable and in such cases a change in the slope will indicate a change from the expected value and the possibility of error. Some of the difficulties with the Cusum plot is that variations are most obvious retrospectively, little information can be gained from a single point and errors are only apparent from several consecutive points. Thus it is debatable whether this type of plot can be classed as true quality control.

Quality assessment

As well as operating quality control programmes, laboratories undertaking a similar range of analyses can cooperate in a group programme. In such a group scheme, the same control samples are analysed by each laboratory and the results compared within the group. This form of assessment is usually a retrospective process enabling overall quality to be maintained or improved. Group schemes do not necessarily demand that the control samples are of known concentration because even using unknown samples, comparisons of single or replicate analyses and of mean values and standard deviation (precision) for the

Procedure 1.8: Construction of a Cusum quality control chart

Data

In monitoring the ascorbic acid content of sample of fruit juice approximately 50 samples are analysed in each batch. For each batch analysed, the mean value is calculated and subtracted from the accepted mean of 100 mg l^{-1}.

Day	Mean	Difference from accepted mean	Cumulative sum
1	104	+4	+ 4
2	99	−1	+ 3
3	104	+4	+ 7
4	100	0	+ 7
5	103	+3	+10
6	99	−1	+ 9
7	103	+3	+12
8	102	+2	+14
9	99	−1	+13
10	103	+3	+16
11	100	0	+16
12	96	−4	+12
13	97	−3	+ 9
14	99	−1	+ 8

Method

A chart is drawn on day 1 to plot the cumulative sum of the differences over the required sequence of days.

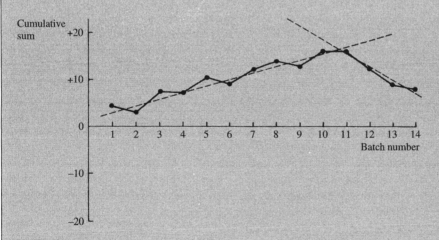

Conclusion

The gradually increasing plot up to day 11 does not indicate that the method is out of control but that the accepted mean of 100 mg l^{-1} is too low. However, the sudden change in direction on day 12 indicates a change in the batch mean, which is probably due to an error in the method. In other circumstances the change may not be so dramatic and not so obvious until several further points have been plotted.

group can be made. Results are frequently published in a coded manner so that each member of the group can identify their result and where they lie in the group profile but cannot identify results from other members in the group.

This type of external scheme is often administered by a commercial organization, which distributes the samples and the results to all participants.

1.2.4　Accreditation of laboratories

Accreditation formally recognizes that a laboratory is competent to carry out its analytical service and is increasingly becoming accepted as a necessary requirement. It is an overall assessment of the performance of a laboratory and covers the quality of management and associated organizational procedures together with the quality of testing. Accreditation is an extensive process requiring a quality audit and review associated with a series of visits from the external accrediting body, to whom a fee is payable.

Various bodies have been established such as NAMAS (National Measurement Accreditation Service) and CPA (Clinical Pathology Accreditation), which operate in the UK. Quality standards appropriate to a wide range of organizational activities, e.g. BS 5750, ISO 9000 and EN 45000 series in Europe, together with the internationally recognized GLP (good laboratory practice) may form part of the accreditation requirement. In the USA, CLIA'88 (Clinical Laboratory Improvement Amendments, 1988) together with the Federal Drug Administration and the US Environmental Protection Agency are involved in accreditation of laboratories.

Each accrediting body produces detailed documents that outline its overall requirements and assessment process. They could include the following aspects: general requirements, organization and management, quality systems, quality audit and review, staff, equipment, measurement traceability and calibration, methods and procedures for tests, laboratory accommodation and environment, handling of test items, records, test reports, handling of complaints and anomalies, sub-contracting of tests, outside support services and supplies.

Several specific aspects of laboratory management which are essential in the process of accreditation are discussed in the following sections. These topics are also very important for laboratories not seeking formal accreditation but concerned about the quality of their work, their credibility and the safety of their employees.

Health and safety

➤ A procedure hazard form should contain all the necessary information regarding potential hazards, mode of disposal and first aid for all the chemicals involved in a particular procedure.

Many of the chemicals and much of the equipment used in laboratories are potentially hazardous. It is essential that these hazards are clearly identified and appropriate working procedures defined together with adequate training of staff and readily available facilities to deal with the effects of any possible accident. These aspects of laboratory work are covered in the UK by COSHH (Control of Substances Hazardous to Health) Regulations and by OSHA (Occupational Safety and Health Administration) in the USA.

Every activity in the laboratory should be assessed for the potential hazards involved and all the relevant information should be presented in a standard format known variously as procedure hazard forms or material safety data

> ➤ A hazard data sheet contains information about a chemical, i.e. its physical properties, precautions for safe handling, mode of disposal, potential hazards and first aid procedures.

sheets (MSDS). There are several steps in producing such a hazard document (Procedure 1.9). The document should then be approved by the laboratory safety officer or equivalent, who should satisfy him- or herself that all the necessary equipment, resources and training are in place for the procedure to be undertaken.

Information on hazards is available from various sources. Chemical manufacturers produce hazard data sheets for their products and some of the major companies produce comprehensive databases. Each data sheet contains information on the physical description of the compound, stability, hazards, first aid measures, storage, transport and disposal requirements.

Procedure 1.9: Preparation of a procedure hazard form

1. List all the chemicals used in the procedure.
2. Search the source information for data sheets for each chemical.
3. From the information on the hazard data sheets, produce a procedure hazard form on which all the relevant information is recorded in an easily accessible format.
    Name and outline of the procedure
    Chemicals involved
    Nature of hazards (if any)
    Handling and disposal procedures
    Essential facilities required
    First aid procedures
4. The procedure hazard sheet, after approval, must be available at the work bench or site of the procedure for use by the analysts.

Chemical hazards are classified under five main headings.

Explosive
In general laboratories there are some compounds that are potentially explosive, e.g. picric acid. It is important that such substances are stored under suitable conditions, e.g. under water, and that they are regularly inspected in order to maintain these conditions. The use of such substances should be carefully controlled and only small amounts used.

Flammable
Flammable liquids and solids are subdivided according to their flash points and many oxidizing substances can cause fire when in contact with combustible materials. Storage, handling and disposal are obviously major features to be considered when using such substances.

Toxic
Substances are graded in toxicity and by the route, e.g. inhalation, swallowing or contact, the latter being particularly important in the design of the working environment. In some instances it is possible to specify exposure limits and in such cases monitoring of the environment may be essential.

Corrosive and irritant
The result of skin contact with corrosive substances is usually obvious but the

effects of irritants are less obvious and hence are potentially likely to be treated more casually. Protective clothing is essential when using toxic or irritant substances.

Radioactive

These substances are subject to very strict control and laboratories must be approved to handle the different categories of radioactivity.

Standard operating procedures (SOPs)

SOPs give written details of the protocol that must be followed for any particular procedure being undertaken. They are not restricted to laboratory operations but are applicable to a wide range of activities in an organization and as such are linked with the overall quality programme. Within a laboratory they are a convenient way of documenting, in a standard, unambiguous format, the information that staff require to be able to carry out their duties in a safe and appropriate manner. Evidence of up-to-date information of this nature is necessary for laboratories seeking accreditation.

SOPs include details of the procedures for collecting and handling the samples; performing the analysis; analysing, storing and retrieving data; and preparing reports. They contain instructions on the use, maintenance, servicing, calibration and repair of equipment and instruments, including computers. The hazards associated with the overall protocol are stated and safe working practices are specified. The quality assurance measures that must be complied with are included together with instructions on the steps to be taken if the method does not perform to specification. The SOP is given a serial number and dated before being signed and stored for future reference. A copy, in a suitable protective cover, must be available at the work bench and the SOP serial number should be quoted on all records relating to its use.

Much information on the mode of operation and verification of performance of laboratory instruments is often available from manufacturers or suppliers in a form that is suitable for incorporation into an SOP.

Computerization

Computers were first used in laboratories to calculate results and generate reports, often from an individual instrument. As automated analysers were developed, so the level of computerization increased and computers now play a major role in the modern laboratory. They are associated with both the analytical and organizational aspects and the term Laboratory Information Management System (LIMS) is often used to describe this overall function. Such systems are available that link the various operations associated with the production of a validated test result, from the receipt of the sample to the electronic transmission of the report to the initiator of the request, who may be at a site removed from the laboratory. Other uses include stock control, human resource management and budgets.

The successful introduction of a system that is appropriate for a particular laboratory is a lengthy process requiring widespread consultation. Current and future needs must be taken into account, as also must the requirements of any external body which may specify the extent of computerization required for accreditation. The introduction of any new computer facilities will necessitate the training of staff and their complete familiarization with the system. As with all other laboratory instruments, SOPs must be written that include hazard assessments and procedures to be followed in the event of systems failure. Regular checks on performance must be carried out and full servicing and maintenance records must be kept.

Good laboratory practice

Good laboratory practice (GLP) is a set of procedures within which the overall performance of a laboratory can be monitored. It is applicable to the organization and functioning of any laboratory but it is particularly relevant to the pharmaceutical industry. Compliance with GLP may be required for accreditation of a laboratory by an external regulating agency.

GLP involves all aspects of the organization which is involved in generating an analytical result, from senior management to the bench workers. The essential features of GLP can be summarized as follows.

Staff. All staff must be adequately trained with designated responsibilities and appropriate qualifications. Full details of all staff must be kept for ten years.

Equipment. This must be of an adequate standard and full records of all maintenance and faults must be kept for ten years.

Procedures. All methods and procedures must be in the form of an SOP.

Data. After the analysis has been completed, all the details of the method, equipment, SOP and the raw results must be stored for ten years.

1.2.5   Samples for analysis

The validity of a laboratory report is affected by additional factors as well as those related to the analytical method used. Of particular importance is the sample itself and the manner in which it is collected and stored prior to analysis.

The collection procedure should be designed to provide a representative sample of the system under investigation and not adversely affect the analytical process. The correct storage conditions for the sample are vital to preserve the integrity of the biological components (Table 1.6); in some situations ensuring that the cellular morphology is not significantly affected. The optimal storage conditions depend on the nature of both the sample and the analysis. The chemical and physical environment are both important and stability studies may indicate the need for the addition of chemicals such as anti-microbial agents, anticoagulants, etc. and storage at a specific temperature for a certain period of time. These conditions should be clearly stated in the laboratory protocol.

In addition to these considerations, the report that is issued is also affected by the procedures that follow the actual analysis (post-analytical) including calculations and transcriptions. The use of computers, particularly in association with bar-coded sample containers, has improved the procedures of ensuring correct sample identity, the production of work lists and the presentation of results. Instructions relating to all the non-analytical aspects should be specified in the SOP associated with the analytical method.

Table 1.6 Examples of storage conditions of biological samples

Possible change in the sample	Examples of methods of prevention
Microbial degradation	Addition of anti-microbial agent, e.g sodium azide
	Store at temperatures below $-20\,°C$
Denaturation of enzymes	Store in 50% glycerol at low temperatures
	Store in liquid nitrogen
Leakage of intracellular components	Separate cells immediately
	Store in isotonic medium
	Usually do not freeze
Oxidation	Add antioxidant, e.g. 2-mercaptoethanol, dithiothreitol
	Store in the dark
	Store under nitrogen or hydrogen
Enzymic conversion of analyte	Add enzyme inhibitor, e.g. fluoride
	Store at temperatures below $-20\,°C$
Coagulation	Add anticoagulant, e.g. heparin, ethylenediamine tetraacetic acid (EDTA)
Gaseous loss	Store under oil, e.g. liquid paraffin

Self test questions

Section 1.2

1. Replicate analyses of a sample gave the following information.
    Mean value 1.75 mg   Standard deviation 0.01 mg
    To which of the following features of analytical methods would this information be most relevant?
    (a) Accuracy.
    (b) Precision.
    (c) Sensitivity.
    (d) Specificity.
2. In a quality assurance programme, the control with a mean value of 10.5 mg and a standard deviation of 0.1 mg was analysed with a batch of test samples and gave a result of 10.0 mg. Which of the following actions should be taken?
    (a) Reject all the test results.
    (b) Accept all the test results.
    (c) Re-analyse the control.
    (d) Re-analyse the test samples.
3. The imprecision of a method can be assessed by determining the standard deviation of replicate analyses
    BECAUSE
    precision falls as random error increases.
4. If a control sample in a quality control programme gives a value that is greater than the mean value for all the control samples by more than 2 SD it suggests that errors have been introduced into the assay
    BECAUSE
    in a normal distribution of replicate results, no more than approximately 2.5% of the values should exceed the mean value by more than 2 SD.

1.3   The production of results

Every analysis is undertaken for a reason and the results of the analysis should be presented in an unambiguous manner to enable valid conclusions to be drawn. There are certain aspects of reporting analytical results which should be considered.

1.3.1   Choice of units

➤ SI units provide a system of universal units that consists primarily of seven base units from which all others may be derived.

Confused and inconsistent use of units of measurement often presents problems and in 1960 the Système International d'Unités (SI units) was introduced. This aimed to produce a universal system of units in which only one unit was used for any physical quantity. It also provides coherence between appropriate measurements and minimizes the number of multiples and sub-multiples in use.

Table 1.7   SI base units

Physical quantity	SI base unit	Symbol
Length	Metre	m
Mass	Kilogram	kg
Time	Second	s
Electric current	Ampere	A
Thermodynamic temperature	Kelvin	K
Luminous intensity	Candela	cd
Amount of substance	Mole	mol

The system defines seven base units (Table 1.7), which are independent of each other but which can be combined in various ways to provide a range of derived units (Table 1.8), each one capable of describing a physical quantity. Coherence is maintained in these derived units because no conversion factors are involved at this stage but in order to provide units of convenient size for different applications a series of standard prefixes may be used. These are multipliers used with coherent units to obtain units of alternative size but only one prefix should be used at a time (Table 1.9).

➤ A derived unit is one that is designed to measure a specific physical quantity and consists of the product of several base units.
➤ A derived unit is said to be coherent if it does not require a numerical factor to define it.

Table 1.8   Derived units

Physical quantity	Derived unit	Symbol	Name
Area	Square metre	m^2	
Volume	Cubic metre	m^3	
Velocity	Metre per second	$m\ s^{-1}$	
Density	Kilogram per cubic metre	$kg\ m^{-3}$	
Force	Kilogram metre per second per second	$kg\ m\ s^{-2}$	Newton
Pressure	Newton per square metre	$N\ m^{-2}$	Pascal
Substance concentration	Mole per cubic metre	$mol\ m^{-3}$	

Table 1.9 Multiples and sub-multiples of units

Factor	Prefix		Factor	Prefix	
	Name	Symbol		Name	Symbol
10^1	deca-	da	10^{-1}	deci-	d
10^2	hecto-	h	10^{-2}	centi-	c
10^3	kilo-	k	10^{-3}	milli-	m
10^6	mega-	M	10^{-6}	micro-	μ
10^9	giga-	G	10^{-9}	nano-	n
10^{12}	tera-	T	10^{-12}	pico-	p
			10^{-15}	femto-	f
			10^{-18}	atto-	a

Certain difficulties arise when attempts are made to use only coherent derived units. In the case of the measurement of mass, for which the name of the base unit is the kilogram, it is unacceptable to use additional standard prefixes, for example millikilogram, etc. It is therefore accepted that in order to prevent confusion, all multiples of mass should be quoted in terms of the gram and not the kilogram while still retaining the kilogram as the base unit for mass. A similar problem occurs in the measurement of volume. Here the unit for volume is the cubic metre (m^3), which is derived from the base unit for length, the metre. The more commonly used litre is a more convenient volume and is the same as a cubic decimetre (dm^3), which, although it is an acceptable derived unit, is not coherent due to the use of the prefix 'deci'. For convenience, the term litre and its symbol (l) are accepted but it should be noted that whatever name and symbol are used, consistency in their use is important.

1.3.2 Calibration

For the majority of quantitative methods, calibration involves analysing solutions that contain a range of known concentrations (standard solutions) of the specified analyte in parallel with the test samples, determining the relationship between the reading and the concentration of the standard analyte, and from this relationship calculating the amount of analyte in the test samples.

In most manual methods of analysis this relationship is most easily expressed in the form of a calibration graph, although in many automated analysis systems, the equation for the relationship is determined by the instrument. One very common and important step in determining this relationship is the preparation of a series of standard solutions.

Standard solutions

The essential starting point in setting up a quantitative analysis is the availability of a sample with a known concentration of the analyte. In many instances this is not very difficult because the substance can be obtained in a pure solid form and an initial or stock standard solution can be prepared by weighing a known amount of the substance and dissolving in a fixed volume

➤ A primary standard is one which is prepared from a pure preparation of the analyte.
➤ A secondary standard is one for which the concentration of the analyte has been determined by analysis using a recognised method.

of solvent. However, in some cases a pure, solid sample of the analyte is not available as a primary standard and it is necessary to use secondary standards. These have been previously analysed using a recognised method and a value attributed to them. This type of standard is frequently available commercially. From the stock standard of either type, a series of dilutions is prepared and analysed.

There are a variety of ways of preparing a set of dilutions but the two most frequently used are percentage dilutions and serial dilutions.

Serial dilutions

Initially a decision needs to be made about the volume of standard that will be required, e.g. 5 ml. A volume of diluent equal to the selected volume, i.e. 5.0 ml, is pipetted into the required number of tubes and an equal volume of the stock standard added to the first tube. The contents are mixed and the same volume, i.e. 5.0 ml, is removed and added to the second tube, mixed and the same volume removed and added to the third tube and so on. This results in a series of dilutions, the concentration of analyte in each one being half that of the preceding tube. They are usually described as 1 in (in a total of) 2, 1 in 4, 1 in 8, 1 in 16, etc.

There are several disadvantages with this method, the prime one being the increasing interval between each successive dilution, which makes the determination of an accurate relationship difficult over the increasing dilution range. Additionally, the units of concentration are less conveniently calculated. If, for instance, the stock solution contained 5.0 mg l^{-1} then the 1 in 16 dilution would contain $5.0/16 = 0.3125$ mgl^{-1}. The technique, however, is useful if it is necessary to cover a wide concentration range, possibly as a preliminary to developing a method for routine use.

Percentage dilutions

As before, a decision about the volume of standard required for the analysis must be made. A convenient volume is selected for the dilutions, ideally one for which 10% intervals are easily measured, e.g. 1.0 ml, 5.0 ml, 10.0 ml, rather than volumes such as 3.0 ml. Having selected a working volume, add the stock standard to a series of tubes in volumes increasing by 10% of the selected volume. If, for instance, the selected volume is 5.0 ml, 10% of this volume of the stock solution is added to the first tube, i.e. 0.5 ml, 20% to the second tube, i.e. 1.0 ml, 1.5 ml added to the third tube and so on. A suitable diluent is then added to each tube to bring the total volume up to the selected volume. For the first tube of the example, 4.5 ml of diluent would be added, 4.0 ml to the second and 3.5 ml to the third tube. The concentration of each dilution is calculated from the initial stock standard concentration and the degree of dilution, e.g. a 30% concentration of a 5.0 ml l^{-1} standard solution has a concentration of $5.0 \times 30/100 = 1.5$ mgl^{-1}.

The method can be used as described, giving a range of standards at 10% intervals or, alternatively, 20% intervals may be used giving a range of five dilutions. It is the most convenient method and gives a regular range of working standards, the value of each being easily calculated. Procedure 1.10 shows this method in practice using 20% intervals.

Procedure 1.10: Preparation of a calibration graph for the determination of salicylate

Method
1. Prepare a stock solution of sodium salicylate containing 200 mgl^{-1}.
2. Set up five tubes and add stock salicylate solution and distilled water as indicated in the table below.
3. Analyse each dilution using the specified assay method and record the absorbance data.
4. Plot the absorbance value against concentration to produce a calibration graph.

Tube number	1	2	3	4	5
Stock salicylate (ml)	1.0	2.0	3.0	4.0	5.0
Distilled water (ml)	4.0	3.0	2.0	1.0	0
Percentage concentration	20	40	60	80	100
Concentration (mg l^{-1})	40	80	120	160	200
Absorbance values	0.18	0.38	0.58	0.73	0.89
(duplicates)	0.20	0.40	0.59	0.75	0.90

Calculation
Plot the absorbance values against the concentration of each dilution.

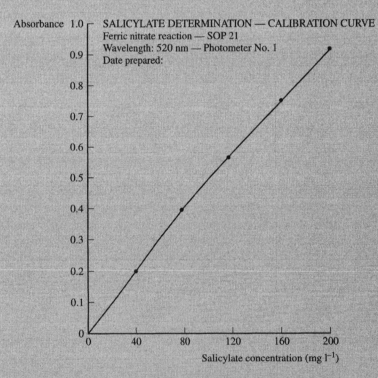

SALICYLATE DETERMINATION — CALIBRATION CURVE
Ferric nitrate reaction — SOP 21
Wavelength: 520 nm — Photometer No. 1
Date prepared:

Conclusion
The resulting calibration graph is almost linear and the concentration of salicylate in any sample can be determined by reading off the concentration corresponding to the test absorbance value.

1.3.3 Graphical presentation of data

Graphs produced for quantitative analytical purposes must be dated and include adequate information on the method of analysis, e.g. the SOP serial number, the analyte and the units of measurement. The scale of each axis should be carefully chosen and not simply fill the available space.

A major difficulty in drawing graphs lies in the nature of the relationship between the two variables and a minimum of five points is essential in order to be able to draw a particular line or curve confidently. In the preparation of a calibration graph, the standard or calibrator solutions are usually analysed either in duplicate or in a limited number of replicates and the mean value is used in the preparation of the curve. Because the mean value from a limited number of replicates is unlikely to be the true mean, it might help in drawing the graph if the mean is plotted and a bar is drawn to indicate the highest and lowest values obtained in the replicates.

In many instances it is known that there should be a linear relationship between two sets of data but the lack of precision makes it difficult to draw a straight line through the points on the graph. In such situations it is advisable to calculate the equation for the straight line that best fits the data using linear regression analysis. This technique can be used to derive the equation describing the relationship between the observed reading and the concentration of the sample. The equation is useful not only to aid the drawing of a graph but is essential in setting up a computer program for calculating test results from a set of calibration values.

> ➤ Linear regression analysis is a statistical technique to determine the equation for the straight line that best describes the relationship between two variables.

Data that give rise to curves rather than straight lines present certain problems not only in their graphical presentation but also in any statistical analysis. A careful examination of the principle of the method may reveal that the variable being measured is not the fundamental one involved and presentation of the data in an alternative form may give a linear relationship. In methods involving radial diffusion, for instance, a plot of distance against concentration often results in a curve, whereas a plot of area (or distance squared) should give a straight line. The use of reciprocal values of either one or both sets of data may result in a straight line plot, as in the Lineweaver–Burk plot for enzyme constants. Many measurements show linear logarithmic relationships as for instance in the determination of relative molecular mass by gel permeation chromatography, while some more complex ones may be linearized by the use of logits:

> ➤ Radial diffusion – see Section 7.3.2.
> ➤ Lineweaver–Burk plot – see Section 8.1.2.

$$\text{logit } x = \log_e \frac{x}{1-x}$$

Because many such relationships are only approximately linear it is often desirable to determine the line statistically rather than visually.

1.3.4 Laboratory report

The report of any measurement must always include enough information to avoid misunderstanding and should contain specific details about the sample as indicated in Table 1.10. In reporting the precision of data, a value for the standard deviation implies that the result is the mean of replicate analyses of the sample. The use of coefficient of variation avoids this problem and is probably more meaningful as a general indication of the quality of the data.

Table 1.10 The laboratory report

Type of information	Examples	
System or sample	Whole blood	Wheat flour
Component	Glucose	Reducing sugars
Method employed	Enzymic (Glucose oxidase)	Copper reduction (Benedict)
Units of measurement	mmol l^{-1}	g l^{-1}
Numerical value	3.50	4.5
Precision: (s)	0.10	0.5
(V)	2%	10%

It may be argued that in practice much of the information is obvious and does not need to be specified. It is important to remember that even this routine data may be required at a future date for reference and lack of this information could then be critical.

Section 1.3
1. Which of the following are acceptable SI units?
   (a) Gram per 100 ml (g/100 ml).
   (b) Mole per second (mol s^{-1}).
   (c) Kilogram per litre (kg l^{-1}).
   (d) Micromole per minute (μmol min^{-1}).
2. Which of the following is the correct term for a millionth of a kilogram?
   (a) Nanokilogram.
   (b) Microgram.
   (c) Milligram.
   (d) Megagram.

1.4   Further reading

Miller, J.N. (1990) *Statistics for analytical chemistry*, 3rd edition, Horwood (Ellis), UK.

Carson, P.A. and Dent, N. (eds) (1994) *Good laboratory and clinical practices: Techniques for the quality assurance professional*, Butterworth-Heinemann, Canada.

Campbell, R.C. (1989) *Statistics for biologists*, 3rd edition, Cambridge University Press, UK.

Callender, J.T. and Jackson, R. (1995) *Exploring probability and statistics with spreadsheets*, Prentice-Hall, UK.

Young, J.A. (ed.) (1991) *Improving safety in the chemical laboratory*, 2nd edition, John Wiley, UK.

Pritchard, E. (1995) *Quality in the analytical chemistry laboratory*, John Wiley, UK.

Gunzler, H. (ed.) (1996) *Accreditation and quality assurance in analytical chemistry*, Springer-Verlag, UK.

2 Spectroscopy

Key topics	• Interaction of radiation with matter
	• Molecular absorptiometry
	• Absorptiometer design
	• Molecular fluorescence techniques
	• Atomic spectroscopy techniques
	• Magnetic resonance spectroscopy

Spectroscopy is the study of the absorption and emission of radiation by matter. The most easily appreciated aspect of the absorption of radiation is the colour shown by substances that absorb radiation from the visible region of the spectrum. If radiation is absorbed from the red region of the spectrum, the transmitted or unabsorbed radiation will be from the blue region and the substance will show a blue colour. Similarly substances that emit radiation show a particular colour if the radiation is in the visible region of the spectrum. Sodium lamps, for instance, owe their characteristic orange–yellow light to the specific emission of sodium atoms at a wavelength of 589 nm.

Measurements of the intensity and wavelength of radiation that is either absorbed or emitted provide the basis for sensitive methods of detection and quantitation. Absorption spectroscopy is most frequently used in the quantitation of molecules but is also an important technique in the quantitation of some atoms. Emission spectroscopy covers several techniques that involve the emission of radiation by either atoms or molecules but vary in the manner in which the emission is induced. Photometry is the measurement of the intensity of radiation and is probably the most commonly used technique in biochemistry. In order to use photometric instruments correctly and to be able to develop and modify spectroscopic techniques it is necessary to understand the principles of the interaction of radiation with matter.

2.1   Interaction of radiation with matter

Radiation is a form of energy and shows both electrical and magnetic characteristics, hence the term electromagnetic radiation (Figure 2.1). It can be considered as being composed of a stream of separate groups of electromagnetic waves and the energy associated with the radiation can be mathematically related to the waveform.

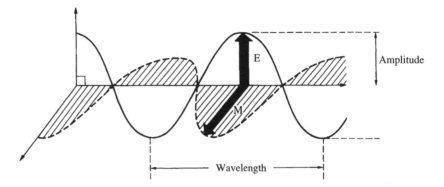

Figure 2.1 Electromagnetic radiation. A representation of electromagnetic radiation with the electric field (E) and the magnetic field (M) at right angles to the direction of the wave movement. Both fields oscillate at the same frequency.

➤ Wavelength is the distance between successive peaks of a waveform.

Waveform can be defined in at least two different ways which are relevant to spectroscopic measurements. Wavelength (λ) is defined as the distance between successive peaks (Figure 2.1) and is measured in subunits of a metre, of which the most frequently used is the nanometre (10^{-9} m). An angstrom unit (Å) is not acceptable in SI terminology but is still occasionally encountered and is 10^{-10} m (i.e. 10 Å = 1 nm). The frequency of radiation (nu, v) is defined as the number of successive peaks passing a given point in 1 second. Hence the relationship between these two units of measurement is:

$$v \propto \frac{1}{\lambda}$$

➤ The visible region of the spectrum is the wavelength range of approximately 400 to 700 nm.
➤ The ultraviolet region of the spectrum is the wavelength range of approximately 200 to 400 nm.
➤ The infrared region of the spectrum is the wavelength range above 700 nm.

The visible region of the spectrum extends approximately over the wavelength range 400–700 nm, the shorter wavelengths being the blue end of the spectrum and the longer wavelengths the red end. Wavelengths between 400 and 200 nm make up the near ultraviolet region of the spectrum and wavelengths above 700 nm to approximately 2000 nm (2 μm) the near infrared region.

The energy associated with a particular waveform is directly related to the frequency of the radiation and therefore inversely related to the wavelength. It may be calculated using the equation:

$$E = hv = \frac{hc}{\lambda}$$

where h = Planck's constant (6.626×10^{-34} J s)
     v = frequency of the radiation
     λ = wavelength of the radiation
     c = speed of light (3.0×10^{8} m s^{-1})

While the energy associated with a waveform is related to its wavelength, the intensity of that radiation is related to its amplitude (Figure 2.1).

The absorption or emission of radiation by matter involves the exchange of energy and in order to understand the principles of this exchange it is necessary to appreciate the distribution of energy within an atom or molecule.

The internal energy of a molecule is due to at least three contributing sources:

1. The energy associated with the electrons.
2. The energy associated with the vibrations between the atoms.
3. The energy associated with the rotation of various groups of atoms within the molecule relative to the other groups.

These energy levels may be altered by the absorption or emission of energy as radiation and, because any given atom or molecule can only exist in a limited number of energy levels, these energy changes must be in definite packets or quanta. The exact amount of energy required to produce a change in the molecule from one energy level to another will be given by the photons of one particular frequency which will be selectively absorbed or emitted.

A study of the wavelength or frequency of radiation absorbed or emitted by an atom or a molecule will give information about its identity and this technique is known as **qualitative spectroscopy**. This information is usually reported as the wavelength of radiation involved and is most easily represented as an absorption or emission spectrum (Figure 2.2). Measurement of the total amount of radiation will give information about the number of absorbing or emitting atoms or molecules and is called **quantitative spectroscopy.**

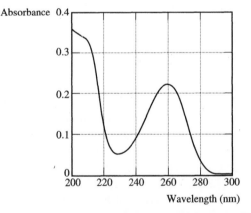

Figure 2.2   **Absorption spectrum of adenosine diphosphate (ADP).**

2.1.1   Absorption of radiation

The increase in energy in a molecule that occurs when radiation is absorbed can be accommodated in the three ways already described. The range of energies involved is characteristic of each type of change and is associated with a well-defined region of the electromagnetic spectrum (Figure 2.3).

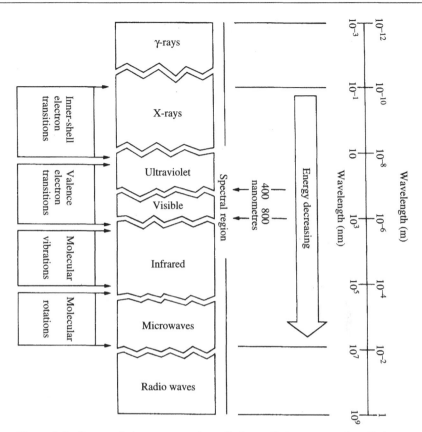

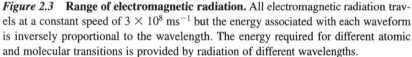

Figure 2.3 **Range of electromagnetic radiation.** All electromagnetic radiation travels at a constant speed of 3×10^8 ms^{-1} but the energy associated with each waveform is inversely proportional to the wavelength. The energy required for different atomic and molecular transitions is provided by radiation of different wavelengths.

Atomic absorption

Individual atoms cannot rotate or vibrate in the same manner as molecules and as a result the absorption of energy is only associated with electronic transitions that are limited in number and associated with very narrow ranges of radiation or absorption lines. The wavelength that is most strongly absorbed usually corresponds to an electronic transition from the ground state to the lowest excited state and is known as the resonance line. The limited number of electronic transitions for any atom results in the fact that excited atoms, when returning to the ground state, emit radiation of the same wavelength as that which was absorbed.

> ➤ An electronic transition involves the raising of an electron to a higher energy level orbital as energy is absorbed.

Molecular absorption – ultraviolet and visible region

This is the area of greatest interest to quantitative biochemistry and is dependent upon the electronic structure of carbon compounds. Absorption of radiation in this region of the spectrum causes transitions of electrons from molecular bonding orbitals to the higher energy molecular antibonding orbitals.

The electronic structure of the carbon atom is designated as $1s^2, 2s^2, 2p_x^1, 2p_y^1$ (Figure 2.4) and this means that in order to fill all the available orbitals

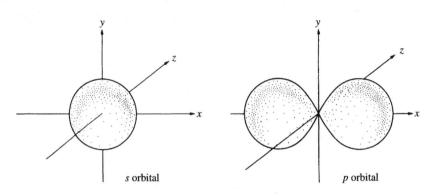

s orbital   *p* orbital

Figure 2.4 Atomic orbitals. An *s* orbital is spherical and can only exist in one orientation but the *p* orbital is directional and can exist in either of three planes (*x*, *y* and *z*), all of equal energy status. All orbitals can accommodate two electrons provided that they are of opposite spin. The number of electrons in a particular orbital is indicated as a superscript to the orbital letter, e.g. $1s^2$.

four further electrons would be required, one in each of the $2p_x$ and $2p_y$ orbitals with two in the unoccupied $2p_z$ orbital. These four vacancies for electrons are not all equivalent in energy or in direction and yet the four valencies of the carbon atom as evidenced in organic compounds are equivalent and symmetrical (Figure 2.5(a)). This symmetry is due to the combination of orbitals with the same basic energy level, i.e. *2s* and *2p*, to give hybrid *sp* orbitals.

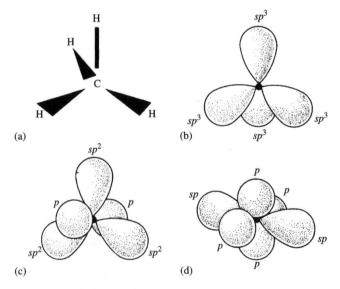

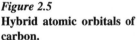

Figure 2.5
Hybrid atomic orbitals of carbon.

Three different combinations are possible depending upon the number of orbitals involved in the hybridization. The four existing outer electrons of the carbon atom (two in the *2s* orbital and one in each of the $2p_x$ and $2p_y$ orbitals) are initially dispersed, one in each of the available orbitals (*2s*, $2p_x$,

$2p_y$ and $2p_z$) and subsequent hybridization of these four orbitals results in the hybrid sp^3 orbital (i.e. one s and three p) which has four symmetrical lobes (Figure 2.5(b)). Combinations that include only two of the p orbitals with the s orbital result in a hybrid sp^2 orbital, which has three symmetrical lobes and a residual $2p$ orbital (Figure 2.5(c)) while a combination of one s and one p orbital results in an sp hybrid (Figure 2.5(d)).

Atoms are linked to form a molecule by covalent bonds, which are a result of the sharing of electrons from two orbitals, one from each atom, and so effectively filling both orbitals with the maximum number of two electrons. The formation of a covalent bond between two carbon atoms which are in the sp^3 hybrid form results in a symmetrical (sigma, σ) bond, which is also known as a single or saturated bond (Figure 2.6(a)). However, if the two carbon atoms are in the sp^2 hybrid form then not only is a bond similar to the sigma bond formed between the two lobes of the sp^2 orbitals but the residual p orbitals combine to produce what is known as a pi (π) bond (Figure 2.6(b)). This bond is not symmetrical about the main sigma bond but is formed by a side-to-side overlap of the two p orbitals resulting in the pi bond electrons being displaced above and below the plane of the two carbon atoms. These are known as delocalized electrons and the overall bond is known as a double or unsaturated

> ➤ A sigma (σ) bond is a single (or saturated) bond between two carbon atoms.
> ➤ A pi (π) bond is a double (or unsaturated) bond between two carbon atoms.

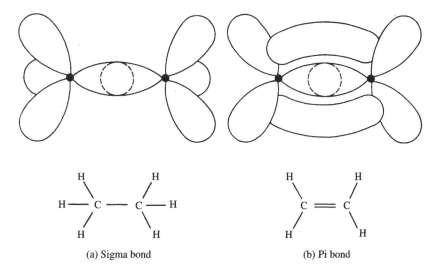

Figure 2.6 **Bonding in carbon compounds.**

(a) Sigma bond   (b) Pi bond

> ➤ A conjugated system is an alternating sequence of single and double bonds.

bond. If an alternating system of sigma and pi bonds is formed (a conjugated system), the delocalized electrons combine to form two molecular orbitals, one above and one below the entire length of the bond sequence (or ring in the case of cyclic compounds).

As indicated earlier, all electrons are capable of existing in one of several energy levels and the change from one level to another involves the exchange of energy. The ease with which an electron can be raised to its next higher energy level is related to the amount of energy required for such a change and, if the energy is provided by radiation, is related to the wavelength of the radiation. Electronic changes can be listed in order of the increasing ease with which the transition may take place, and this parallels the increasing wavelength causing the transition.

Table 2.1 Absorption maxima of hydrocarbons

Compound	Structure	Type of transition	Absorption maximum (nm)
Ethane	C—C	$\sigma \rightarrow \sigma^*$	< 180
Ethylene	C=C	$\pi \rightarrow \pi^*$	190
Benzene		Cyclic $\pi \rightarrow \pi^*$	256
Naphthalene		Cyclic $\pi \rightarrow \pi^*$	290
Anthracene		Cyclic $\pi \rightarrow \pi^*$	360

Molecules that show increasing degrees of conjugation require less energy for excitation and as a result absorb radiation of longer wavelengths (Table 2.1). In addition, the introduction of saturated groups (methyl, hydroxyl, etc.) into a compound that already absorbs radiation may result in an increase in the wavelength of radiation absorbed. The effect of conjugation is most dramatically demonstrated by certain dyes which in the reduced forms are colourless but in the oxidized forms show a characteristic colour. The absorption characteristics of the reduced form of methyl red (Figure 2.7) are due to two separate conjugated systems which absorb radiation in the ultraviolet region of the spectrum and hence are not visible. The effect of oxidation is to join the two systems in a way that increases the extent of conjugation and shifts the wavelength of the absorbed radiation into the visible region of the spectrum, giving a colour to the compound.

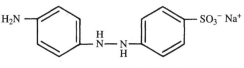

Reduced form – colourless

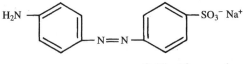

Oxidized form – red

Figure 2.7
The effects of increased conjugation in the dye methyl red.

The absorption peaks found in ultraviolet and visible spectroscopy are much broader than those found in infrared spectroscopy. This is due to the additional effects of vibrational and rotational transitions being superimposed on the basic electronic transition (Figure 2.8). Molecules existing in an electronic ground state may be at different vibrational and rotational energy levels and as a result will require different total amounts of energy to undergo an electronic transition. Similarly the final excited molecule may exist at various vibrational and rotational energy levels. The overall effect is that a range of wavelengths is absorbed and the spectrum shows the characteristic broad peak which is useful in quantitative studies but not so useful in qualitative work.

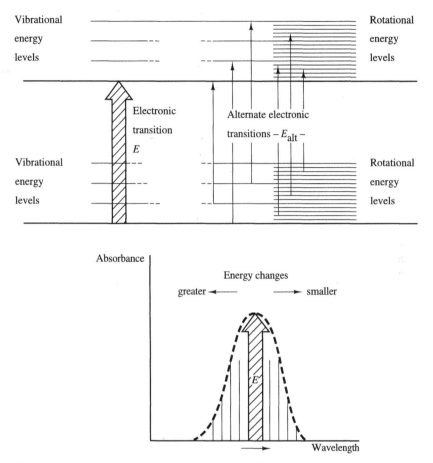

Figure 2.8   **Range of energy levels associated with electronic transitions.**
Not all the molecules in a sample show the same energy change (E) for a particular electronic transition because although they all exist in the same electronic energy level, they may have different rotational and vibrational energies. They will therefore require different amounts of energy (E_{alt}) in order to be raised to the next electronic energy level. Each different amount of energy will be provided by a different wavelength of radiation, resulting in the characteristic absorption peak.

Molecular absorption – infrared region

Not all organic compounds absorb radiation in the ultraviolet and visible regions of the spectrum but they do show the absorption of infrared radiation due to vibrational changes. Vibrational transitions within a molecule may be achieved with energy levels associated with the near infrared region of the spectrum while rotational changes in molecular energy are associated with the far infrared and microwave regions (Figure 2.3).

The wavelength range of infrared radiation lies between 700 nm and 1×10^6 nm (1 mm) but because of the large numerical values involved in quot-

➤ Wavenumber is the number of wavelengths in 1 centimetre.

ing wavelength, absorption band positions are usually quoted as the wavenumber, which is the reciprocal of the wavelength in centimetres:

$$\text{Wavenumber (cm}^{-1}) = \frac{1}{\text{wavelength (cm)}} = \frac{1 \times 10^7}{\text{wavelength (nm)}}$$

➤ A vibrational transition is the bending or stretching of bonds caused by the absorption of energy.

The two main types of molecular vibrations associated with the absorption of infrared radiation involve the stretching and bending of bonds. Stretching is a vibrational movement in which the bond length alters and the two nuclei move harmonically relative to each other. For a compound involving two atoms there is only one bond that is capable of being stretched and in general terms it can be said that for a compound containing n atoms there are $(n - 1)$ possible stretching effects. Bending vibrations present many more possible variations, the total number being defined as $(2n - 4)$ for linear molecules and $(2n - 5)$ for non-linear molecules.

Although a large number of vibrational changes are possible for any given molecule, not all of them will result in the absorption of radiation. An absorption maximum in the infrared spectrum of a compound can be demonstrated only when a vibration results in a change in the dipole of the molecule, a process for which energy is required. Carbon dioxide, for example, has a linear structure and can undergo two stretching vibrations $(n - 1)$ and two bending vibrations $(2n - 4)$ (Table 2.2). An examination of the four vibrational changes possible will show that one stretching effect does not result in a change in the relative distribution of the charge associated with the molecule (dipole) and as a result will not show an absorption maximum. The other three vibrational changes do result in a change in the overall distribution of the charge and give absorption maxima in the infrared. However, the two bending vibrations are identical in energy level, although in a different plane, and therefore show the same absorption maximum.

➤ A dipole is an unequal distribution of charge in a molecule.

Table 2.2 Fundamental vibrations of carbon dioxide

Structure	Vibrations	Absorption maximum (cm^{-1})
O—C—O	None	Normal structure
O→C←O	Stretching	None
O➤◄C—O	Stretching	2349
⌄C⌄ O O (bent)	Bending (horizontal plane)	667
O◄—C►—O	Bending (vertical plane)	667

Infrared spectra show not only absorption bands due to fundamental vibrations but also additional and usually weaker bands due to multiples of the fundamental frequencies (overtones) and to the combination of fundamental vibrational effects and their multiples (combination bands). The interpretation of such spectra is further complicated by the fact that the spectrum for a given substance may vary depending upon the solvent used or the method of sample preparation. Infrared absorption involves a large number of quite different transitions, both vibrational and rotational, which result in a spectrum which is composed of a large number of separate, narrow peaks and is more useful in qualitative than in quantitative analysis.

> A rotational transition is the turning of one group about a bond relative to the whole molecule.

> An atom or molecule is said to be excited when it has undergone a transition of some form.

2.1.2 Emission of radiation

An atom or molecule is said to be in an excited or unstable state when it absorbs energy which results in either electronic or vibrational transitions. It will return to its ground state very rapidly and the energy may be lost in one of three ways:

1. As a result of a chemical reaction.
2. By dissipation as heat.
3. By emission as radiation.

If the atom or molecule loses all or part of this energy as radiation, photons of energy will be emitted which correspond to the difference between the energy levels involved. Since these levels are clearly defined for any given atom or molecule, the radiation emitted will be of specific frequencies and will show up as bright lines if the emitted light is dispersed as a spectrum.

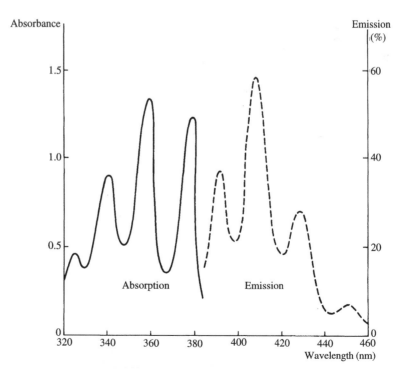

Figure 2.9
Absorption and emission spectra of anthracene in solution in dioxane.

Atomic emission is displayed by many elements, particularly the metals, which after being excited either thermally or electrically emit a discontinuous spectrum, the strongest lines of which are due to transitions ending in the ground state (resonance lines). Molecular emissions are more complex than atomic emissions, the radiation emitted consisting of broad bands of radiation rather than the narrow lines associated with atomic emission and resulting in a spectrum which is an approximate mirror image of the absorption spectrum of the compound (Figure 2.9).

Molecular emissions are due to electronic transitions within the molecule but are modified by variations in bond length. The bond between two atoms assumes a particular length as a result of the various forces acting upon the atoms involved. The attractive forces between the electrons of one atom and the nucleus of the other atom are balanced by the repulsive forces of the like-charges carried by both nuclei. The Morse curve (Figure 2.10) describes

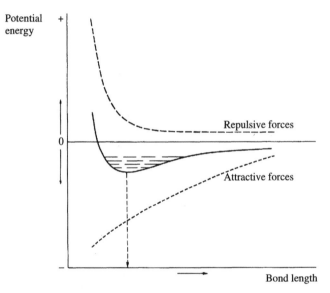

Figure 2.10   **Morse curve.** Inter-atomic attractive and repulsive forces result in the formation of a bond length with a minimum energy level.

the relationship between these forces and illustrates the fact that at a particular bond length the energy associated with the bond is minimal and the molecule is said to be in its ground state. However, this only describes the most stable and hence the most frequent form of the molecule. More energetic forms will also exist and these are indicated by the higher energy levels in the Morse curve diagram.

A new Morse curve is required to describe the energy levels associated with an excited molecule and this is displaced to the right and above the original curve. The excitation of a molecule with a particular bond length results in an excited molecule with the same bond length but with a higher internal energy as illustrated by the new Morse curve (Figure 2.11). An excited molecule initially will revert to the minimum energy state for the excited molecule

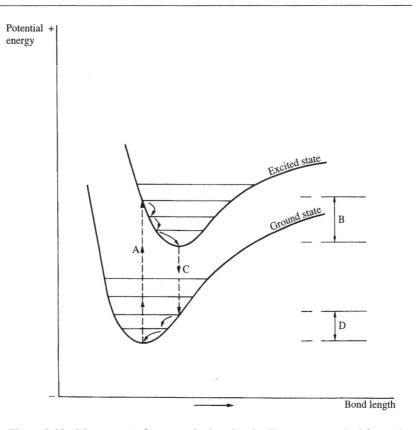

Figure 2.11 **Morse curve for an excited molecule.** The energy required for excita-tion (A) is lost as the molecule returns to the ground state but only the energy lost between states (C) may be emitted as radiation. Energy losses due to internal rearrangements (B and D) are non-radiative.

before reverting to the ground state, with the result that the energy originally absorbed is lost in at least three stages (B, C + D) as the molecule returns to its ground state. Some energy losses are due to internal rearrangements of each of the two molecular forms and as such involve a non-radiant loss of energy: it is only the energy lost during the interchange between the two molecular forms that may possibly be emitted as radiation. Any energy lost by radiation (C) must be less than the total energy absorbed (A) and hence the wavelength emitted will be greater than that which was absorbed in the first place.

A molecule in the ground state has a pair of electrons with opposite spin in each molecular orbital. On excitation, one electron is elevated to an anti-bonding orbital and because it is not restricted by the presence of another electron it can exist with either its original spin (a singlet transition) or reversed spin (a triplet transition). If the return to the ground state involves a fall from the singlet transi-tion and radiation is emitted, the compound is said to fluoresce. However, in a return to the ground state from a triplet transition, fluorescence will not occur and the energy loss will probably be non-radiative. The intensity of fluorescence depends upon the proportion of the total molecules which undergo singlet transi-tions and it is mainly for this reason that different compounds show different degrees of fluorescence. Some compounds do show radiative emission during triplet transitions and these compounds are said to phosphoresce. The most demonstrable difference between these two types of emission is the fact that fluorescence occurs within about 1×10^{-8} seconds and persists for only about 1 to 1×10^3 nanoseconds and is much faster than phosphorescence, which persists for up to 1×10^{-3} seconds due to the time taken for the spin change to occur.

> Luminescence – see Section 8.3.4.

Chemiluminescence is another form of molecular emission in which the initial electronic transition is caused by an exergonic reaction rather than the absorption of radiant energy. Most chemiluminescence reactions are of the oxidative type and those involving hydrogen peroxide are particularly useful biochemically. Luminol (5-amino-2,3-dihydrophthalazine-1,4-dione), for instance, will emit light when reacting with hydrogen peroxide and may be used in monitoring many oxidative enzymes such as glucose oxidase, amino acid oxidase, etc. Bioluminescence is a special type of chemiluminescence in which the process of light emission is catalysed by an enzyme. The enzyme luciferase, extracted from the firefly, uses ATP to oxidize the substrate luciferin with the emission of radiation and can be used to measure ATP concentrations. The enzyme extracted from bacterial sources can oxidize long chain aliphatic aldehydes in the presence of oxygen, FMN and NADH with the emission of radiation.

> Scintillation counter – see Section 5.2.2.

Luminescence methods are very sensitive, quantities as little as 1 femtomole of ATP being detectable and while measurements may be made using scintillation counters, much simpler equipment that requires neither a radiation source nor a monochromating system is satisfactory.

Self test questions

Section 2.1
1 What describes radiation with a wavelength of 340 nm?
   (a) Ultraviolet radiation.
   (b) Visible radiation.
   (c) Near infrared radiation.
   (d) Far infrared radiation.
2 Ultraviolet radiation will frequently induce which of the following molecular transitions?
   (a) Inner shell electronic transition.
   (b) Valence electronic transition.
   (c) Molecular vibrational transition.
   (d) Molecular rotational transition.
3 An effect of an increased level of conjugation in a molecule is to shift the absorption maximum to a longer wavelength
   BECAUSE
   increasing the level of conjugation in a molecule reduces the amount of energy required to induce an electronic transition.
4 The process in which energy is emitted as radiation after a chemically induced electronic transition is known as fluorescence
   BECAUSE
   in fluorescence the emitted radiation is always of a longer wavelength than that which initially induced the transition.

2.2   Molecular absorptiometry

Photometric measurements provide the basis for the majority of quantitative methods in biochemistry and are related to the amount of radiation absorbed rather than the nature of such radiation. This relationship is expressed in two experimental laws, which provide the mathematical basis for such quantitative methods.

2.2.1   Beer–Lambert relationship

Lambert's law states that the proportion of radiant energy absorbed by a substance is independent of the intensity of the incident radiation. Beer's law states that the absorption of radiant energy is proportional to the total number of molecules in the light path. Beer's law describes the basic relationship between the concentration of the absorbing substance and the measured value of absorbed radiation. Lambert's law is of major significance in the manner in which measurements are made and the fact that a given sample always absorbs the same proportion of the incident radiation regardless of its intensity greatly simplifies the design of instruments.

The amount of radiation absorbed by a substance cannot be measured directly and it is usually determined by measuring the difference in intensity between the radiation falling on the sample (incident radiation, I_0) and the residual radiation which finally emerges from the sample (transmitted radiation, I) (Figure 2.12).

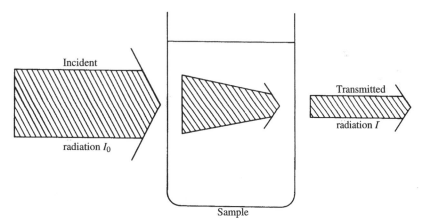

Figure 2.12   **Absorption of radiation.**

> ➤ The Beer–Lambert law expresses the amount of light absorbed by a sample in terms of the concentration of the sample and the length of the light path.

Using such measurements, the Beer–Lambert law can be expressed as an equation:

$$\log_{10} \frac{I_0}{I} = \epsilon c l$$

where c is the concentration of the substance in gram molecules per litre, l is the light path in centimetres and ϵ (epsilon) is known as the molar absorption coefficient for the substance and is expressed in litres per mole per centimetre $(1 \ \mathrm{mol}^{-1} \ \mathrm{cm}^{-1})$.

> Transmittance is the amount a light passing through a sample expressed as a percentage of the light incident on the sample.

The values for I and I_0 cannot be measured in absolute terms and the measurements are most conveniently made by expressing I as a percentage of I_0. This value is known as the percentage transmittance (T) and only shows a linear relationship with the concentration of the test substance if the logarithm of its reciprocal is used. It is therefore more convenient to report the measurements initially as this logarithmic function of I and I_0, a parameter which is known as absorbance (A):

$$\text{Percentage transmittance } (T) = \frac{I}{I_0} \times 100$$

$$A = \log_{10} \frac{100}{T}$$

> Absorbance is a measure of the amount of light absorbed by a sample expressed in terms of the logarithm of the ratio of the transmitted and incident radiation.

In many instruments the meter read-out is calibrated in absorbance units using a logarithmic scale while other instruments retain the convenience of a linear scale but convert the signal from the detector to a logarithmic one by electronic or mechanical means. It is essential when using a photometric instrument to know if it is calibrated in absorbance or transmittance units.

2.2.2 Deviations from the Beer–Lambert law

The assumption that the difference between the incident and the transmitted radiation is a measure of the radiation absorbed by the analyte is not completely true because this may not be the only reason why the incident radiation does not appear in the transmitted form. A certain amount of radiation will be reflected from the surface of the sample holder, usually a glass or plastic cell, or absorbed by the material of which the cell is composed. The sample may also be dissolved in a solvent which itself may also absorb or reflect radiation:

$$\text{incident } (I_0) = \text{absorbed} + \text{transmitted } (I) + \text{other losses}$$

For this reason the radiation transmitted by a blank sample is measured and taken to be the effective incident radiation. This blank should be identical to the test sample in all aspects except the presence of the test substance:

$$\text{blank reading} = I_0 - \text{other losses}$$

Hence:

$$\text{absorbed} = \text{blank} - \text{transmitted } (I)$$

> Monochromatic radiation is radiation of only one wavelength.

An important factor which influences the Beer–Lambert relationship is the wavelength range of the incident radiation. Ideally the radiation should only provide a specific unit of energy and not a range of energy levels. Such specific radiation is known as monochromatic radiation and can only be produced under very restricted conditions. The best that can often be achieved is radiation with a very limited range of wavelengths. If the incident radiation contains wavelengths that will not be absorbed by the test substance, the difference between the intensities of incident and transmitted radiation will not be proportional to the concentration of the test substance. In such cases a nonlinear relationship will exist as illustrated in Figure 2.13. This problem is particularly relevant to instruments that use simple glass filters rather than monochromating systems such as prisms or diffraction gratings.

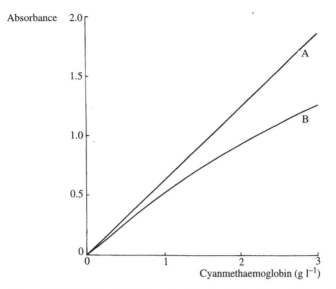

Figure 2.13 **Validity of the Beer–Lambert relationship for different monochromating systems.** The absorbance of varying concentrations of cyanmethaemoglobin was measured at 540 nm using a spectrophotometer (A) and a simple photometer (B) with a glass filter.

Alterations that may occur in the molecular nature of the sample due to changes in concentration may also result in deviation from the Beer–Lambert relationship. Molecules may tend to associate with one another when the concentration is high or, conversely, complexes may tend to dissociate in low concentrations. Both types of change may possibly affect the absorption characteristics of the compound and result in non-linear graphs.

Quantitative measurements rely on the assumption that the only radiation to reach the detector has passed through the sample but in practice it is very difficult to design instruments that are capable of effectively eliminating all extraneous radiation. Much of this unwanted radiation arises from the scattering of the incident radiation by irregularities in surfaces caused by faults in manufacture or scratches, etc. Certain general precautions are usually taken to reduce this extraneous light in photometric instruments, such as the use of matt black interior surfaces to give maximal absorption and the insertion of baffles and windows to compartmentalize the instrument. Optical light paths should be designed to produce parallel beams of light and sharply focused spectra by the use of collimating lenses and mirrors.

Imperfections in monochromators result in the presence of a small proportion of unwanted wavelengths in the incident radiation. Such **stray light** results in a deviation from a Beer–Lambert relationship (Figure 2.14) and the effect is that absorbance measurements are lower than they should be.

It is possible to assess the proportion of stray light by measuring the amount of radiation transmitted by samples that are optically opaque at the wavelength to be assessed but that transmit radiation of other wavelengths. The instrument is set to zero and 100% transmittance in the normal way and the opaque substance introduced into the sample compartment. The amount of light transmitted by the sample, measured in percentage transmittance, is

➤ Stray light is the term given to unwanted radiation that is not removed by the monochromating system of a photometric instrument and is transmitted with the selected wavelengths.

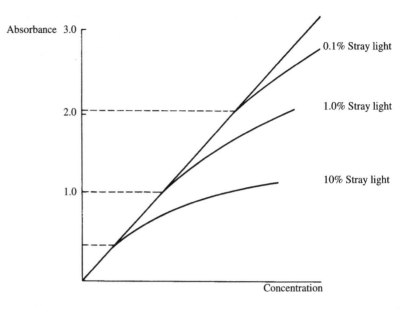

Figure 2.14 **Effect of stray light.** In the absence of stray light samples of increasing concentration result in a linear plot against the measured absorbance value. Increasing proportions of stray light result in deviations from this linear relationship above the absorbance values indicated (----).

Table 2.3 Stray light measurements for various instruments measured at 370 nm

Optical system	Stray light(%)
Single beam prism monochromator	<0.02
Single holographic monochromator	<0.02
Double holographic monochromator	<0.0002

quoted as the stray light at a specified wavelength. At 220 nm a solution of sodium iodide (10 g l^{-1}) transmits less than $1 \times 10^{-10}\%$ of incident radiation but transmits more than 50% at wavelengths greater than 265 nm. The stray light at 340 nm can be measured using a solution of sodium nitrite (50 g l^{-1}). Alternatively, various filters are available for use in a similar manner, a Vycor filter being opaque to radiation below 220 nm. Examples of stray light measurements for various instruments are given in Table 2.3. Errors are most significant near the wavelength limit of an instrument and may result in the recording of 'false' absorption peaks in addition to incorrect absorbance values.

2.2.3 Quantitative measurements

The measurement of the intensity of radiation is indirect, involving the generation and measurement of an electric current, and it is necessary to standardize instruments before test readings are made. The method of standardization is in principle the same for all instruments but does vary in practice depending upon the design of a particular instrument.

The basic procedure involves setting the minimum and maximum conditions of transmitted radiation and adjusting the metering system to give appropriate readings. To set maximum transmittance a blank sample is used and the instrument is adjusted to give either a reading of 100% transmittance or zero absorbance. Zero transmittance is set when all light to the detector is cut off using an opaque shutter and the meter is adjusted to give a transmittance reading of zero. The corresponding value for absorbance is infinity and because it is technically difficult to set to this value, methods of adjustment vary from instrument to instrument but are always described in the instrument manual.

Although the Beer–Lambert law describes the quantitative relationship between the absorption of radiation and the concentration of the absorbing substance, it does not specify the precise wavelength of radiation used, although we have noted that ideally the radiation should be monochromatic. In order to achieve maximum sensitivity from the method, the absorbance measurements for any given concentration should be as great as possible and preferably should be made at the absorption maximum. Examination of the absorption spectrum of adenosine diphosphate (ADP) (Figure 2.2) shows that at the absorption maximum at 258 nm an absorbance value of 0.22 is given while at 270 nm the absorbance value is only 0.14. In theory either wavelength would be suitable for the quantitation of ADP but obviously measurement made at 258 nm would offer greater sensitivity.

Absolute methods

It is possible to measure the absorbance of a sample of a known compound at its absorption maximum and to calculate the actual concentration of the compound in the sample using a known value for the molar absorption coefficient (often obtainable from published spectral tables). In Figure 2.2 the absorption spectrum of ADP shows an absorbance of 0.22 at 258 nm. The quoted value for the molar absorption coefficient of ADP at this wavelength is 1.54×10^4 l mol^{-1} cm^{-1} and hence the concentration of ADP in the sample used can be calculated from the Beer–Lambert equation:

$$A = \epsilon c l$$

Hence:

$$c = \frac{A}{\epsilon} = 1.42 \times 10^{-5} \text{ mol l}^{-1}$$

> ➤ Molar absorption coefficient is the mathematical constant in the Beer–Lambert equation and is the absorbance at a specified wavelength of a 1 mol l^{-1} solution of the analyte in a light path of 1 cm.

In practice, however, this is not always possible owing to a variety of reasons. If the sample is a mixture of several organic compounds, the measured absorbance value will be the cumulative effect of all the substances that absorb at the selected wavelength. In addition the precise value of the molar absorption coefficient depends to some extent on the particular instrument used and ideally values for the constant should be determined rather than accepted from the literature. The use of the molar absorption coefficient assumes that the Beer–Lambert relationship is valid over the range of absorbance values measured and, because this is not always true, it is essential that this relationship is always confirmed experimentally using the particular instrument in question.

Determination of the molar absorption coefficient
In order to determine the molar absorption coefficient, a pure, dry sample of the compound must be available. Purity is often difficult to check in a routine analytical laboratory but dryness may be achieved by desiccation. Care must be exercised in the choice of desiccant and the temperature used, owing to the potential instability of the compound at even ambient temperatures or the effect of light. It is advisable to determine the value for the coefficient using different samples of the compound and subsequently compare the results.

A small amount of the substance is accurately weighed, dissolved and made up carefully in a volumetric flask to a definite volume, e.g. 100 ml. From the known relative molecular mass (RMM) of the compound it is possible to calculate the molar concentration of the solution:

$$\text{Molar concentration} = \frac{\text{weight}}{\text{RMM}} \times \frac{1000}{\text{volume (ml)}} \text{ mol l}^{-1}$$

Concentrations in the region of 0.1 mol l^{-1} are often convenient but it obviously depends upon such factors as the amount of substance available, the cost, the solubility, etc. From this stock solution, a series of accurate dilutions are prepared using volumetric glassware and the absorbance of each dilution measured in a 1-cm cuvette at the wavelength of maximum absorbance for the compound. A plot of absorbance against concentration will give an indication of the validity of the Beer–Lambert relationship for the compound and a value for the molar absorption coefficient may be calculated from these individual measurements or from the slope of the linear portion of the graph:

$$\text{Molar absorption coefficient} = \frac{\text{absorbance}}{\text{molar concentration}} \text{ 1 mol}^{-1} \text{cm}^{-1}$$

Comparative methods
If the use of the molar absorption coefficient is inappropriate then it may be sufficient to use a single reference solution of known concentration (known as a standard or calibrator solution) and to compare the test absorbance with that of this standard. The principle of quantitation is exactly the same as before except that the molar absorption coefficient is eliminated from the calculation by measuring the absorbance of both the test and standard solutions at the same wavelength and comparing their absorbance values.

Procedure 2.1: Determination of the molar absorption coefficient of NADH at 340 nm

Method

A series of dilutions were made from an accurately prepared solution of NADH and the absorbance of each dilution was measured at 340 nm.

NADH (mmol l^{-1})	Absorbance (at 340 nm)
0.1	0.60
0.2	1.25
0.3	1.75
0.4	2.20
0.5	2.40

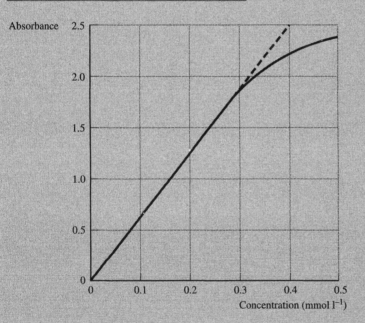

Calculation

The molar absorbance can be calculated from the slope of the straight line section of the graph:

$$\epsilon = \frac{0.95}{0.15} = 6.333$$

Result

Because the concentration is quoted in millimoles the molar absorption coefficient is 6.333×10^3 l mol^{-1} cm^{-1}.

The graph indicates that for the instrument used, the Beer–Lambert relationship for NADH is only valid up to a concentration of 0.3 mmol l^{-1} (an absorbance value of 1.75).

A standard solution of concentration c_s gives an absorbance value of A_s while the test solution gives an absorbance of A_t. Hence:

$$A_s = \epsilon c_s l$$

and

$$A_t = \epsilon c_t l$$

Therefore

$$\frac{A_s}{A_t} = \frac{\epsilon c_s l}{\epsilon c_t l} = \frac{c_s}{c_t}$$

and the concentration of the test

$$c_t = c_s \times \frac{A_t}{A_s}$$

The use of a single standard in this way assumes that the Beer–Lambert relationship is valid over the absorbance range measured and again it is necessary to confirm the relationship before using the method.

The analysis of a range of known concentrations of the test substance is necessary to validate the Beer–Lambert relationship. If the plot of absorbance values against concentration results in a straight line then either of the two previously outlined methods may be used. If, however, the resulting graph shows a curve instead of a straight line then the implication is that the actual value for the molar absorption coefficient is dependent to some extent upon the concentration of the compound and as a result invalidates both methods. In such circumstances a graphical plot (calibration curve) will be needed.

➤ Calibration – see Section 1.3.2.

Care must be taken in preparing the standard solutions because a pure preparation of the test compound will not be subject to the same interference effects of a crude or mixed test sample. To compensate for these differences it may be necessary to prepare the standard by adding a known amount of the pure component to the actual sample and using the resulting increase in absorbance as the absorbance value for the standard.

Analysis of mixtures

When there are several compounds which absorb radiation of the same wavelength, although not necessarily having the same absorption maxima, the absorbance value measured is the cumulative effect of all the absorbing compounds and no longer reflects the concentration of any one compound. The most frequent approach to such a problem involves the use of a chemical reagent to modify one of the compounds chemically and to produce a new compound with significantly different absorption characteristics. This technique is particularly useful if it results in a shift of the absorption maximum from the ultraviolet to the visible region of the spectrum (Figure 2.15).

Alternatively the absorbance of the mixture can be measured at different wavelengths (usually one for each compound present) and, provided that the molar absorption coefficients for each compound at each wavelength are known, the concentration of each compound can be calculated. Procedure 2.2 illustrates such a calculation.

Procedure 2.2: Spectrophotometric analysis of mixtures

A mixture of two compounds A and B gives an absorption spectrum which is the cumulative effect of both compounds. A knowledge of the molar absorption coefficients of each compound would permit the concentration of each in the mixture to be determined.

Data

Compound	Molar absorption coefficient ($l \, mol^{-1} \, cm^{-1}$)	
	at 250 nm	**at 350 nm**
A	120	30
B	15	40

Method
The absorbance of the mixture was measured at the same two wavelengths:

Absorbance at 250 nm   1.10

Absorbance at 350 nm   1.65

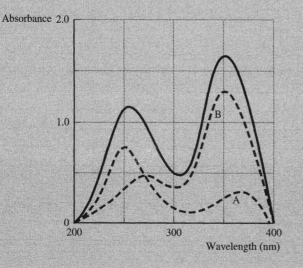

Calculation
Using the Beer–Lambert equation, $A = \epsilon c l$

At 250 nm $1.10 = 120 \, Conc_A + 15 \, Conc_B$ [1]

At 350 nm $1.65 = 30 \, Conc_A + 40 \, Conc_B$ [2]

In order to solve the two simultaneous equations, it is necessary to eliminate one of the unknown values. Hence if equation [2] is multiplied by four

$$6.60 = 120 \, Conc_A + 160 \, Conc_B \qquad [3]$$

Subtracting equation [1] from equation [3] gives

$$5.50 = 0 + 145 \, Conc_B$$

Therefore the concentration of compound B is

$$\frac{5.50}{145} = 0.038 \text{ mol l}^{-1}$$

Substituting in equation [1] gives

$$1.10 = 120 \text{ Conc}_A + 15 \times 0.038$$

Therefore the concentration of compound A is $0.0044 \text{ mol l}^{-1}$.

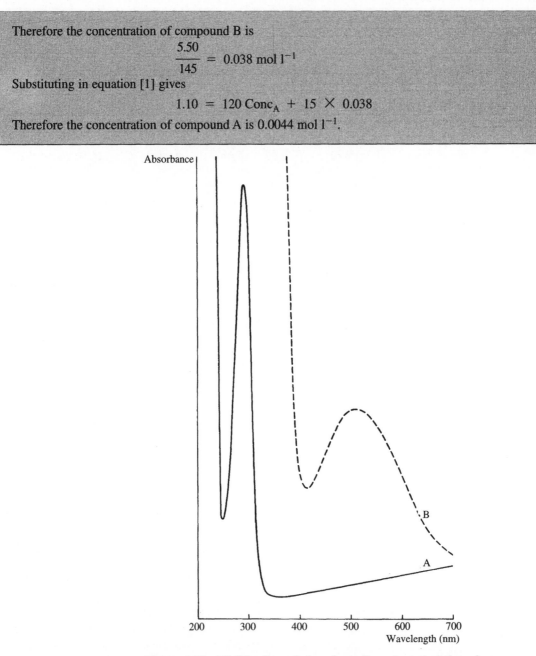

Figure 2.15 **Modification of the absorption characteristics of a compound.** Salicylate has an absorption maximum at 290 nm (A). Complexing with ferric nitrate in an acid results in a shift of the absorption maximum into the visible region at 510 nm (B) and provides the basis for a method of quantitation.

Difference spectroscopy

Some samples may contain different forms of the same substance which show very similar absorption spectra and hence make the detection or measurement of one rather difficult. This situation arises, for instance, in the measurement of the oxidized and reduced forms of cytochrome c and in the detection of various haemoglobin derivatives in blood. In these situations the use of difference spectroscopy may be useful.

Difference spectroscopy requires a blank solution which includes the interfering substance. This gives an absorption spectrum which is a plot of the difference in absorption between the blank and the sample at each wavelength. Compounds that are common to both blank and sample will give a zero difference in absorbance and any slight spectral differences in the sample will be shown as either peaks or troughs (Figure 2.16). The technique will demonstrate small differences in absorption characteristics but the sensitivity of the method is obviously significantly affected by the quality of the spectrophotometer.

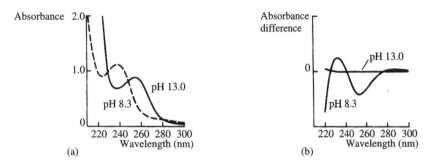

Figure 2.16   **Difference spectroscopy.** Phenobarbitone (a) shows different absorption maxima at pH 13.0 (keto form) and pH 8.3 (enol form).
A difference spectrum of two solutions of phenobarbitone with identical concentrations but at different pH (b) (pH 13.0 as the reference and pH 8.3 as the sample) has a positive peak at 232 nm and a negative peak at 254 nm.
The difference in absorbance between the peak and the trough may be used as a direct measure of the concentration even in the presence of many interfering substances.

Self test questions

Section 2.2

1. Which of the following statements are true about the absorption of radiation?
   (a) The amount of radiation absorbed is proportional to the incident radiation.
   (b) Absorbance is the reciprocal of transmittance.
   (c) The Beer-Lambert law is only valid at the absorption maximum.
   (d) Absorbance is dependent on the concentration of the analyte.
2. Calculate the molar concentration of a solution of compound X, which has an absorbance of 1.2 at 350 nm. The molar absorption coefficient of compound X is $0.5 \times 10^3 \, 1 \, mol^{-1} \, cm^{-1}$ at 350 nm. Which of the following is the correct answer?
   (a) $2.4 \, mmol \, 1^{-1}$.
   (b) $4.2 \, mmol \, 1^{-1}$.
   (c) $6.0 \, mmol \, 1^{-1}$.
   (d) $2.4 \, mol \, 1^{-1}$.

3. It is preferable to measure the absorbance of an analyte at its absorption maximum
   BECAUSE
   absorptiometric methods are most sensitive when used at the absorption maximum of the analyte.
4. Stray light may cause deviations from the Beer–Lambert relationship
   BECAUSE
   the wavelength range of the radiation incident on a sample does not affect the validity of the Beer–Lambert relationship.

2.3   Absorptiometer design

> Photometric instruments are designed to measure the intensity of radiation.

> Instruments prefixed by the term 'spectro' are capable of splitting radiation into a spectrum.

There is a wide range of instruments available for the measurement of absorbed radiation and it is important to appreciate the characteristics of the different instruments and the nature and quality of the results that can be obtained. A **photometer** is an instrument that is designed to measure the intensity of a beam of light and usually does so by comparing it with the intensity of some reference source of radiation. A **spectrometer** is an instrument that is capable of splitting the incident radiation into a spectrum and is used to identify the individual wavelengths that are present. A **spectrophotometer** is an instrument that combines both of these two functions. It is capable of producing radiation of defined wavelength characteristics from a mixed source of radiation and subsequently measuring the intensity of that radiation.

The basic design of a spectrophotometer is illustrated in Figure 2.17. Light from the lamp passes through a monochromating device and the selected wavelength, after passing through the sample, is measured by a suitable photoelectric detector. The intensity of the initial radiation is controlled either by an attenuator, which is often a simple shutter system, or by varying the operating voltage of the lamp. The geometry of the light path is controlled by a series of lenses or focusing mirrors. The degree of sophistication of the instrument (and hence its price) and the quality of the results depend upon the individual components and the optical design but the basic method of operation is similar for all instruments.

The simplest types of photometric instrument are designed for measurements in the visible region of the spectrum only and rely on coloured filters and simple photoelectric detectors. The name colorimeter is often used to describe such instruments although this is not necessarily correct and the word should probably be reserved for visual comparators rather than photoelectric instruments.

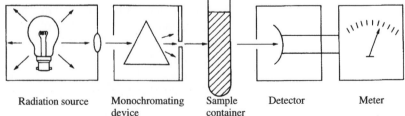

Figure 2.17 **Components of a spectrophotometer.**

| Radiation source | Monochromating device | Sample container | Detector | Meter |

2.3.1 Radiation sources

The tungsten lamp is a reliable and cheap source of radiant energy for use mainly in the visible region of the spectrum. It is not a satisfactory source of ultraviolet or infrared radiation owing to the nature of the emission and the absorbing characteristics of the glass envelope. Although it may be used in the near ultraviolet and infrared regions (330 nm to 3 μm) the large amount of visible light also emitted causes technical problems, particularly in the infrared region.

The usual sources of ultraviolet radiation are hydrogen or deuterium discharge lamps (the latter usually being preferred) or the mercury vapour lamp. All ultraviolet sources must be fitted with quartz or silica glass windows and none of the lamps named emits any significant amounts of radiation above 400 nm.

Black body radiators are used as sources of infrared radiation in the range 2–15 μm, e.g. the Nernst glower, which consists of a hollow rod made of the fused oxides of zirconium, yttrium and thorium. For use it is preheated and, when a voltage is applied, it emits intense continuous infrared radiation with very little visible radiation.

2.3.2 Monochromators

The ability to produce monochromatic radiation is a very desirable feature of photometric instruments because the Beer–Lambert relationship is only strictly true for monochromatic radiation. A good spectrophotometer may provide radiation of a specified wavelength with a range or bandwidth of as little as 0.1 nm, but it can be appreciated that even this is still not monochromatic when it is considered that the bandwidth of the sodium emission line is about 1×10^{-5} nm. This fact provides one of the major problems in the design of photometric instrumentation, namely the production of so-called monochromatic radiation.

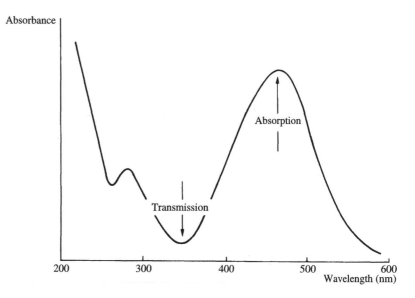

Figure 2.18
Absorption spectrum of methyl orange.

There are various devices designed to produce radiation of limited wavelength range, some being extremely effective while others transmit a considerable range of wavelengths.

Coloured filters

The absorption spectrum of a dye shows not only an absorption maximum in the visible region of the spectrum but also an absorption minimum which indicates the wavelengths of radiation that are transmitted by the dye (Figure 2.18). Such dyes, in the form of stained glass or gelatin filters, provide the simplest approach to the design of monochromating systems.

Filters obviously do not provide monochromatic radiation and in many cases the bandwidth may be as great as 100–150 nm (Figure 2.19). Increasing the concentration of the dye may reduce the bandwidth to some extent but will also reduce the amount of radiation transmitted and this then will impose a strain on the detecting system and the sensitivity of the instrument.

Interference filters

Interference filters modify the intensity of the radiation transmitted by the filter, enhancing the required wavelength and suppressing the unwanted wavelengths.

> ➤ Bandwidth refers to the range of wavelengths of radiation transmitted by a monochromator and is measured at half maximum transmittance.

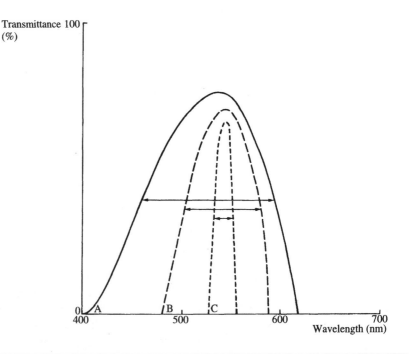

Figure 2.19
Transmission characteristics of various filters.

Spectrum	Filter	Colour or wavelength	Bandwidth (nm)
A	Glass OGRI	Light green	135
B	Glass 625	Medium green	68
C	Interference	540 nm	14

Bandwidth is defined as the range of wavelengths transmitted at half maximum transmittance.

They give transmission spectra with considerably reduced bandwidths (10–30 nm) compared with coloured filters (Figure 2.19) and can be used over the spectral range varying from the near ultraviolet (350 nm) to the near infrared (5 μm).

Interference filters consist of two pieces of glass each mirrored on one side in such a manner that half of the incident radiation is reflected and the remainder transmitted. They are separated by a thin layer of transparent material, often

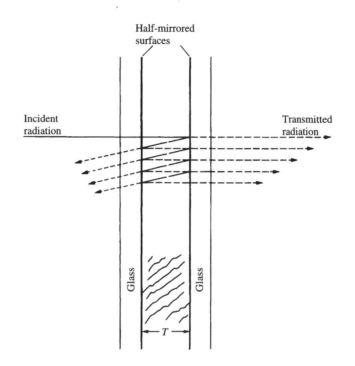

Figure 2.20
An interference filter.

calcium or magnesium fluoride, the thickness of which is the critical factor in determining the wavelength of the radiation transmitted by the filter. As a result of internal reflections, the beam of light that is incident upon the filter will emerge as a series of superimposed beams of light, each one retarded from the preceding beam by twice the thickness of the filter (Figure 2.20). The wavelength composition of the transmitted beam will be the same as the incident beam but those wavelengths for which the distance $2T$ is a whole multiple will be in phase in all transmitted beams. These wavelengths will be reinforced in the beam by constructive interference while all other wavelengths will be out of phase either completely or partially and will suffer destructive interference to a greater or lesser extent (Figure 2.21).

The refractive index of the material used to separate the two pieces of glass also affects the process and the wavelength reinforced will be that radiation which in the material used has a wavelength of $2T$. The wavelength of radiation in air which will be reinforced by such a filter will be:

$$\lambda = 2Tn$$

where n is the refractive index of the material and λ is the wavelength.

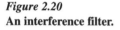

> Interference between beams of radiation can only occur with radiation of the same wavelength.

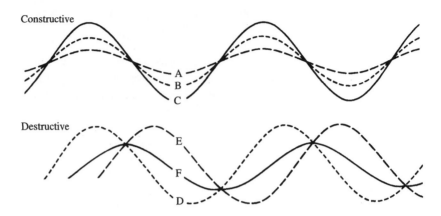

Figure 2.21 **Interference effects.** Combination of the waveforms (A) and (B) which, despite having different amplitudes, are of the same wavelength and in phase with each other, results in a constructive effect and the final waveform (C) has an increased amplitude. Combination of waveforms (D) and (E), which are out of phase with each other, results in destructive interference and the final waveform (F) has a reduced amplitude.

Because any wavelength for which $2Tn$ is a whole multiple will also be reinforced, these filters will also transmit second-, third- and fourth-order bands in which the wavelength transmitted will be one-half, one-third and one-quarter of the original. In some cases second- and third-order bands are preferred to the first-order band despite a decrease in the intensity of the transmitted radiation, because the bandwidth of higher order transmissions becomes progressively narrower. It is relatively simple to eliminate the unwanted bands of radiation using additional coloured filters because of the large differences in wavelength.

Wedge-shaped interference filters which have an increasing distance between the two pieces of glass produce an apparent spectrum due to the range of wavelengths constructively enhanced. These are extremely useful in the design of variable wavelength photometers but the range of radiation produced is not a true spectrum and there is still considerable overlap between wavelengths.

Prisms
The refractive index of a material is different for radiation of different wavelengths and hence if a parallel beam of polychromatic white light is passed through a prism the various wavelengths of which it is composed will be bent through different angles to produce a spectrum. The introduction of a variable slit into the optical light path will permit the selection of a particular section of the spectrum (Figure 2.22). Silica or quartz prisms are necessary for the ultraviolet region of the spectrum. While glass may be used in the visible region, rock salt prisms are necessary for work in the infrared.

Prisms are aligned so that the beam of light passing through the prism is parallel to its base with the result that the angle of deviation of the emerging beam is minimal and gives maximum resolution between the individual wavelengths. A Littrow mounting is designed so that the spectrum is focused on a

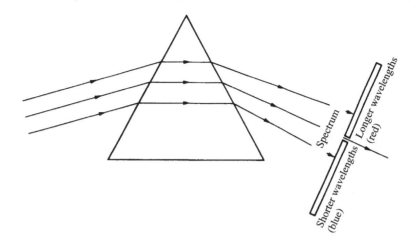

Figure 2.22
A prism monochromator.

plate which incorporates a slit and permits the position of the prism to be adjusted so that radiation from the selected section of the spectrum passes through the slit. The width of the slit is obviously a major factor in determining the bandwidth of the transmitted radiation but the geometry and refractive index of the prism also have a significant effect.

The dispersion of spectra produced by prisms varies depending upon the design of the prism and also on the region of the spectrum involved. Dispersion is less at shorter wavelengths and greater at longer wavelengths. Some typical values are:

> ➤ The dispersion of a monochromator is the linear spread of the spectrum which it produces and is expressed as nanometres per millimetre.

Wavelength	Linear dispersion
250 nm	0.5 nm per mm
450 nm	5.5 nm per mm
600 nm	12.0 nm per mm

This means that if a fixed slit width is used, the radiation transmitted will have different bandwidths depending upon the region of the spectrum. The mechanical efficiency of the mobile slit is therefore extremely important in the ability to select specific wavelengths.

Diffraction gratings
In its simplest form a diffraction grating consists of a series of parallel opaque lines drawn on a piece of glass. When the grating is held in a parallel beam of light each clear space acts as a source of radiation (Figure 2.23(a)). When viewed from any angle (θ) the overlapping radiation from all the slits is composed of all the wavelengths of the original light but the radiation from each slit is retarded or advanced relative to the radiation from an adjacent slit by the distance $d \sin \theta$, where d is the distance between each slit (Figure 2.23(b)). This means that only wavelengths for which $d \sin \theta$ is a whole multiple will be in phase with each other and will be constructively reinforced while all other wavelengths will suffer destructive interference to a greater or lesser extent. Hence the wavelength of radiation produced at an angle is

$$n\lambda = d \sin \theta$$

where n is a whole number.

At any specified angle the wavelengths that are whole multiples of the value $d \sin \theta$ will be enhanced and are known as first-, second- and third-order spectra. The unwanted wavelengths can be eliminated by the use of simple cut-off filters as previously described for interference filters. If a particular grating, for instance, has 1000 slits per millimetre and is viewed at an angle of 8.6°, the wavelengths of radiation enhanced will be 150 nm, 300 nm, 450 nm, 600 nm, etc.

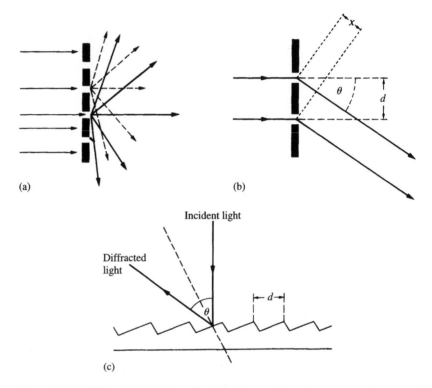

Figure 2.23   Diffraction gratings. Each line in a grating acts as a separate source of radiation (a) but radiation transmitted at any angle θ is retarded relative to radiation from the preceding line (b) by the distance x. In the transmitted radiation some wavelengths will undergo constructive interference while the majority will suffer destructive effects. Reflectance gratings (c) are frequently used and the principles of monochromation are the same as for transmission gratings.

Diffraction gratings are usually of the reflectance type rather than the transmission type (Figure 2.23(c)) mainly because of the loss of ultraviolet radiation that occurs during transmission through glass. Gratings are sometimes made by ruling a series of parallel grooves with a diamond in a layer of aluminium prepared on a glass blank, a process that is difficult because there may be as many as 2000 lines per millimetre. A holographic grating, however, is much simpler to produce and is made by coating a glass blank with a thin layer of photoresist which is then exposed to the parallel interference fringes produced when two beams of light from lasers intersect. The photoresist is developed in an etching solution to yield a surface pattern of parallel grooves and is subsequently coated with a thin layer of aluminium to produce a reflectance grating.

Because the dispersion is due to the geometry of the grating and is not a property of the material as it is with prisms, gratings may be produced with optimal characteristics for specific spectral ranges. Such spectra show an equal distribution of wavelengths compared with those produced by prisms and this not only simplifies instrument design but also means that values for linear dispersion hold true for the whole wavelength range of a grating, often being as low as 1 nm per mm. The efficiency of monochromation can be considerably improved by the use of a double monochromator, in which the selected part of a spectrum from the first grating can be further resolved by a second grating, resulting in bandwidths as low as 0.1 nm without excessive reduction in the intensity of radiation.

2.3.3   Detectors

> Photoelectric detectors produce a current which is proportional to the intensity of the light falling on them.

The materials and design of the various photoelectric detectors available are such that the absorption of radiation results in the displacement of electrons and hence in the development of a potential difference between two electrodes. The main types of photoelectric detectors may be classified as either photovoltaic or photoconductive (Figure 2.24).

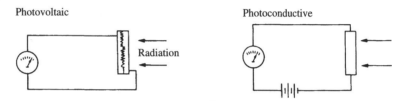

Figure 2.24   **Photoelectric detectors.** Photovoltaic detectors measure the flow of electrons displaced by the absorption of radiation. Photoconductive detectors measure the changes in conductivity caused by the absorption of radiation.

Ultraviolet and visible region of the spectrum

Photovoltaic devices are the most useful in this region of the spectrum and the barrier layer or solar cell (Figure 2.25) may be used in simple photometers. These cells are only effective in the visible region of the spectrum and while giving a linear response to the intensity of the radiation over a limited range, they show a fatigue effect in which the signal decreases as the radiation continues to fall on the cell. The current generated is very small and is difficult to amplify because of the low internal resistance of the cell. As a result such cells are not very sensitive.

Developments in photoelectric cells have resulted in the production of photodiodes. These in principle are very similar to the barrier layer cell described above but the selenium is replaced by silicon products. Silicon is a semiconductor and with traces of boron or gallium added the trace element can accept electrons displaced from the silicon leaving it positively charged (p-silicon). Conversely, trace amounts of arsenic will act as electron donors, giving the silicon a negative charge (n-silicon). The displacement of electrons from silicon can be caused by the absorption of radiation and if the diode is designed in much the same manner as the barrier layer cell, with p-silicon as the face plate and n-silicon as the back plate, the absorption of radiation will result in a small current flowing between the plates of silicon.

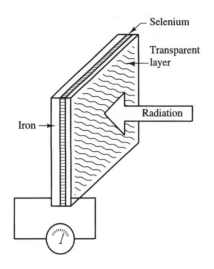

Figure 2.25 **Barrier layer cell.** Radiation absorbed by the semiconductor, often selenium, causes electrons to be released and a small current flows which can be measured using a microammeter.

➤ Diode arrays – see Section 3.2.2.

Photodiodes can be produced as very small, solid-state devices which are effective over the wavelength range 200–1000 nm and the signals produced are amenable to computer handling. They are frequently produced in units in which a large number of diodes are arranged side by side, giving what is known as a diode array and as such can be used in spectrophotometers and monitoring systems for chromatography.

Photoemissive tubes are necessary for work in the ultraviolet range and they show greater sensitivity and precision than photoelectric cells. A simple photoemissive tube consists of two electrodes in a vacuum. A silver cathode coated with an alkali metal is maintained at a potential difference of about 100 V from the anode, which is a plain silver wire and serves to collect the electrons (Figure 2.26(a)).

A photomultiplier consists of a series of electrodes (usually ten), the first one being a photoemissive electrode and the remainder being electron multiplier electrodes (Figure 2.26(b)). Radiation falling on the first electrode results in the release of electrons which are collected by the next electrode, which is held at a slightly more positive charge. These electrons stimulate the release of more electrons from this electrode, increasing the electron flow by a factor of four or five. The successive electrodes are operated at voltages increasing in steps of about 100 V and the overall device requires a very stable high voltage power supply. The resulting amplification of the initial current is several million fold, making the photomultiplier tube a very sensitive photoelectric device.

The spectral response of photoemissive tubes depends upon the composition of the cathode and the use of various mixtures of elements permits the production of a wide range of tubes of varying responsiveness (Figure 2.27). Photomultiplier tubes can be used to detect low intensity radiation and even in the absence of any light will still generate a small current due to various emissions from the material of the tube, etc. This 'dark current' has to be compensated for in any measurements that are made.

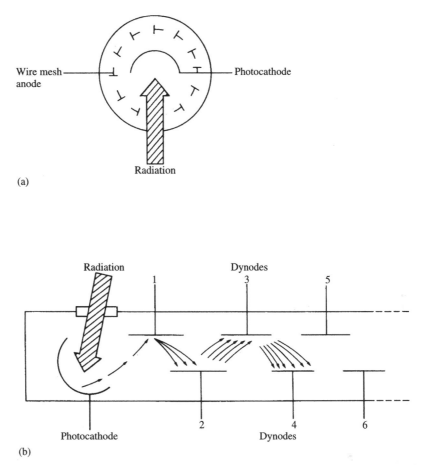

Figure 2.26 **Photoemissive tubes.** Light enters a simple phototube (a) and causes the release of electrons from the photoemissive alloy of the cathode. Owing to the potential difference between the anode and the cathode, the electrons are captured by the anode and the resulting current can be amplified and measured. Photomultiplier tubes (b) are a development of simple phototubes and result in internal amplification of the current initially developed at the photocathode.

Infrared region of the spectrum

Photoconductive detectors, such as lead sulphide, are useful in the spectral region of 800–4000 nm (i.e. the near infrared). They consist of a thin layer of a semi-conductor deposited on glass or another insulating material and protected against atmospheric damage by either a vacuum or a thin film of lacquer. The absorption of radiation increases the conductivity of the layer by displacing electrons. If a constant potential difference is applied across the layer, the current flowing will increase significantly when radiation is absorbed.

Detection of the middle and far range of infrared radiation requires thermal detectors, the simplest of which is a thermocouple, in which the change in temperature at one junction of the thermocouple results in a small voltage being produced. Although simple in design, thermocouples lack sensitivity. Bolometers are more sensitive and are based on the fact that as the temperature of a conductor varies so does its resistance. A suitable conductor (nickel plate or a thermistor) is incorporated in a Wheatstone bridge circuit through which a steady current is passed. When radiation strikes the sensing element, the electrical resistance changes and a corresponding change in the current can be measured.

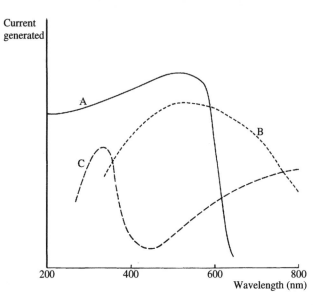

Figure 2.27 **Spectral response of cathode elements in photoemissive tubes.** Various combinations of elements respond in different ways to specific wavelength ranges and are only suitable for the specified spectral regions: calcium–antimony (A); sodium–potassium–caesium–antimony (B); silver–caesium (C).

2.3.4   Optical materials

Normal glass will only transmit radiation between about 350 nm and 3 μm and, as a result, its use is restricted to the visible and near infrared regions of the spectrum. Materials suitable for the ultraviolet region include quartz and fused silica (Figure 2.28). The choice of materials for use in the infrared region presents some problems and most are alkali metal halides or alkaline earth metal halides, which are soft and susceptible to attack by water, e.g. rock salt and potassium bromide. Samples are often dissolved in suitable organic solvents, e.g. carbon tetrachloride or carbon disulphide, but when this is not possible or convenient, a mixture of the solid sample with potassium bromide is prepared and pressed into a disc-shaped pellet which is placed in the light path.

2.3.5   Optical systems

A basic photometer (Figure 2.17) is a single-beam instrument in which there is only one light path and the instrument is standardized using a blank solution, which is replaced with the sample to obtain a reading. The major problem with a single-beam design lies in the fact that the absorbance value recorded for the

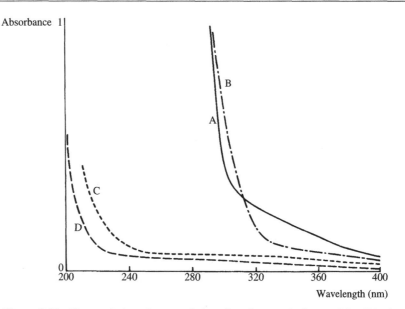

Figure 2.28. **Transmission characteristics of various optical materials:** (A) glass, (B) plastic, (C) silica glass, (D) quartz.

sample may be a result of other factors as well as the absorbing properties of the sample. Fluctuations in the intensity of the light source or instability in the detection system will result in errors, while the fact that the absorption of the blank will vary at different wavelengths means that the setting-up procedure must be undertaken every time the wavelength is changed. In order to mini-mize these potential errors and to permit automation or the measurement of absorbance at varying wavelengths it is necessary to introduce a constant ref-erence signal by using a double-beam optical system (Figure 2.29).

► Double-beam instruments involve two light paths, one to measure the sample and the other a reference or blank.

In a double-beam instrument the radiation is split into two beams, one passing to the detector through the sample and one through the blank. A series of half-mirrors may be used, permitting both light paths to function simultane-ously or alternatively the beam may be 'chopped' mechanically, deflecting a beam of radiation alternately along each light path. The latter method is the most frequently used because the maximum intensity of the incident beam is maintained and it is usually accomplished by means of a rotating sector mirror. The detector receives alternate pulses of radiation which have passed through the blank and the sample and it produces alternating electrical signals which are directly proportional to the intensities of the radiation transmitted by both blank and sample. In a ratio-recording photometer system this signal is resolved elec-tronically into two voltages corresponding to the blank and sample signals and the former is used as the standardizing potential for a recording potentiometer. In an optical null balancing photometer, the alternating current is used to power a servo motor, which drives an attenuator into the blank light path until the intensity of the beam is exactly the same as the sample beam. The attenuator is designed so that the distance moved is proportional to the absorbance of the sample and is usually mechanically coupled to a pen recorder system.

In some situations measurement of the reflected, rather than the transmitted, radiation may be made to assess the amount of radiation that has been absorbed by the sample. There are two main ways by which radiation might be reflected. Specular reflection is similar to the reflection by a mirror and, for quantitative work, the angles of the incident and the reflected radiation are important. Diffuse reflection is from within the layers of the material and the reflected light is disbursed over a range of 180°. This type of reflection is measured in the thin films used in dry chemistry systems. The term reflectance density is often used, which is defined in a manner comparable to absorbance: the logarithm of the ratio of incident to reflected light.

➤ Reflectance photometry – see Section 6.1.2.
➤ Dry chemistry analyser – see Section 6.1.2.

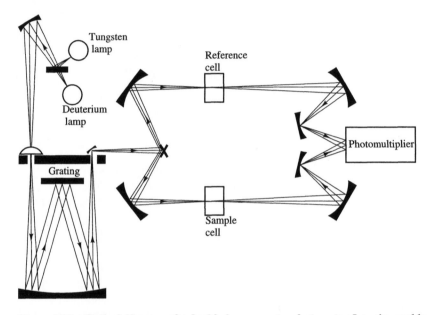

Figure 2.29 **Optical diagram of a double-beam spectrophotometer.** Interchangeable lamps are available for work in the visible and ultraviolet regions and monochromation is achieved with a reflectance diffraction grating.

The components of the instrument are similar to absorptiometry but their relative positions are changed in order to catch the reflected radiation. In most of these instruments, rather than incorporating a monochromator, a light-emitting diode (LED) of a specific wavelength is used to produce the incident radiation.

The principles of absorptiometry have been applied to the measurement of turbidity. Suspensions of particles scatter incident radiation and, while there is no absorption of radiation by the analyte, the reduction in the transmitted radiation can be used as a measure of the degree of turbidity. Because absorption is not involved, there is no requirement for monochromation but the fact that the extent of light scattering increases as the wavelength of the incident radiation decreases explains the fact that some instruments do incorporate a simple monochromation system.

The sensitivity of the technique is considerably improved by measuring the amount of light scattered rather than the light transmitted. In such instruments, known as nephelometers, the optical system is more similar to reflectance systems or fluorimeters in order to detect the scattered light rather than the transmitted light. The use of lasers appreciably enhances the performance of these techniques by allowing much more critical design of instrumental light paths and reducing the unwanted reflections from cuvettes, etc., which reduce the sensitivity of simpler instruments.

➤ Nephelometry – see Section 7.3.1.

Self test questions

Section 2.3
1. Which of the following factors influence the wavelength of radiation selected when using a reflectance diffraction grating?
   (a) The distance between the lines.
   (b) The angle between the incident and reflected radiation.
   (c) The material of the grating.
   (d) The thickness of the grating.
2. Which of the following materials are suitable as sample holders for ultraviolet absorptiometry?
   (a) Plastic.
   (b) Glass.
   (c) Silica glass.
   (d) Rock salt.
3. A reflectance grating is the most suitable monochromation system for ultraviolet radiation
   BECAUSE
   a reflectance grating will have a narrower bandwidth than an interference filter.
4. Infrared spectrophotometers use photovoltaic-type detectors to measure the intensity of radiation
   BECAUSE
   electrons are displaced in a photovoltaic cell when radiation is absorbed.

2.4   Molecular fluorescence techniques

Molecular fluorescence involves the emission of radiation as excited electrons return to the ground state. The wavelengths of the radiation emitted are different from those absorbed and are useful in the identification of a molecule. The intensity of the emitted radiation can be used in quantitative methods and the wavelength of maximum emission can be used qualitatively. A considerable number of compounds demonstrate fluorescence and it provides the basis of a very sensitive method of quantitation. Fluorescent compounds often contain multiple conjugated bond systems with the associated delocalized pi electrons, and the presence of electron-donating groups, such as amine and hydroxyl, increase the possibility of fluorescence. Most molecules that fluoresce have rigid, planar structures.

2.4.1   Instrument design

Fluorescent radiation is emitted in all directions and use is made of this fact to avoid difficulties that may be caused by the transmission of incident radiation by the sample. By moving the detection system at right angles to the cell (Figure 2.30) only fluorescent radiation is detected. Some instruments are designed to measure 'front face' fluorescence, i.e. the radiation that is emitted along a light path at an acute angle to the incident radiation. Such fluorescence measurements are comparable to reflectance measurements.

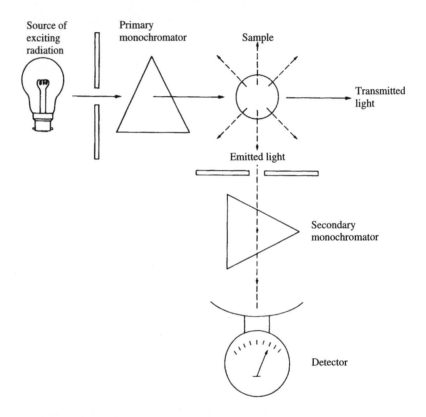

Figure 2.30
Fluorimeter design.

Despite the measurement of the emitted radiation by these means it is still possible for scattered or reflected incident radiation to reach the detector. To prevent this, fluorimeters require a second monochromating system between the sample and the detector. Many simple fluorimeters use filters as both primary and secondary monochromators but those instruments that use true optical monochromators for both components are known as spectrofluorimeters. Other instruments incorporate a simple cut-off filter system for the emitted radiation while retaining the optical monochromator for the excitation radiation. Because the wavelengths of both excitation and emission are characteristic of the molecule, it is debatable which monochromator is the most important in the design of a fluorimeter.

2.4.2 Fluorimetric methods

In order to select the operating conditions for any fluorimetric method, the excitation and emission spectra of the analyte must be determined. Figure 2.9 illustrates the fact that an emission spectrum is an approximate mirror image of the excitation spectrum, the latter being similar to the absorption spectrum of the compound.

The main advantage of fluorescence techniques is their sensitivity and measurements of nanogram (10^{-9} g) quantities are often possible. The reason for the increased sensitivity of fluorimetry over that of molecular absorption spectrophotometry lies in the fact that fluorescence measurements use a non-fluorescent blank solution, which gives a zero or minimal signal from the detector. Absorbance measurements, on the other hand, demand a blank solution which transmits most of the incident radiation and results in a large response from the detector. The sensitivity of fluorimetric measurements can be increased by using a detector that will accurately measure very small amounts of radiation.

Fluorescence measurements are quoted relative to a standard maximum intensity of fluorescence and it is usual to standardize the instrument on zero fluorescence (blank) and 100% fluorescence (highest standard). The intensity of fluorescence of all intermediate standards and test samples is then measured and expressed as a percentage of the highest standard. Over a limited concentration range, the intensity of fluorescence is directly proportional to the concentration of the compounds but in practice it is advisable to produce a calibration curve for each assay. Fluorimetric techniques often demonstrate a greater degree of specificity than comparable absorption techniques for several reasons. Many compounds that absorb in the ultraviolet do not fluoresce and the presence of such substances would possibly interfere in the absorptiometric analysis of a sample but would not interfere in a fluorimetric assay. The converse is also true that the presence of a fluorescing substance in an absorptiometric assay would reduce the apparent absorbance by the amount of the fluorescence which reached the detector. This is a problem in spectrophotometers in which the monochromator is in front of the sample but in practice many instruments position the monochromator after the sample and before the detector. Probably the main reason for the improved specificity of fluorimetric techniques is the ability to identify and select the emission wavelength and this is particularly significant in situations where compounds absorb at the same wavelength but show different emission spectra.

A major difficulty in fluorimetry is the fact that a wide range of substances and conditions can suppress or quench the emission of fluorescent radiation. Many excited molecules tend to lose their energy by molecular collisions rather than by fluorescence and the proportion doing so increases as the temperature rises. Frozen samples often show considerably more intense emission than do liquid samples and the control of temperature during measurements is an important consideration. Similarly a decrease in the pH of the sample often reduces the intensity of fluorescence owing to the binding of protons by non-bonding electrons. Some compounds, particularly those containing electrophilic groups (e.g. carboxylic, azo and halides), can quench fluorescence, and their presence even in trace amounts can significantly reduce the intensity of emission. Because of this all equipment should be scrupulously clean and all reagents should be of the highest purity.

2.5   Atomic spectroscopy techniques

The atoms of certain metals when heated emit radiation and methods of analysis have been developed which use the wavelength of emission for qualitative analysis and the intensity of emission in quantitative work (Table 2.4). Emission spectroscopy is most frequently encountered as flame emission photometry, which is almost entirely restricted to the visible region of the spectrum and to those elements that are easily excited at the temperature of a flame. Alternatively, excitation may be achieved by means of a high voltage arc struck between two electrodes. More recently, temperatures as high as 10000 °C, produced by inductively coupled plasma (ICP) discharges, have been used to cause the excitation of atoms and offer improved levels of sensitivity. Atomic fluorescence is a variant of flame photometry in which atoms are excited by the absorption of radiation rather than thermal energy. The instruments used for such measurements require optical systems similar to those used in molecular fluorimetry. However, despite the great potential increase in sensitivity of such a method, it is limited to a few elements for which the necessary intense excitation sources of radiation are available.

Table 2.4   Emission lines of various elements

Element	Wavelength most suitable for:		Other lines	Flames suitable for:	
	Atomic absorption	Flame emission		Atomic absorption	Flame emission
Calcium	**422.7**	**422.7**	239	A–Ac	N–Ac
Copper	**324.7**	324.7	216, 222, 249, etc.	A–Ac	N–Ac
Iron	**248.3**	372.0	248, 252, 373	A–Ac	N–Ac
Lanthanum	550.1	**579.1**	418, 495, etc.	N–Ac	N–Ac
Lead	**283.3**	405.8	217, 261, 368	A–Ac	N–Ac
Lithium	670.8	**670.8**	323, 610	A–Ac	N–Ac
Magnesium	**285.2**	285.2	202	A–Ac	N–Ac
Mercury	**253.6**	253.6	—	A–Ac	N–Ac
Phosphorus	**213.6**	526.0	213, 214	N–Ac	A–H
Potassium	766.5	**766.5**	769, 404	A–Ac	A–Ac
Sodium	589.0	**589.0**	589.6, 330	A–Ac	A–Ac
Zinc	**213.9**	213.9	307	A–Ac	N–Ac

The emission line in bold type indicates the method of choice.
Key to gases: A, air; Ac. acetylene; N. nitrous oxide; H. hydrogen.

Whereas flame emission photometry relies on the excitation of atoms and the subsequent emission of radiation, atomic absorption spectrophotometry relies on the absorption of radiation by non-excited atoms. Because the proportion of the latter is considerably greater than that of the excited atoms, the potential sensitivity of the technique is also much greater.

2.5.1 Flame emission photometry

A flame photometer (Figure 2.31) is designed to cause atomic excitation of the analyte and subsequently to measure the intensity of the emitted radiation. A monochromating system is essential to distinguish between the emission of the test element and other radiation from the flame.

The **flame** combines both the source of radiation and the atomized sample and hence must be very stable if steady readings are to be obtained. The flame temperature must be high enough to excite the atoms under investigation; the hotter the flame, the greater the proportion that will be excited (Table 2.5). If it is too hot, however, the atoms may be raised to higher energy levels and electrons may be removed altogether, resulting in ionization.

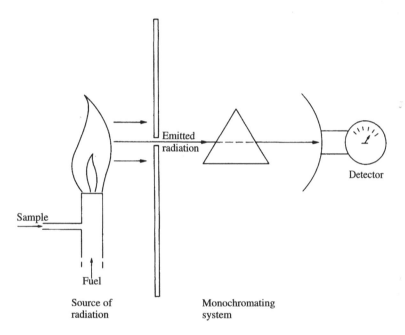

Figure 2.31 **Components of a flame photometer.**

> ➤ An atomizer is a device designed to produce an aerosol, i.e. a very fine liquid spray.

The sample is converted into an aerosol in an **atomizer**. It then passes through an expansion chamber to allow a fall in the gas pressure and the larger droplets to settle out before passing to the burner, where the solvent evaporates instantly, the atoms remaining as a finely distributed gas. Atoms in the sample that are bound in molecules should be decomposed at the flame temperature so rapidly that the same effect is achieved. In practice only a small proportion of the sample (approximately 5%) is effectively atomized because the drop size of the remaining 95% is so large that the water is never effectively stripped away. In low temperature flames, for instance, only one sodium atom in about 60000 is excited but despite this apparently low efficiency the technique is very sensitive.

Any of the **monochromating systems** described for absorptiometers may be used, although the cheaper models of flame photometer usually employ filter systems. In these cases interference from other elements at wavelengths near to the test wavelength may be a real problem but owing to the intensity of emission it is possible to reduce the bandwidth by using very dense filters. In many instruments, the use of interference filters improves the specificity of the analysis.

Table 2.5   Fuels for flame spectroscopic techniques

Fuel	Oxidant	Remarks	Temperature (°C)
Propane Butane	Air Air	Low temperature flames suitable for easily excited atoms, such as Na, Li, K and Ca	1900
Acetylene	Air	Medium temperature flame suitable for most emission analyses, e.g. Mg, Mn and Sr	2300
Acetylene	Nitrous oxide	High temperature flame necessary for the more refractile elements, e.g. P	3000

➤ The effect of overall composition of the sample on the analysis of one component is known as the matrix effect.

The presence of high concentrations of other elements in the sample might not only stress the monochromating system but also cause suppression of the emission by the atoms under investigation, a phenomenon known as the matrix effect. In such situations it is often necessary to prepare standard solutions that contain equivalent amounts of the interfering elements. Similar effects can also result from significant differences in the viscosity or density of the sample and standards, often caused by the presence of protein. Various anions, such as phosphate and aluminate, may complex the cations being measured and give falsely lowered readings. This can be countered by using organic chelating agents, e.g. EDTA or citric acid, which, although binding the cations themselves, are readily decomposed in the flame compared with the very stable phosphates and aluminates. The use of lanthanum chloride and, less effectively, strontium chloride, is perhaps the most convenient method of overcoming the effect of such anionic interference. These cations bind the interfering anions and are possibly precipitated when the aerosol is formed, leaving the test cations available for excitation. In the absence of such protecting cations it is possible to decompose the phosphate and aluminate complexes thermally by using flames of higher temperature but care has to be exercised using this approach as unwanted ionization may occur. Such anionic interferences are often relevant to atomic absorption spectroscopy.

Quantitative methods using flame emission photometry cannot be absolute because an unknown, although relatively constant, proportion of the sample will reach the flame of which only a further small proportion of atoms will actually be excited and subsequently emit radiation. Hence it is essential to construct calibration curves for any analysis. The radiation emitted by the flame when pure solvent is sprayed is used to zero the instrument and the maximum reading set when the standard with the highest concentration is sprayed.

➤ Calibration – see Section 1.3.2.

A range of standard solutions is then sprayed and a calibration graph is drawn. As with fluorescence measurements, the intensity of emission is reported as a percentage of the emission of the highest standard solution, and over a limited concentration range there is a linear relationship between percentage emission and the concentration of the atoms (Figure 2.32).

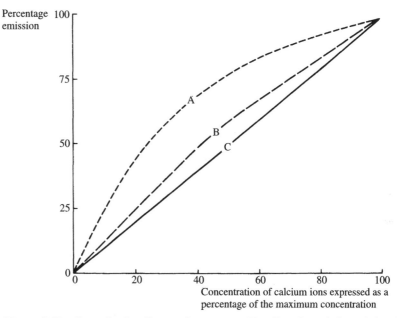

Figure 2.32 **Quantitative flame photometry.** The linearity of the relationship between the concentration and the percentage emission becomes better as the concentration range becomes smaller. The maximum concentrations for the calibration curves are (A) 25 mmol l^{-1}, (B) 2.5 mmol l^{-1} and (C) 0.25 mmol l^{-1}.

In addition to the emission due to the test element, radiation is also emitted by the flame itself. This background emission, together with turbulence in the flame, results in fluctuations of the signal and prevents the use of very sensitive detectors. The problem may be appreciably reduced by the introduction into the sample of a constant amount of a reference element and the use of a dual-channel flame photometer, which is capable of recording both the test and reference readings simultaneously. The ratio of the intensity of emission of the test element to that of the reference element should be unaffected by flame fluctuations and a calibration line using this ratio for different concentrations of the test element is the basis of the quantitative method. Lithium salts are frequently used as the reference element in the analysis of biological samples.

2.5.2 Plasma emission spectroscopy

Atoms can be excited using the high energy levels associated with inductively coupled plasma (ICP) instead of a flame. Such a method of excitation is far more effective and permits the analysis of elements beyond the scope of simple flame emission techniques, such as the refractory elements of boron, phosphorus and tungsten. The high temperature eliminates many of the interference effects and the instrumentation is designed with a series of photomultiplier tubes set for the different emission wavelengths of specific elements, permitting multi-element analysis of samples. The method is very sensitive and specific.

The ICP discharge is caused by the effect of a radio-frequency field on argon gas flowing through a quartz tube. The high power frequency causes a changing magnetic field in the gas and this in turn results in a heating effect. Temperatures of 9000–10000 °C can be produced in this way.

2.5.3 Atomic absorption spectrophotometry

When atoms are dispersed in a flame, the vast majority remain in the ground state and only a very small proportion are thermally excited. If a beam of polychromatic radiation is passed through such a flame, atoms in the ground state will absorb the appropriate wavelength and, for each element, a series of absorption bands may be demonstrated in the transmitted radiation. These absorption bands will correspond to the excitation of the valence electrons and although the energy required for such excitation varies considerably from one element to another, the energy levels for a single element are very limited and result in very narrow absorption bands (e.g. 0.001 nm). For atoms of metals or semi-metals, the energy can be supplied by radiation in the range of 200–900 nm but for all non-conducting elements (insulators) the energy required is very large and radiation of the far ultraviolet and X-ray regions would be required.

The proportion of the incident radiation which is absorbed by the atoms in the flame (or vapour) is measured and related to the number of atoms in the flame (Figure 2.33) in a manner directly comparable to molecular absorption spectrophotometry.

Instrument design
The absorption bands due to atoms are very narrow and the use of white light as the incident radiation would swamp even the best monochromating system with unabsorbed radiation on either side of the absorption band. It is fundamental, therefore, to the technique of atomic absorption spectrophotometry

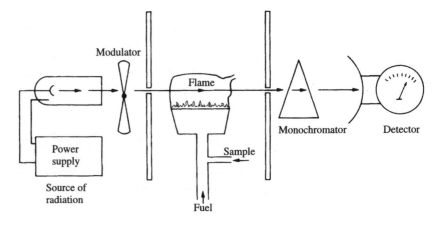

Figure 2.33 **Components of an atomic absorption spectrophotometer.**

that the incident radiation is of the correct wavelength, bandwidth and intensity. This radiation is produced by a lamp in which the cathode is coated with atoms of the element under investigation and which emits radiation of precisely the same wavelength as that which will be absorbed by non-excited atoms of the same element in the flame.

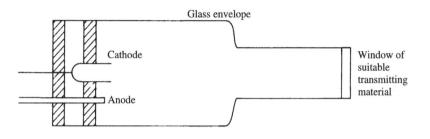

Figure 2.34
A hollow cathode lamp.

> ➤ Hollow cathode lamps produce monochromatic radiation for atomic absorption spectroscopy.

A **hollow cathode lamp** (Figure 2.34) consists of two electrodes sealed in a glass envelope filled with an inert gas, usually argon or neon. The end window of the lamp must be of an appropriate material in order to transmit the emitted radiation and is either quartz or silica. The cathode of the lamp, usually cup shaped, is either made of the element whose spectrum is required or coated with the element, and the application of a potential of between 300 and 400 V is usually required to cause excitation of the atoms and discharge of the appropriate radiation. Lamps are available with cathodes which contain two or more elements with emission lines that are easily distinguishable. These are used for the analysis of several elements without changing lamps, although such multi-element lamps do tend to give less satisfactory performance than single-element lamps. For certain elements, electrodeless discharge lamps (EDLs) have been designed in which the excitation of atoms is achieved by radio frequencies that induce resonance effects and the energy liberated causes vaporization and excitation of the element.

> ➤ Electrodeless discharge lamps (EDLs) emit radiation as a result of radio frequencies providing the exciting energy.

Double-beam atomic absorption spectrophotometers are designed to control variations which may occur in the radiation source but they are not as effective as double-beam molecular absorption instruments in reducing variation because there is no blank sample in flame techniques.

The **flame**, as well as containing the unexcited atoms of the element, will also emit radiation due to the thermal excitation of a small proportion of atoms, and it is essential that the detector is capable of distinguishing between the identical radiation that is transmitted by the flame and that emitted from the flame. This is achieved by introducing a characteristic signal or modulation into the incident radiation by means of a rotating segmented mirror (chopper) or an electrically induced pulse. This pulsed beam is detected as an alternating signal, which is superimposed on the relatively constant signal generated by the emission from the flame. The difference between the two signals is automatically measured by the instrument and only the light emitted from the flame is recorded.

The design of the burner head and the method of atomization of the sample both influence the sensitivity which can be achieved. The burner head is designed to give a long narrow flame so that as many atoms as possible are presented in the light path. It needs to be kept spotlessly clean to minimize background emission and the position in the light path has to be adjusted to give maximum sensitivity, the precise position depending not only on the gases being burnt but also on their flow rate. The proportion of fuel to oxidant alters the characteristics of the flame, a high proportion of fuel resulting in a flame with reducing properties while an excess of oxidant gives an oxidizing flame. For each element analysed the optimum proportions must be determined.

The basic design of an atomizer is the same as that for flame emission spectroscopy (Figure 2.35). The method of producing an aerosol involves spraying the sample in air or oxidant gas. The larger drops precipitate on the baffles of the

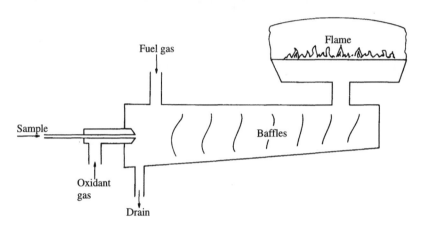

Figure 2.35
A spray atomizer.

expansion chamber and flow to waste. The fuel gas is introduced and the components (fuel, oxidant and aerosol sample) are mixed before passing to the burner.

High quality **monochromating systems** are necessary to isolate the required emission line of the element from those emission lines due to the gases that are also present in the lamp. Owing to the very narrow bandwidth of atomic emission lines, it is not adequate simply to select the required wavelength using the monochromator scale, and a procedure known as 'peaking up' has to be undertaken. The wavelength of the emission line required can be obtained from the literature and the monochromator is initially set to this value. With the lamp operating, the monochromator is adjusted to give maximum response from the detector. In some instances, owing to the presence of interfering elements, it may be necessary to use an emission line other than the resonance line for the test element, despite the loss in sensitivity which this causes. As distinct from molecular absorption spectrophotometry, the onus for specificity in atomic absorption spectrophotometry lies with the radiation source rather than with the monochromating system.

Conventional flame techniques present problems when dealing with either small or solid samples and in order to overcome these problems the **electrothermal atomization** technique was developed. Electrothermal, or flameless, atomizers are electrically heated devices which produce an atomic vapour (Figure 2.36). One type of cuvette consists of a graphite tube which has a small injection port drilled in the top surface. The tube is held between electrodes, which supply the current for heating and are also water-cooled to return the tube rapidly to an ambient temperature after atomization.

➤ A resonance line is the wavelength of the most intense emission of an element.
➤ Electrothermal atomizers produce an atomic vapour by rapid heating of the sample.

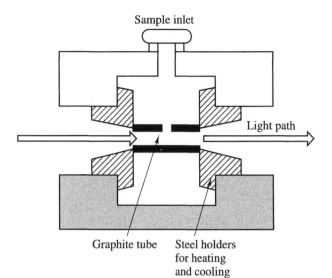

Sample inlet

Light path

Figure 2.36
An electrothermal atomizer.

Graphite tube   Steel holders
for heating
and cooling

The sample is introduced into the cuvette either using a micropipette or as a weighed portion. It is obvious that these manipulations are a potential source of error and considerable care must be taken at this stage. The first step in the analysis is the heating of the sample (ashing) to remove the solvents and to destroy the matrix. The presence of any organic material in the sample will produce smoke or carbon particles and interfere with any absorbance measurements being made at the same time. Care must be taken in the choice of the ashing temperature to ensure that the sample matrix is destroyed and the resulting smoke eliminated without the loss of any of the test element before the analytical stage. This is particularly important with the more volatile elements such as mercury and lead.

When ashing is complete, the temperature of the cuvette is rapidly raised to the atomizing temperature, which is usually about 1000°C higher than the ashing temperature. This causes vaporization of the sample and a resulting transient increase (2–5 seconds) in the absorbance. Atomic absorption spectrophotometers must be able to detect and record this maximum absorbance value (Figure 2.37). The cuvette is cleaned after the analysis by a further increase in temperature of several hundred degrees. When samples contain a large amount of protein or other organic material, repeated heatings or mechanical cleaning may be necessary.

Although electrothermal atomizers have certain advantages, they are slower than flame techniques particularly when large numbers of samples have to be analysed, and the transient readings which result from such methods may show poorer precision than do the steady readings obtained by sample aspiration.

A microsampling system known as the Delve's cup is a hybrid of flameless and flame techniques. The sample is placed in a small crucible, which is held in the flame by means of a wire loop. The sample is ashed in a cooler part of the flame and then moved to the hotter part in order to cause the rapid vaporization of the element. The cup is held beneath an opening in a nickel or aluminium tube which is in the light path of the instrument. The atomic vapour is trapped in the tube for a short period of time (up to 10 seconds) while the absorbance measurements are made.

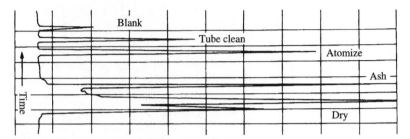

Figure 2.37 **Analysis of chromium by atomic absorption spectrophotometry using electrothermal atomization.**

Analysis sequence (monitoring at 357.9 nm)

Dry	200 °C	25 s
Ash	1400 °C	30 s
Delay		10 s
Atomize	2950 °C	5 s
Delay		10 s
Tube clean	3100 °C	5 s
Delay		10 s
Blank	2950 °C	5 s

(Reproduced by permission of G. Murray, Clinical Chemistry Department, Doncaster Royal Infirmary, UK.)

Interference effects

Atomic absorption spectroscopy is highly specific and there are very few cases of interference due to the similar emission lines from different elements. General interference effects, such as anionic and matrix effects, are very similar to those described under flame emission photometry and generally result in reduced absorbance values being recorded. Similarly, the use of high temperature flames may result in reduced absorbance values due to ionization effects. However, ionization of a test element can often be minimized by incorporating an excess of an ionizable metal, e.g. potassium or caesium, in both the standards and samples. This will suppress the ionization of the test element and in effect increase the number of test atoms in the flame.

Any ions that need to be introduced into a sample either to prevent ionization or suppression effects (for example, a lanthanum and caesium mixture) or as an internal reference (for example, lithium) are usually incorporated in the diluting fluid in which the samples are prepared.

Quantitative methods

The relationship between absorbance and the concentration of the sample as specified in the Beer–Lambert equation is only true over a limited concentration range and, in practice, a calibration graph may show some curvature. The main reasons for this lie in the interference effects and the stray light present with flame techniques due to the narrow bandwidth of emission lines and the broad bandwidth of the monochromator. Obviously other factors such as flame instability and sample distribution within the flame play a part.

It is usual to record data in absorbance units and, although a straight line relationship is theoretically valid, the effective linear range does not usually exceed 1.0 absorbance unit and in many cases may be only up to 0.5 absorbance units. In order to increase the versatility of an instrument, some manufacturers incorporate a concentration mode in which it is possible to alter the sensitivity of the instrument and so work over various concentration ranges. If a direct read-out of concentration is used, the problem of non-linearity becomes more serious and in order to try to overcome this, some instruments incorporate a curvature correction device.

2.6   Magnetic resonance spectroscopy

The absorption of electromagnetic radiation resulting in molecular transitions is not restricted to the ultraviolet, visible and infrared regions of the spectrum; energy associated with the longer wavelengths, i.e. microwaves and radiowaves, can be absorbed and result in molecular transitions. Nuclear magnetic resonance (NMR) spectroscopy and electron spin resonance (ESR) spectroscopy are examples of techniques that depend upon the magnetic properties exhibited by nuclei and electrons in certain molecular situations. They are useful tools in the study of molecular structures and biochemical reactions such as enzyme catalysis.

Magnetic properties are only exhibited by molecules that have either atoms with an odd mass number or an uneven number of electrons. These two features provide the basis for NMR and ESR spectroscopic techniques respectively. Nuclei with an equal number of protons and neutrons will have an even distribution of charge, as will an electron orbital occupied by the full complement of two electrons with opposite spins. An imbalance in either will give the atom or orbital a magnetic moment, i.e. it will have a positively charged section and a negatively charged section. Under normal conditions there will be no preferred orientation but if the molecule is placed in a magnetic field it will tend to occupy a position involving the least energy and will, like a compass needle, align itself with the magnetic field. In most instances the only other possible orientation is in a position directly opposite to the magnetic field, i.e. at 180° to the original position. The transition between these two orientations, which is due to the electric field effects, will require specific quanta of energy, which can be provided by the magnetic aspects of electromagnetic radiation in a comparable manner to absorption in the ultraviolet and visible region of the spectrum. The energy is provided by radiation in the microwave and radiowave regions of the spectrum with wavelengths in centimetres or metres, corresponding to frequencies of gigahertz (GHz) and megahertz (MHz). These are associated with electron (ESR) and nuclear (NMR) transitions respectively.

A nucleus with its magnetic moments aligned with the magnetic field is said to show 'parallel' alignment, while the higher energy transition state is said

> ➤ A magnetic moment is the tendency of a nucleus or electronic orbital with an unequal distribution of charge to align itself in a magnetic field.

to be 'antiparallel'. For electron magnetic moments the terms used are 'up-spin' and 'down-spin', the latter being the higher energy state.

Table 2.6   Elements important in NMR studies of biological compounds

Group of compounds	Relevant atoms	Natural occurrence (%)
All	^1H	100
	^{13}C	1
Nucleotides		
Phospholipids	^{31}P	100
Phosphorylated compounds		
Amino acids	^{14}N	99.9
Peptides	^{15}N	0.4
Proteins	^{33}S	0.7
Substitute for hydrogen	^2H	0.2
	^{19}F	

There are an appreciable number of elements with odd atomic mass numbers although some are rare and have only a low natural occurrence. Those that are useful in biological applications of NMR are given in Table 2.6. These include the hydrogen atom (^1H) and the isotope of carbon (^{13}C), both of which are applicable to all biochemical compounds. The isotopes of phosphorus (^{31}P) and nitrogen (^{15}N) are useful in the study of nucleotides and amino acids or their derivatives. The hydrogen nucleus is particularly useful but does usually require the NMR studies to be performed in a solvent that is free from protons, deuterated water, D_2O, being most frequently used. However, it is possible to analyse aqueous samples by saturating the signal due to water by applying a long pulse (1–2 s) of radiation at the water frequency. This equilibrates the hydrogen nuclei between their various energy levels and after the saturating radiation is stopped, but before the hydrogen atoms revert to their ground state (relax), the sample is test scanned with a short pulse of radiation.

The number of molecules with single electron orbitals, and therefore suitable for ESR, is limited due to the electron-sharing feature of the usual covalent bond. This tends to restrict its use to compounds containing transition metals and reactions involving free radicals. However, this does make ESR very useful for monitoring reactions involving metallo-enzymes or free radicals.

2.6.1   Instrumentation

The basic technique of magnetic resonance spectroscopy involves placing the sample in a magnetic field generated between the poles of either a permanent magnet or an electromagnet. The energy for the transition is produced by a radio-frequency generator and the strength of the signal is measured by a radio receiver (Figure 2.38). The detection of radio signals does not demand an optical light path as in ultraviolet/visible spectroscopy and in some instruments the same probe (antennae/aerial) is used both as a transmitter and as a receiver, with very rapid changes from one function to the other.

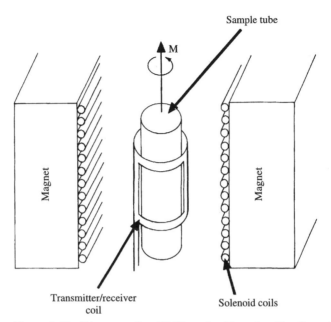

Figure 2.38 **Diagram of an NMR probe.** The solenoid coils on the pole faces of the magnet produce the variable magnetic field. The direction of the magnetic field through the sample is indicated by M.

The energy required to cause a transition depends not only on the molecular/atomic features of the test substance but also on the strength of the magnetic field that has to be overcome during the transition. If the technique was directly comparable with ultraviolet/visible spectroscopy, the absorption characteristics would be monitored as the radio frequency was increased. However, in practice, because of technical difficulties in the generation and detection of a signal of varying frequency, the frequency is held constant and as the magnetic field is changed the absorption is monitored.

The magnets give a constant magnetic field and the variation required is introduced by coils mounted on the pole faces of the magnet; the current applied to the coils provides the basis for the *x* axis of the resulting spectrum, i.e. comparable to wavelength in a visible absorption spectrum. All magnets show some variation in the magnetic field across the face of the magnet and in order to average out the effects of these differences, the sample is rotated rapidly in the field.

The NMR scan is basically a series of pulses and delays. A short pulse of radiation is applied to the sample as the magnetic field is scanned. If the signal due to water is to be suppressed there is an initial saturation signal at the water frequency as described earlier. Before another test scan can be taken, the sample must be allowed to relax and during this period a further saturation of the water signal can be done and the process repeated. The sequence of events and collection of data are automated.

2.6.2 The spectrum

> ➤ The resonance frequency is that radio frequency which is absorbed by a nucleus in a specific magnetic field.

It is technically possible, but very difficult, to measure the exact frequency of a radio signal, and in practice the frequency of the energy absorbed by a test compound (usually called the resonance frequency) is measured relative to that of a reference compound. This reference may be mixed with the sample (direct referencing), or if contamination of the sample is undesirable it may be placed in a separate container within the sample tube (external referencing). In proton and ^{13}C NMR, the reference compound usually used is TMS (tetramethyl silane) or its water-soluble derivative DSS (2,2-dimethylsilapentane 5-sulphonic acid). These compounds give a sharp proton peak at the right-hand side of a typical NMR spectrum (Figure 2.39).

> ➤ Chemical shift is a value derived from the resonance frequency of the nucleus as a function of the reference frequency.

The difference between the test peak and the reference peak is known as the 'chemical shift' of the test. This is a scale of frequency normalised to give the reference peak a value of zero.

$$\text{Chemical shift} = \frac{\text{sample frequency} - \text{reference frequency}}{\text{reference frequency}} \times 10^6$$

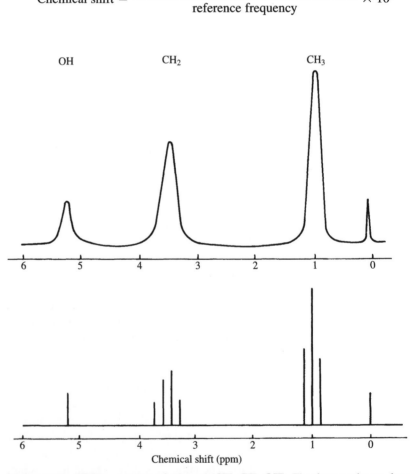

Figure 2.39 **NMR spectrum of ethanol ($CH_3\ CH_2\ OH$).** The three peaks are due to the three groups in the compound, which can be split into multiple peaks using a higher resolution. The peak at zero chemical shift is due to the reference compound, DSS.

This value, being a ratio, has no units and the factor 10^6 makes it a manageable figure. As a result chemical shift values are reported as being 'parts per million' (ppm).

The absorbing property of a nucleus is affected by the magnetic effects of adjacent nuclei. This results in what is known as 'spin–spin splitting' of the initial resonance peak. In general, if there are n adjacent similar nuclei, the peak will be split into $n + 1$ components and this can give information about the manner in which the nuclei are arranged in the molecule.

Figure 2.39 shows a proton NMR spectrum of ethanol, a compound that has three different kinds of hydrogen nuclei, those in the CH_3, those in the CH_2 and the one in the OH group. These three kinds of nuclei absorb energy at different frequencies and as a result give three resonance peaks in the spectrum. The integrated area under each peak is proportional to the number of protons in the group and the position of the resonance peak on the scale (the chemical shift) is characteristic of the particular molecular environment in which the protons are found, i.e. the specific group. The peak on the right-hand side of the spectrum with a chemical shift of zero is the reference peak of DSS.

When a higher resolution spectrum is produced some peaks are split into multiple peaks. The methyl protons, each giving the same chemical shift, are all affected by the two methylene protons, resulting in three peaks from the original one (i.e. $n + 1$). In a similar manner, the two protons of the methylene group are affected by the three protons in the methyl group, which results in the single methylene peak splitting into four.

2.6.3 Applications

The biological applications of NMR include the study of the structure of macromolecules such as proteins and nucleic acids and the study of membranes, and enzymic reactions. Newer methods and instruments have overcome, to a large extent, the technical difficulties encountered with aqueous samples and the analysis of body fluids is possible, permitting the determination of both the content and concentration of many metabolites in urine and plasma. NMR is not a very sensitive technique and it is often necessary to concentrate the sample either by freeze drying and dissolving in a smaller volume or by solid phase extraction methods.

The principles of ESR spectroscopy are very similar to NMR spectroscopy but the technique gives information about electron delocalizations rather than molecular structure and it enables the study of electron transfer reactions and the formation of paramagnetic intermediates in such reactions. In some situations, information regarding molecular structure can be obtained when suitable prosthetic groups are part of a molecule, e.g. FMN (flavin mononucleotide) in certain enzymes or the haem group in haemoglobin. Sometimes it is possible to attach suitable groups to molecules to enable their reactions to be monitored by ESR techniques. Such 'spin labels' as they are called, are usually nitroxide radicals of the type

Self test questions

Sections 2.4/5/6

1. Which of the following spectroscopic techniques involve the measurement of absorbed radiation?
   (a) Fluorescence.
   (b) Flame emission spectroscopy.
   (c) Atomic absorption spectroscopy.
   (d) Nuclear magnetic resonance spectroscopy.
2. In which of the following techniques is the Beer–Lambert relationship of significance?
   (a) Fluorescence.
   (b) Flame emission spectroscopy.
   (c) Atomic absorption spectroscopy.
   (d) Nuclear magnetic resonance spectroscopy.
3. A fluorimeter must have a source of ultraviolet radiation
   BECAUSE
   fluorescence always involves the emission of ultraviolet radiation.
4. The technique of atomic absorption spectroscopy needs a source of monochromatic radiation such as a hollow cathode lamp
   BECAUSE
   the flame in atomic absorption spectroscopy only contains unexcited analyte atoms which absorb radiation.

2.7   Further reading

Willard, H.H., Merritt, L.L., Dean, J.A. and Settle, F.A. (1988) *Instrumental methods of analysis,* 7th edition, Wadsworth Publishing Co., USA.

Metcalfe, E. (1987) *Atomic spectroscopy and emission spectroscopy*, John Wiley, UK.

Hollas, J.M. (1996) *Modern spectroscopy*, 3rd edition, John Wiley, UK.

Abraham, R.J., Fisher, J.P. and Loftus, P. (1992) *Introduction to NMR spectroscopy*, John Wiley, UK.

Harris, D.A. and Bashford, C.L. (eds) (1987) *Spectroscopy and spectrofluorimetry – a practical approach*, IRL Press, UK.

Thomas, M.J.K. (1996) *Ultraviolet and visible spectroscopy*, John Wiley, UK.

3 Separation methods

Key topics	• Principles of separation techniques • Methods based on polarity • Methods based on ionic nature • Methods based on size • Methods based on shape

One of the major difficulties in the analysis of biological samples is the presence of substances that may be confused with the one under investigation. This means that it is often necessary either to isolate the test substance or to remove the interfering substance before the analysis can proceed. Even the presence of substances that are grossly different from the substance under investigation can often affect the quality of the analytical results. For this reason many methods include a preliminary separation step. Proteins are frequent interfering substances in biological samples and these may be removed by techniques such as dialysis, ultrafiltration and gel permeation chromatography. Alternatives include precipitating the protein with high concentrations of salts, organic acids or solvents or by heating (Table 3.1). The precipitated protein can be subsequently removed by either filtration or centrifugation. Because many small non-polar molecules are solubilized by being bound to proteins, they will also be removed with the protein and in order to quantitate these substances it is necessary to break these complexes, often by extraction into an organic solvent, before removing the precipitated protein.

➤ Protein separation – see Section 11.3.

Although the removal of interfering substances is an important application of separation methods, this chapter is more concerned with the use of these procedures to isolate, identify or quantify a particular substance or group of substances in the presence of other very similar substances. The quantitation of one amino acid in the presence of other amino acids is only one example of such an analytical problem.

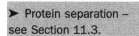

➤ Amino acid separation – see Section 10.5.

Table 3.1 Protein precipitants

Precipitating agent	Final concentration in mixture
Anionic	
Perchloric acid	1.0 mol l^{-1}
Picric acid	10 g l^{-1}
Salicylsulphonic acid	25 g l^{-1}
Trichloroacetic acid	10 g l^{-1}
Tungstic acid	Sodium tungstate 10 g l^{-1}, sulphuric acid 0.05 mol l^{-1}
Cationic	
Zinc hydroxide	Zinc sulphate 10 g l^{-1}, sodium hydroxide 0.05 mol l^{-1}
Copper tungstate	Cupric sulphate 5.0 g l^{-1}, sodium tungstate 2.5 g l^{-1}
Neutral	
Sodium sulphate (and other salts)	250 g l^{-1}
Organic solvents	
Acetone	63% (v/v)
Chloroform	25% (v/v)
Ethanol	70% (v/)
Methanol	75% (v/v)

3.1   Principles of separation techniques

All separation procedures depend primarily upon some physical characteristic of the compounds. It is relatively easy to separate substances that have significantly different physical characteristics by simple techniques, such as solvent extraction. However, if the various compounds are similar to each other then it is essential that any slight differences between them are exploited in order to achieve separation. Table 3.2 lists the major physical properties that form the basis of separation procedures.

It is possible to exaggerate any differences by repeating a manipulation many times, for instance, in sequential solvent extractions. Paper chromatography is an example of a separation procedure in which a large number of partition equilibria occur between an organic solvent and the water in the cellulose.

The polarity of a molecule is a characteristic that significantly affects many of its properties, in particular its solubility and volatility. Molecules are said to be polar if they have a dipole. This is defined as an unequal distribution of electrons within the molecule, caused by differences in the electronegativity of the atoms involved. Water is a polar substance because oxygen shows greater electronegativity than hydrogen and tends to develop a slight negative charge, leaving the hydrogen atoms with a slight positive charge.

➤ A polar molecule is one that has an uneven distribution of charge even though, overall, it might be uncharged.

Owing to this partial ionic or polar nature, water molecules tend to associate with one another to form a lattice structure stabilized by these ionic interactions. Other molecules that also show this partial ionic nature will fit more readily into the lattice structure than non-polar molecules and will dissolve in water. This attraction between two polar molecules has other effects besides affecting the solubility of the substance. If one of the substances is a solid, the binding of the other polar substances is known as adsorption and the strength of the adsorptive bonds reflects the ionic interactions between the two molecules. Separation techniques, such as gas–liquid chromatography, liquid chromatography and adsorption chromatography, all depend to a large extent on polar interactions, and slight differences in polarity can be used to facilitate separation.

The possible formation of a dipole is a feature of covalent bonding but it is obvious that an ionic bond results in a definite unequal distribution of electrons within a molecule and such molecules (or ions) are extremely polar. However, the fact that they carry a definite charge enables additional separation techniques to be applied. The rate of migration in an electric field (electrophoresis) and the affinity for ions of opposite charge (ion-exchange chromatography) are extremely valuable techniques in the separation of ionic species.

Molecules that vary significantly in their size can be separated by ultrafiltration or dialysis, while molecules that are only slightly different in size can often be separated by gel permeation chromatography. Ultracentrifugal techniques, while apparently separating on the basis of size, are strictly speaking more influenced by the mass and density of the molecule and to a lesser extent by its shape.

Many molecules show a definite affinity for another molecule based on a shape relationship, e.g. enzymes and their substrates, antibodies and their

Table 3.2 Classification of separation techniques

Molecular characteristic	Physical property	Separation technique
Polarity	Volatility	Gas−liquid chromatography
	Solubility	Liquid−liquid chromatography
	Adsorptivity	Liquid−solid chromatography
Ionic	Charge	Ion-exchange chromatography
		Electrophoresis
Size	Diffusion	Gel permeation chromatography
(mass)		Dialysis
	Sedimentation	Ultracentrifugation
Shape	Ligand binding	Affinity chromatography

antigens, and such relationships form the basis of extremely useful and specific methods of separation known generally as affinity chromatography.

The classification of separation techniques as shown in Table 3.2 is concise and easy to remember but it is also simplistic because it appears to imply that only one factor is involved in each technique. In practice, the effectiveness of any method is a composite of many factors, the one indicated in the table usually being the most significant. Some of the developments in separation procedures exploit this range of factors involved in any separation technique by using conditions or reagents designed to minimize one or maximize another. As a consequence, the techniques and instrumentation of separation methods are constantly changing but the fundamental principles remain the same and need to be understood in order to appreciate the usefulness and limitations of any particular technique.

3.1.1   General methods of separation

The movement of the analyte is an essential feature of separation techniques and it is possible to define in general terms the forces that cause such movement (Figure 3.1). If a force is applied to a molecule, its movement will be impeded by a retarding force of some sort. This may be as simple as the frictional effect of moving past the solvent molecules or it may be the effect of adsorption to a solid phase. In many methods the strength of the force used is not important but the variations in the resulting net force for different molecules provide the basis for the separation. In some cases, however, the intensity of the force applied is important and in ultracentrifugal techniques not only can separation be achieved but various physical constants for the molecule can also be determined, e.g. relative molecular mass or diffusion coefficient.

There are basically three ways in which separation methods can be performed. They are most easily described in relation to chromatography but they are also relevant to other methods such as electrophoresis and ultracentrifugation.

Zonal methods

In zonal methods (Figure 3.2) the sample is applied as a band or zone and separation takes place in an appropriate solvent system, resulting in the separation of the individual components of the sample into separate bands. The separation process can take place entirely under uniform conditions (normal development) but it may be desirable to alter the conditions during the procedure in order to change either the impelling or retarding forces on a predetermined basis. Such gradient developments may involve gradual changes in the conditions, such as pH or temperature. In zonal separation, the individual components of the sample are separated from each other and it is often possible to use the method in a preparative manner to obtain pure preparations of each component as in some forms of chromatography, electrophoresis and ultracentrifugation.

➤ Gradient separations involve a gradual change in the eluting characteristics of the system.

Frontal methods

Frontal analysis does not involve the use of separate solvent systems but separation takes place within the liquid sample usually as a result of one compo-

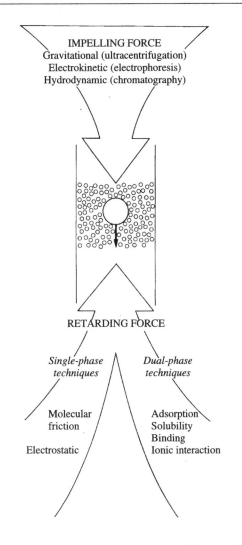

Figure 3.1
**A representation of some
of the factors involved in
separation techniques.**

nent moving faster than another (Figure 3.2). The result is that a series of concentration boundaries develop, each one being due to one component of the mixture. While only a small fraction of the fastest component can actually be obtained in a pure form, the movement of each boundary is characteristic of the component and it is feasible to study a component despite never having a pure preparation. This technique is used in ultracentrifugation for the determination of relative molecular mass and sedimentation coefficient.

Displacement methods
In some chromatographic techniques the components of a mixture show different affinities for a solid phase and can only be displaced from that phase by other molecules that have a greater affinity for it. They are zonal techniques because the sample is applied as a band or zone but separation is achieved using a solution of the displacing solute rather than the pure solvent (Figure 3.2). It is the solute that is important rather than the solvent. These techniques

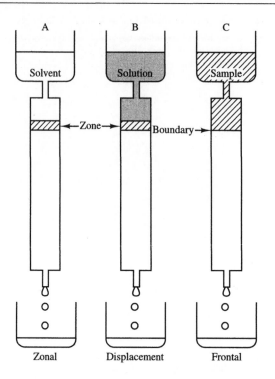

Figure 3.2 **Three major methods in chromatography.** The commonest form of chromatography involves the introduction of a small volume of sample onto a column and is known as zonal chromatography. Movement down the column is effected by the mobile phase, which may be simply a solvent (A) allowing partition of the test molecules between the stationary and mobile phases. Alternatively, the mobile phase may be a solvent containing solute molecules (B), which actively displace test molecules from the stationary phase. A less frequently used method known as frontal separation (C) does not involve a separate mobile phase but a large volume of the sample is allowed to pass through the column and as the various components separate, concentration fronts develop and their movement can be monitored.

may be used in a normal development mode (i.e. without changing the composition of the eluting solution) or in a gradient mode in which the composition of the eluting solution is progressively changed.

While the individual components of the mixture may be separated from each other, they will all be contaminated to some extent with the eluting solute. Techniques such as ion-exchange chromatography and various types of adsorption and affinity chromatographies are examples of displacement methods.

3.1.2   General methods of detection

Some detection procedures involve the loss of the sample: for example, in the flame ionization detectors used in gas chromatography the sample is burnt, while in liquid chromatography the addition of a colour reagent will also result in the loss of the sample. For such methods to be used in a preparative manner, it is necessary to split the effluent stream from the column, testing one part in the normal manner and collecting the remainder until the identity or position of each component is known. A fraction collector is normally used for this purpose. In thin-layer and electrophoretic techniques, destructive locating methods such as colour reagents and stains can be used in a preparative manner by either masking part of the chromatogram or cutting it into slices and only staining part of it, the remainder being kept for further use.

A range of non-destructive physical methods are available to monitor the effluent from a column. Many compounds absorb radiation in either the visible or ultraviolet region of the spectrum, while some are fluorescent. In the absence of specific monitoring methods, changes in the composition of a solution can often be detected potentiometrically because of changes in resistance, degree of ionization, etc. Changes in the concentration of a solution also tend to alter various general characteristics, e.g. refractive index.

▶ Chemical modification – see Section 1.1.1.

If the compound shows no convenient intrinsic feature, it may be possible to develop a detectable characteristic by chemical modification and, particularly in thin-layer chromatography and electrophoresis, many locating reagents which are specific for various groups of compounds are available.

The inclusion of a fluorescent dye into thin-layer plates can be used to detect substances that quench its fluorescence and so result in dark zones when the chromatogram is examined under ultraviolet radiation. Autoradiography can also be used in thin-layer chromatography and electrophoresis when samples are radio-labelled.

▶ Radioisotope label – see Section 5.3.

Self test questions

Section 3.1

1 Which properties does the polarity of a molecule affect?
   (a) Solubility.
   (b) Volatility.
   (c) Diffusion.
   (d) Adsorption.

2 Which of the general methods of separation can be used in a preparative way?
   (a) Zonal.
   (b) Displacement.
   (c) Frontal.

3 Methyl alcohol (CH_3OH) is a polar compound
   BECAUSE
   the hydrogen atom of an OH group is more electronegative than the oxygen.

4 Chromatographic techniques that use antibodies to bind the analyte are known as affinity chromatography
   BECAUSE
   antibodies bind specifically to their corresponding antigen.

3.2   Methods based on polarity

Figure 3.3 outlines the similarities in those methods in which the polarity of the various test substances is an important consideration. All of these methods are essentially chromatographic, a word coined originally by Tswett in 1906 which now implies the separation of the components of a mixture by a system involving two phases, one of which is stationary and the other mobile.

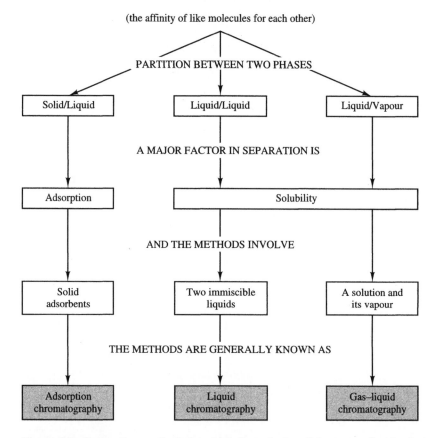

Figure 3.3 **Separation methods in which the polarity of the test molecule plays an important part.**

The range of applications of the different chromatographic methods depends mainly upon the number of alternative phases available. Thin-layer chromatography is restricted to very few stationary phases but has a wide range of mobile phases available. In gas–liquid chromatography (GLC) there are a large number of stationary phases available, while in high performance liquid chromatography (HPLC) there are a wide range of both stationary and mobile phases.

3.2.1 Liquid–solid chromatography (adsorption)

> ➤ Adsorption is the process by which some substances physically bind to the surface of a solid polar substance.

When a solution of a polar compound is in contact with a finely divided solid such as charcoal or silica, fairly extensive adsorption takes place on the surface of the solid. The majority of adsorbents are polar, either acidic (e.g. silica) or basic (e.g. alumina), and a large surface area is necessary for a significant degree of adsorption to take place.

Charcoal is a non-polar adsorbent that will bind large or non-polar molecules from an aqueous solution, but its effects are not very predictable. However, several synthetic non-polar adsorbents have been developed, known as XAD resins, which are synthetic polymers, often polystyrene based. They are used mainly as preparative media for extracting substances from samples which, after washing the resin, can be eluted from it with a polar organic solvent.

> ➤ Hydrogen bond — see Section 11.1.

The adsorptive effects of the polar adsorbents are often due to the presence of hydroxyl groups and the formation of hydrogen bonds with the solute molecules. The strength of these bonds and hence the degree of adsorption increases as the polarity of the solute molecule increases. In order to separate the solute from the adsorbent it is necessary to use a solvent in which the solute will dissolve and which also has the ability to displace the solute from the adsorbent. Solvents that are too polar will overwhelm the adsorptive effects and result in the simultaneous elution of all the components of a mixture. Table 3.3 lists various classes of compounds and some common solvents in order of polarity.

While only a small number of adsorbents are commonly used, which are all relatively non-specific (Table 3.4), the versatility of adsorption chromatography is due to the range of solvent mixtures which can be used.

Table 3.3 Polarity of selected solutes and solvents

Solute	Adsorption energy		Solvent	Solvent strength
Hydrocarbons	0.07	↑ Increasing polarity ↓	Hexane	0.01
Halogen derivatives	1.74		Benzene	0.32
Aldehydes	4.97		Chloroform	0.40
Esters	5.27		Acetone	0.55
Alcohols	6.50		Pyridine	0.71
Acids/bases	7.60		Methanol	0.95

Adsorption energy is a measure of the affinity of a solute for an adsorbent and it varies from one adsorbent to another. The above data are relevant to silica as the adsorbent.

Solvent strength is a measure of the affinity of a solvent for the adsorbent. Complete lists are known as eluotropic series and will include mixtures of solvents as well as the pure solvents above.

Methods of adsorption chromatography

Separations may be achieved using columns with zonal development, a technique that is very useful for preparative and qualitative purposes and in HPLC offers a high level of selectivity. However, thin-layer techniques are particularly popular because of their convenience and speed and they are particularly useful in qualitative analysis.

Table 3.4 Some examples of adsorbents and possible applications

Adsorbent	Strength	Applications
Silicic acid (silica gel)	Strong	Steroids, amino acids, lipids
Charcoal	Strong	Peptides, carbohydrates
Aluminium oxide	Strong	Steroids, esters, alkaloids
Magnesium carbonate	Medium	Porphyrins
Calcium phosphate	Medium	Proteins, polynucleotides
Cellulose	Weak	Proteins

Thin-layer plates consist of a layer of the adsorbent spread over an inert, flat support, usually glass, aluminium or rigid plastic film. The layer is usually stabilised by incorporating a binding agent, such as plaster of Paris (10%) in the mixture. Plates are prepared by shaking a mixture of the adsorbent and the binding agent with an appropriate volume of water. A uniform layer of this slurry is spread over clean plates and allowed to set. The plates are then dried in an oven and stored in a desiccator. Variations in layer uniformity result in differences in separation patterns, and commercially prepared plates are normally used because of their quality and convenience.

Samples are applied as discrete spots or as streaks with capillary or micropipettes as volumes of between 1 and 20 μl. Damage to the surface must be avoided in order to achieve a regular pattern of the separated components. Appropriate reference mixtures must be applied to the same plate as the samples to assist in identifying the test components.

After the samples have been applied and dried, the plate is placed in a tank that contains the solvent (0.5–1 cm deep). It is particularly important when using volatile solvents that the atmosphere of the tank is saturated with solvent vapour. This prevents both evaporation from the surface of the plate and alteration to the composition of the solvent due to the loss of volatile components.

When the solvent has moved to the top of the plate (10–20 cm) or, preferably, to a line that has been scored across the plate about 1 cm from the top edge (Figure 3.4) the plate is removed and dried quickly. The separated bands can be visualized by using a suitable reagent or by viewing under ultraviolet light and their positions marked.

Test spots can be identified by comparing their migration with that of reference samples, together with additional evidence, e.g. using a specific colour reagent. The R_F value (rate of flow) is a measure of the movement of a compound compared with the movement of the solvent:

➤ The rate of flow (R_F) is a qualitative parameter used in thin-layer chromatography.

$$R_F = \frac{\text{distance compound has moved from the origin}}{\text{distance solvent has moved from the origin}}$$

Ideally, this value should be the same for a given compound using the same solvent regardless of the distance moved by the solvent. However, it is difficult to standardize chromatographic conditions, and variations that may

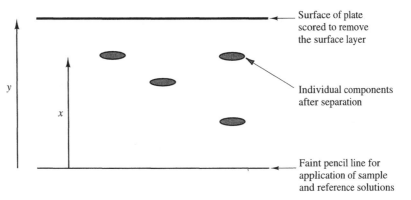

Figure 3.4 **Thin-layer chromatography.** Rate of flow (R_F) for the analyte indicated is x/y. The R_F values for the other components can be calculated in a similar manner, the value y being the same in each case.

occur in the structure and thickness of the layer, the amount of water remaining and the effect of the binding agent all contribute to the non-reproducibility of R_F values.

Demonstration of identical R_F values is not really conclusive proof of the test identity. The evidence may be strengthened by demonstrating identical R_F values for the test and reference compounds in a range of solvents. In addition, a mixture of the sample and the reference compound can be run and if the two substances are identical, a single spot should result after chromatography. This latter technique is known as co-chromatography.

> ➤ Co-chromatography – the chromatography of a mixture of an analyte and the reference compound to demonstrate identical separation characteristics.

3.2.2 Liquid–liquid chromatography

The partition of a solute between two immiscible liquid phases provides the basis for simple solvent extraction techniques. The polarity of both solute and solvent are important factors in determining the solubility of the solute, and polar solutes will dissolve more readily in polar solvents than in non-polar solvents.

Liquid–liquid chromatography in its simplest form involves two solvents that are immiscible. However, many recently developed media consist of a liquid (the stationary phase) that is firmly bound to a solid supporting medium. As a result, it is possible to use a second solvent (the mobile phase) which under normal conditions would be miscible with the first solvent. The second solvent is permitted to move in one direction across the stationary phase to facilitate the separation process. The presence of a supporting medium introduces some problems in the system and, in theory, it should be completely inert and stable, showing no interaction with the solutes in the sample. However, this is not always the case and sometimes it affects the partitioning process, resulting in impaired separation.

Paper chromatography
The simplest form of liquid–liquid chromatography is paper chromatography (or commercially prepared cellulose thin-layer plates) in which the

water bound to the cellulose acts as the stationary phase (polar) and the mobile phase (less polar) can be a mixture of various organic solvents, butanol, chloroform, etc.

The efficiency of any chromatographic technique depends upon the number of sequential separations or equilibria that take place, which in the case of paper chromatography are due to the large number of 'compartments' of cellulose-bound water. The test solutes are carried up the paper dissolved in the mobile phase and encounter successive compartments of water. At each one, rapid partition between the two phases occurs leaving the mobile phase to carry up the residual solute to the next water compartment and another partitioning effect. The solute, which is dissolved in the water and hence not carried up the paper, is now presented with fresh solvent rising up the paper and again is redistributed between the two phases.

The extremely large number of partitioning phenomena greatly exaggerates the differences in the relative solubilities of the solutes and results in the effective separation of the components. Those solutes that show greater solubility in the stationary phase will move less rapidly up the paper than those solutes that show a greater solubility in the mobile phase.

In a typical chromatographic system the samples are spotted onto a strip of filter paper, dried and then placed in a tank containing a suitable solvent, which is allowed to run up the paper by capillary action. For long chromatographic runs (exceeding 20 cm), a descending technique may be used in which the paper is hung from the reservoir, the solvent flowing down the paper.

> Tailing refers to a separation pattern in which the desired discrete sample zone is spoiled by a trail of analyte behind the initial zone.

Cellulose is itself polar in nature and can cause some adsorption, which may result in the 'tailing' of zones. However, this adsorptive effect may contribute to the separation process in some instances and the use of a polar mobile phase can enhance this effect further, e.g. the separation of amino acids using an aqueous solution of ammonia as the mobile phase. The combination of partition and adsorption generally influences separations on cellulose thin-layer plates, which have superseded paper chromatography in most instances and offer increased speed and resolution.

High performance liquid chromatography

Column chromatographic techniques were originally used as preparative tools but over the years major advances have taken place. HPLC is now a highly developed technique and a wide range of stationary phases are available. These enable partition, adsorption, gel permeation, affinity and ion-exchange chromatography to be performed.

For efficient separations it is essential to have very small and regular-shaped support media, a supply of mobile phase pumped at a pressure that is adequate to give a suitable constant flow rate through the column and a convenient and efficient detection system. For all types of stationary phases the apparatus required consists of five basic components (Figure 3.5), each available with varying degrees of sophistication.

The solvent delivery system includes a pump that delivers the appropriate solvent from a reservoir at a preselected rate. The use of fine particle columns, and the high flow rates demanded, require relatively high pressures

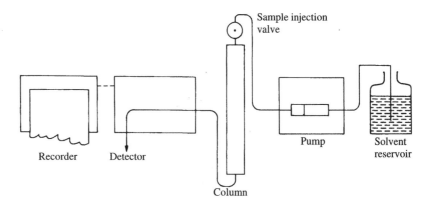

Figure 3.5
A representation of the components of an HPLC system.

of solvent, sometimes up to 35 MN m^{-2} (5000 lb in^{-2}) although more usually about 10 MN m^{-2} (1500 lb in^{-2}). In addition to being able to produce these pressures and in order to obtain a steady baseline signal from the detector, it is necessary to eliminate gross fluctuations in pressure which often result from simple pumps. To achieve this 'pulse-free' pumping it is necessary either to incorporate pulse dampening systems or to use more sophisticated dual-piston pumps.

> Isocratic liquid chromatography involves the use of only one solvent mixture.

The use of a single solvent is known as an isocratic system (with equal power), but it is often desirable to have the ability to alter the composition of the solvent mixture during the chromatographic run, a process known as gradient elution. The gradient may be as simple as a stepwise change from one previously prepared solvent mixture to another at any given time in the separation. However, more frequently and more satisfactorily, the production of the gradient is microprocessor controlled.

The fact that the solvent is delivered under pressure demands a suitable **sample injection device**, usually in the form of a sampling valve, which will give good precision. The typical six-port valve (Figure 3.6) includes a loop of specified volume (10 μl to 2 ml), which is filled using a normal low pressure syringe before the sample is introduced into the column by switching the solvent flow through this loop.

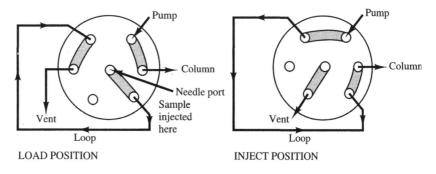

Figure 3.6 **Diagram of a six-port sample injection valve.** In the load position the loop can be filled with the sample using a simple syringe. When the valve is turned to the inject position, the sample is washed into the column.

A **continuously monitoring detector** of high sensitivity is required and those that measure absorption in the ultraviolet are probably the most popular. These may operate at fixed wavelengths selected by interference filters but the variable wavelength instruments with monochromators are more useful. Wavelengths in the range of 190–350 nm are frequently used and this obviously means that a mobile phase must not absorb at those wavelengths.

▶ A diode array is a spectroscopic device for the continuous monitoring of absorbance over a specified wavelength range.

Conventional spectroscopic detectors can only monitor at a single selected wavelength but the development of diode array detection permits the continuous, simultaneous monitoring of the column effluent over any selected wavelength range in the ultraviolet and visible region of the spectrum. A diode array consists of a large number of microdiodes (sometimes up to 500) arranged so that the spectrum from a holographic grating is focused on them. Each diode then will record variations in the intensity of radiation from a particular section of the spectrum (Figure 3.7). The data from all the diodes are handled by a computer and can be presented as a traditional chromatogram but with the advantage that an absorption spectrum can be produced for any point on the chromatogram or the data can be manipulated in any appropriate manner by the computer.

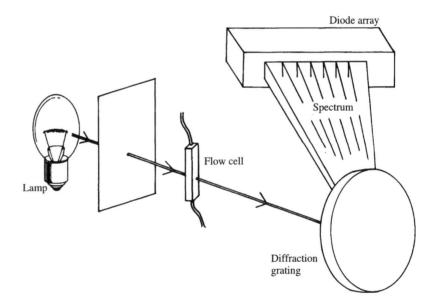

Figure 3.7 **Diode array detector.** Light from the lamp passes through the flow cell and to a holographic reflectance grating and the resulting spectrum is focused on the diode array. Detectors frequently have a spectral range of 190–800 nm and can offer bandwidths as low as 1.0 nm.

Fluorescence detectors can also be used and while their sensitivity may be greater, they are less widely applicable owing to the smaller number of fluorescent compounds. Differential refractometers will detect changes in the refractive index of the solvent due to the presence of solutes and, while they are less sensitive than the other detectors and often cannot be used effectively with gradient elution techniques, they are capable of detecting the presence of any solute.

➤ Electrochemical
detectors – see Section
4.4.2.

Electrochemical detection systems are becoming more popular. Although most electrochemical methods have been investigated, e.g. potentiometric, conductometric, coulometric, etc., the most frequently used are amperometric (Figure 3.8). In these detectors a working and a reference electrode are held at a selected potential difference determined by the discharge potential of the analyte. The current resulting from an oxidation/reduction reaction of a small proportion of the test molecules at the working electrode is measured and, under certain conditions, is proportional to their molar concentration. Careful selection of the working voltage can permit the selective determination of one analyte in the presence of another requiring a higher working voltage. Some detectors offer increased selectivity by incorporating multiple working electrodes, each one set at a different working voltage.

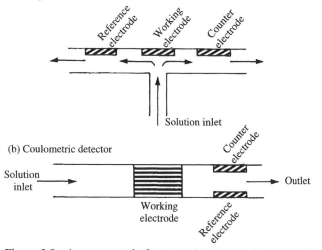

Figure 3.8 **Amperometric detectors** (a) measure the current that flows between the working electrode, usually a glassy carbon electrode, and a reference electrode, at a fixed voltage, usually close to the discharge potential for the compound. **Coulometric detectors** (b) are less common and are designed with a porous carbon flow cell so that all the analyte reacts in the cell, the amount of current consumed during the process being proportional to the amount of the substance.

When electrochemical detection systems are used, it is essential that the mobile phases used are electrically conductive but only result in a low background current at the voltage selected. Cell volumes are of the order of 1 μl and the control of flow rate, pH and temperature is critical for reliable detection.

The **flow cell** for spectroscopic detection has to be made of material that transmits radiation of the wavelength required and it is important that the volume of the cell is small enough to give good resolution between two sample components that are close together. Generally speaking the volume of a flow cell should be approximately one-tenth that of the peak volume, which can be calculated from the solvent flow rate and the time base of the peak. Problems occur when bubbles of air become trapped in the flow cell and it is important that it should be designed to clear bubbles if they occur. However, this occurrence may be substantially reduced if the solvents have been previously 'degassed', a process which may be required to improve the quality of separation. The simplest method of degassing is to shake the solvent vigorously in a large flask under a vacuum. Alternative methods include ultrasonication and bubbling nitrogen through the solvent.

Columns are most frequently made of internally polished stainless steel tubing of approximately 4 mm internal diameter and 25 cm in length, although for preparative work the dimensions may be increased. It is important that the space not occupied by the supporting medium (the dead volume) is kept to an absolute minimum to prevent excessive dilution of the sample by the mobile phase. Microbore columns with internal diameters ranging from 0.5 to 2 mm may be used to reduce considerably consumption of solvent and to improve resolution. However, specially designed detection systems are needed as well as special pumps designed to deliver volumes in the order of 10 μl per minute. Such systems have an advantage in that they can be coupled directly to a mass spectrometer.

The dimensions of the **packing material** are important in achieving efficient separations and the flow pattern around particles causes mixing of solute and solvent and increases the dilution effects, contributing to the phenomenon known as 'band broadening' (Figure 3.9). Pellicular packing materials are

➤ Band broadening is the blurring and spreading of the boundaries of a sample zone due to diffusion.

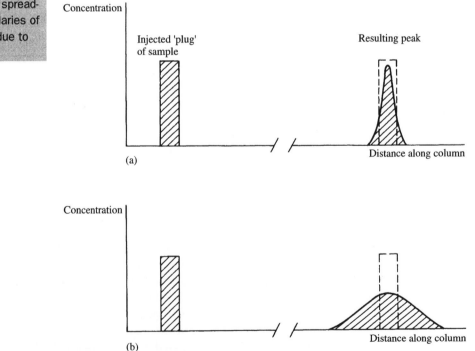

Figure 3.9 **Band broadening.** The sharp boundaries of the injected sample are blurred due to diffusion as it is carried down the column (a). If there is a large amount of dead space, the diffusion effect becomes excessive, resulting in a very broad peak which is not very useful analytically (b).

solid cores of approximately 40 μm in diameter with a porous surface layer approximately 2 μm thick and are most useful when high pressures or extremes of solvent polarity are required. Microparticulate packings consist of totally porous particles usually only 5–10 μm in diameter and result in substantially better performance than the pellicular materials. Columns of pellicular materials may be prepared by dry packing techniques whereas the microparticulate materials require the pumping of a slurry in an appropriate solvent under high pressures in order to achieve uniform packing.

Assessment of column efficiency

Many factors contribute to the quality of the separation of the individual components in a sample and it is necessary to have a defined means of assessing and comparing the efficiency of the process under different conditions.

The ability of a chromatographic process to separate or resolve two similar compounds (Figure 3.10) is measured as the resolution index (R_s).

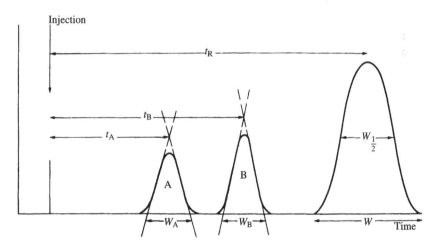

Figure 3.10 **Assessment of column efficiency**

$$\text{Resolution index } (R_s) = \frac{2(t_B - t_A)}{W_A + W_B}$$

$$\text{Number of theoretical plates } (N) = 5.54 \left(\frac{t_R}{W_{\frac{1}{2}}} \right)^2$$

where $W_{\frac{1}{2}}$ is the peak width at half peak height.

> ➤ The resolution index (R_s) is a measure of the ability of a separation process to separate two similar substances.

Regardless of any quantitative considerations, the presence of two separate peaks in the chromatogram indicates the extent of the resolution and may be quantified using the following equation:

$$\text{Resolution Index}(R_s) = \frac{\text{twice the distance between the two peaks}}{\text{sum of the base width of the two peaks}}$$

$$= \frac{2(t_B - t_A)}{W_A + W_B}$$

The greater the value for R_s, the better the resolution of the two compounds. However, large values of R_s indicate a significant time difference between the two peaks and R_s values of about 1.5 are ideal.

The injection of the sample will theoretically result in a square-sided zone which will be broadened by mixing with the solvent to produce a trace which approximates to a Gaussian distribution about the mean. Assuming that the base of the original sample zone is small, the extent of this peak broadening may be expressed as the variance (σ^2) and the base of the curve (W) will equal 4σ (Figure 3.10).

The concept of a theoretical plate is based on the number of equilibria that may have taken place during the separation process and this number is related to the number of times the effective volume of a column is greater than the peak volume. The variance (σ^2) for the peak is a measure of the broadening of the injection volume while the square of the retention time (t_R^2) is a measure of the effective column volume for that compound. Hence the number of theoretical plates (N) may be calculated from the following equation:

$$N = \frac{t_R^2}{\sigma^2} = \left(\frac{t_R}{\sigma}\right)^2$$

Using information from the dimensions of the peak (Figure 3.10) this equation can be converted to various forms, e.g.:

$$N = 16\left(\frac{t_R}{W}\right)^2$$

Because it is often difficult to measure the base width (W) accurately, it is probably better to use the peak width at half peak height, which is known to equal 2.355σ (Figure 3.10) for a Gaussian distribution. Hence:

$$N = \left(\frac{2.355 \times t_R}{\text{peak width at half peak height}}\right)^2$$

$$= \left(\frac{\text{retention distance}}{\text{width at half peak height}}\right)^2 \times 5.54$$

> ➤ Theoretical plate number is a measure of the efficiency of the separation process.
> ➤ The height equivalent to a theoretical plate (HETP) is the length of the column divided by the theoretical plate number.

The **theoretical plate number** (N) is inversely related to the amount of zone broadening occurring in a column. The greater the value for N, the more efficient is the column but differences of less than 25% are not very significant.

The **height equivalent to a theoretical plate** (HETP) may be calculated from the value for N and the column length:

$$\text{HETP} = \frac{\text{length of the column}}{N}$$

While the value for N is useful for comparing the relative efficiencies of different columns, the HETP is useful in assessing the varying efficiency of the same column under different conditions. The value decreases as the efficiency of the column increases, a characteristic that is generally better for small particle supports and less viscous fluids. In order to assess column efficiency independently of the particle size, the reduced plate height (h) is sometimes used.

$$h = \frac{\text{HETP}}{\text{particle diameter } (\mu m)}$$

Figure 3.11 illustrates the effect of varying the flow rate of the mobile phase on the efficiency of the separation process and provides a standard method of determining the optimum flow rate for a specific column and mobile phase system.

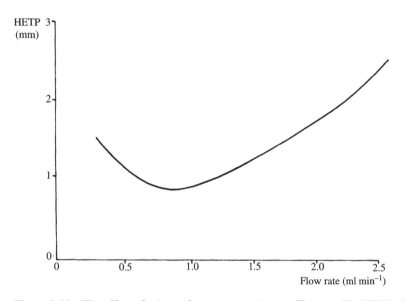

Figure 3.11   **The effect of solvent flow rate on column efficiency.** The HETP of the column for the test compound is determined at different solvent flow rates. The flow rate resulting in the lowest value for HETP gives the most efficient separation.

Qualitative analysis

Retention distance (or time) is normally used to aid the identification of a component of a mixture, provided that a known sample of the component has been subjected to separation under identical conditions. Because of the variations that can occur in the retention time due to technical factors, e.g. fluctuations in flow rate, condition of the column, the **relative retention** or **selectivity factor** (α) is sometimes used. This expresses the test retention time as a ratio of the retention time of another component or reference compound when both are injected as a mixture:

$$\text{Relative retention} = \frac{\text{test retention time}}{\text{reference retention time}}$$

For instance, the relative retention time for phenobarbitone to barbitone in procedure 3.1 is calculated as follows:

$$\text{Relative retention} = \frac{4.4}{3.4} = 1.3$$

Quantitative analysis

Although the response of the detector is usually proportional to the concentration of the test substance, this relationship can vary and it is essential that the response to a series of standard solutions is measured and a calibration factor or curve determined.

The relationship between the concentration of the solute and the peak produced in the chromatogram is, strictly speaking, only valid for peak area measurements, but in most instances it is more convenient to measure peak height. Such peak height measurements should only be used when all the peaks are very narrow or have similar widths. The tedium and lack of precision associated with non-automated methods of peak area measurements may be overcome using electronic integrators, which are features of most modern instruments.

Variation in sample volume is the factor that most affects the precision of quantitative measurements and the use of an injection valve may overcome this and permit the use of external standards. However, it is still often desirable to use an internal standardization procedure as this will reduce the effects of any variation in the detector responsiveness over a period of time.

> ➤ An external standard is measured in a separate analysis from the test sample.

For **external standardization**, replicate standards of known concentration of the pure substance are injected and the height of the resulting peaks measured. The sample is then injected and the peak height compared with those of the standards to calculate the concentration of the test. In using this technique, replicate injections of both the standards and the test must be made.

> ➤ An internal standard is added to the sample before analysis.

If the precision offered by the external standard technique is not adequate, it is necessary to use an **internal standard**. In this procedure, a known amount of a reference substance, not originally present, is introduced into the sample. This will result in the appearance of an additional peak in the chromatogram of the modified sample and any variations in the injections volume will equally affect the standard and the test compounds.

Because a detector may respond differently to the test substance and the internal standard, it is necessary to determine the response factor (R) of

---

Procedure 3.1: Quantitation of barbiturates by HPLC using an internal standard

Equipment and reagents

Column:	ODS
Solvent:	Methanol (60 parts)
	Water (40 parts)
	Flow rate 1.5 ml min^{-1}
Detection:	Absorbance at 224 nm
Standard:	Internal standard of barbitone (5.0 mmol l^{-1})

Method

To 1.0 ml of sample
0.1 ml of the internal standard was added
20 μl of the mixture was injected.

Results

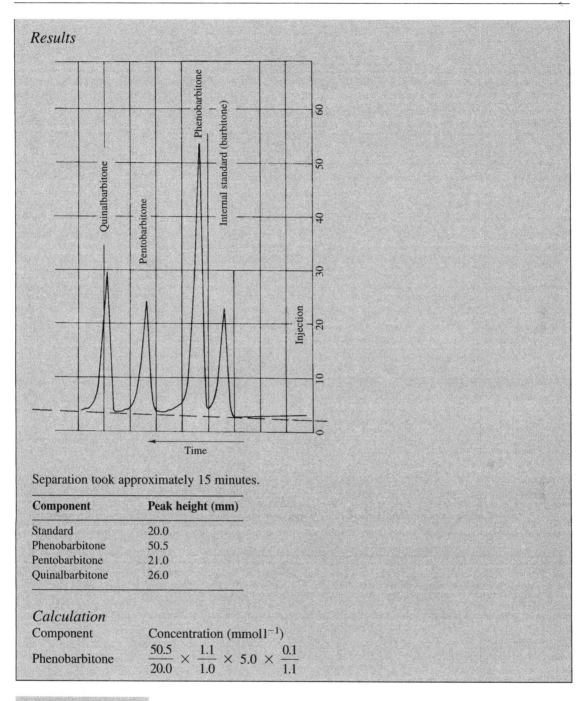

Separation took approximately 15 minutes.

Component	Peak height (mm)
Standard	20.0
Phenobarbitone	50.5
Pentobarbitone	21.0
Quinalbarbitone	26.0

Calculation

Component   Concentration (mmol l^{-1})

Phenobarbitone $\dfrac{50.5}{20.0} \times \dfrac{1.1}{1.0} \times 5.0 \times \dfrac{0.1}{1.1}$

➤ The response factor (*R*) is the ratio of the detector response to equal concentrations of the standard and the test substance.

the detector for each test substance relative to the internal standard. This can be done by injecting solutions that contain known amounts of the test substance and the internal standard:

$$\frac{\text{test peak height}}{\text{standard peak height}} \times R = \frac{\text{test concentration}}{\text{standard concentration}}$$

An internal standard can be used in two ways, usually depending upon the nature of the sample. In its simplest form the method uses the equation for calculating the response factor. A known constant amount of the internal standard is introduced into the sample and an aliquot of the mixture is injected. Knowing the concentration of the internal standard (C_s) and the response factor of the test substance (R) the concentration of the test compound is:

$$C_T = \frac{\text{test peak height}}{\text{standard peak height}} \times C_s \times R$$

This method assumes that the response factor is constant over a range of concentrations and it is often more acceptable to determine the response factor for a range of test concentrations. In this method, a calibration curve is produced by incorporating a fixed amount of the internal standard in samples that contain known amounts of the test compound. For each concentration the ratio of peak heights is determined and plotted against concentration (Procedure 3.2). For quantitation of a test sample, the same amount of the internal standard is introduced in its usual way and the ratio of peak heights for the standard and unknown is used to determine the concentration of the unknown from the calibration curve.

➤ Amino acid analyser – see Section 10.6.

This latter method is mainly used when a single substance in a sample is being determined but where the analysis involves the quantitation of many or all of the components of the sample, e.g. in an amino acid analyser, the former method is the more suitable.

An internal standard must always be introduced into the sample before any extraction or purification procedures are undertaken as this will compensate for losses in the analysis as a result of these processes.

Procedure 3.2: Quantitation of ethanol by GLC using an internal standard

Equipment and reagents

Column	PEG200
Temperature	70°C
Detector	Flame ionization
Reference	Propanol (1.0 g l^{-1})
Standard	Ethanol (20 g l^{-1})

A series of dilutions in water were prepared.

Method

Standardization

To 1.0 ml of the propanol reference solution
0.1 ml of the ethanol standard was added
And 1.0 μl of the mixture was injected.

Analysis of sample

To 1.0 ml of propanol reference solution
0.1 ml of the sample was added
And 0.1 μl of the mixture injected.

Results

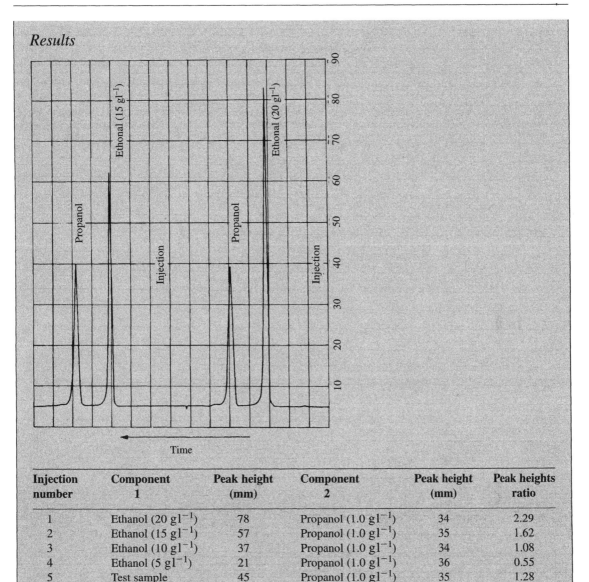

Injection number	Component 1	Peak height (mm)	Component 2	Peak height (mm)	Peak heights ratio
1	Ethanol (20 gl^{-1})	78	Propanol (1.0 gl^{-1})	34	2.29
2	Ethanol (15 gl^{-1})	57	Propanol (1.0 gl^{-1})	35	1.62
3	Ethanol (10 gl^{-1})	37	Propanol (1.0 gl^{-1})	34	1.08
4	Ethanol (5 gl^{-1})	21	Propanol (1.0 gl^{-1})	36	0.55
5	Test sample	45	Propanol (1.0 gl^{-1})	35	1.28

Calculation

For each injection the ratio of ethanol peak height to propanol peak height was calculated and a graph of the peak heights ratio against ethanol concentration was drawn.

The ethanol concentration corresponding to the peak heights ratio for the test sample was read off the graph giving a concentration of 12.1 gl^{-1}.

Choice of column materials

HPLC techniques were initially developed as liquid–liquid chromatographic methods and difficulties in maintaining the stationary phase were resolved by chemically bonding it to the particulate support. Subsequently a whole range of column materials have been developed that enable the basic HPLC instrumentation to be used for the major chromatographic techniques.

In selecting a column material for the separation of a specific substance, it is necessary to decide which physical characteristic of the molecule may be useful (Table 3.5). The initial major consideration for small molecules is usually the polarity of the molecule. A choice can then be made from ion-exchange chromatography for ionic species, adsorption chromatography for molecules showing moderate degrees of polarity and partition chromatography, which can be applicable to most molecules. For large molecules, gel permeation chromatography should be considered, using non-compressible gels which are effective at the pressures used.

The bonding of cyclodextrins to silica has provided a range of media known as chiral stationary phases (CSPs), which are capable of

Table 3.5 Chromatographic media

Name of medium	Surface group	Chromatographic applications
Adsorption		
Partisil	Silica	Steroids
LiChrosorb Si60	Silica	Alcohols
Nucleosil	Silica	Organic acids
LiChrosorb Alox	Alumina	Vitamins
Spherisorb A	Alumina	Pesticides
Partition (normal phase)		
Partisil PAC	Cyano-amino	
LiChrosorb NH$_2$	Amino	Sugars
Nucleosil NO$_2$	Nitrite	Food preservatives
LiChrosorb Diol	Hydroxyl	
Partition (reverse phase)		
Partisil ODS	Octadecylsilane	Barbiturates
LiChrosorb RPB	Octyl	Esters, ethers
LiChrosorb RP18	Octadecyl	Aromatics
LiChrosorb RP2	Silane	Steroids
Ion-exchange		
Partisil SCX	Sulphonate	Amino acids, nitrogenous bases
LiChrosorb KAT	Sulphonate	Nucleosides
Partisil SAX	Quaternary ammonium	Nucleotides
Nucleosil N(CH$_3$)$_2$	Dimethyl amine	Organic acids
Gel permeation		
LiChrospher Si	Rigid porous silica	
Macrogel	Semi-rigid polystyrene	Synthetic polymers
Sephacryl S	Dextran-acrylamide	

separating some diastereoisomers. Cyclodextrins have rigid, highly defined structures and are able to form inclusion complexes with a range of compounds. The structure of the cyclodextrin is often such that one stereoisomer can fit easily into the molecule but the other isomer cannot. The latter isomer is therefore eluted from the column first. Such CSPs provide an alternative method to the production of FLEC derivatives for separating stereoisomers (Figure 3.12). Usually the method is only appropriate for compounds that contain an aromatic group near the chiral centre but development of new chiral media is continually extending the range of applications.

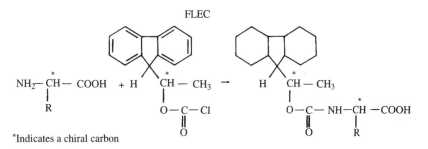

*Indicates a chiral carbon

Figure 3.12 **FLEC** reacts with alcohols and amines without any racemization of the original sample to produce highly fluorescent diastereoisomers which can be separated by reverse phase chromatography.

Chemically bonded media provide both polar and non-polar stationary phases for partition methods. Techniques using a non-polar stationary phase with a polar mobile phase are known as reverse-phase systems in contrast to normal-phase chromatography, which uses a non-polar mobile phase. Because the stationary phase is chemically bonded to a supporting medium, the elution of the liquid by the mobile phase is prevented. It is still necessary to avoid extremes of pH to prevent losses due to hydrolytic or other cleaving reactions. The most common supporting medium for partition methods is silica, to which the stationary phase is linked by either an Si–C bond, or more frequently, by a reaction with an organochlorosilane (R_3SiCl) and a surface silanol group (SiOH) to give $-Si-O-SiR_3$. Variations in the nature of this R group result in a range of stationary phases with considerable differences in polarity.

> ➤ Reverse-phase chromatography describes a system with a non-polar stationary phase and a polar mobile phase.

The most popular and versatile bonded phase is octadecylsilane (ODS), $n\text{-}C_{18}H_{37}$, a grouping that is non-polar and used for reverse phase separations. Octylsilane, with its shorter chain length, permits faster diffusion of solutes and this results in improved peak symmetry. Other groups are attached to provide polar phases and hence perform normal phase separations. These include cyano, ether, amine and diol groups, which offer a wide range of polarities. When bonded stationary phases are used, the clear distinction between adsorption and partition chromatography is lost and the principles of separation are far more complex.

The mobile phase

The great versatility of HPLC lies in the fact that the stability of the chemically bonded stationary phases used in partition chromatography allows the use of a wide range of liquids as a mobile phase without the stationary phase being lost or destroyed. This means that there is less need for a large number of different stationary phases as is the case in gas chromatography. The mobile phase must be available in a pure form and usually requires degassing before use. The choice of mobile phase (Table 3.6) is influenced by several factors.

Table 3.6 Solvents for high performance liquid chromatography

Solvent	Solvent strength	Viscosity	Ultraviolet cut-off (nm)
Pentane	0	0.24	200
Petroleum ether	0.01	0.30	226
Hexane	0.01	0.31	200
Carbon tetrachloride	0.18	0.97	263
Toluene	0.29	0.59	284
Benzene	0.32	0.65	278
Diethyl ether	0.38	0.24	218
Chloroform	0.40	0.58	245
Dichloromethane	0.42	0.43	245
Ethylene dichloride	0.49	0.79	228
Methylethyl ketone	0.51	0.42	329
Dioxane	0.56	1.44	215
Acetone	0.56	0.30	330
Acetonitrile	0.65	0.34	190
Ethanol	0.88	1.2	210
Methanol	0.95	0.55	210
Water	Large	1.00	200

Compatibility with the stationary phase

The mobile phase must not react chemically with the stationary phase or break the bond linking it to the supporting material. For this reason extremes of pH and strong oxidizing agents should normally be avoided. The working pH range of a medium will be quoted by the supplier.

Compatibility with the detection system

In any chromatographic analysis the method of detection is determined by the nature of the analyte and the mobile phase used must not interfere with this system. The use of ultraviolet absorption detection systems is very common but the solvents used must not absorb significantly at the wavelength used. For instance, absorption at 280 nm is frequently used to detect protein but some solvents, e.g. acetone, absorb at this wavelength. Similarly the use of concentration gradients in the mobile phase may present problems with refractive index and electrochemical detection systems.

➤ Absorption spectra of proteins – see Section 11.2.1.

Pressure considerations

Small particle media with their desirable large surface areas require relatively high pressures in order to achieve a realistic solvent flow rate. As the viscosity of the solvents will appreciably affect this pressure-to-flow rate relationship a solvent must be chosen that achieves the desired separation without requiring pressures too high for the system.

Polarity

> ➤ Solvent strength is a measure of the molecular polarity of a liquid.

The polarity of a solvent is expressed as **solvent strength** (Table 3.6), which is a measure of the ability of the solvent to break adsorptive bonds and elute a solute from an absorbent. High values indicate high polarity. The major factor in selecting a mobile phase for separations based on partition or adsorption is the polarity of the analyte. The polarity of the mobile phase should be such that there is an effective partition of the analyte between the two phases, i.e. the stationary and the mobile phase, and not a complete affinity for only one. If a test sample contains more than one component, as is usually the case, it may not be possible to achieve complete separation using a single mobile phase (isocratic separation). In such cases gradient elution is required in which the solvent strength of the mobile phase is gradually changed during the separation process by altering the proportion of solvents in the mixture.

Reverse-phase chromatography is used mainly for the separation of non-ionic substances because ionic, and hence strongly polar, compounds show very little affinity for the non-polar stationary phase. However, ionization of weak acids (or weak bases) may be suppressed in solvents with low (or high) pH values. The effect of such a reduction in the ionization is to make the compound more soluble in the non-polar stationary phase but the pH of the solvent must not exceed the permitted range for bonded phases, i.e. pH 2–8.

> ➤ The chromatographic technique of ionic suppression involves the use of a mobile phase which reduces the charge carried by the analyte.
> ➤ Ion-pair chromatography involves the introduction of ions into the mobile phase which will bind to the analyte and suppress its charge.

While the technique of **ionic suppression** (or **ionization control**) is only effective with weakly ionic species, **ion-pair chromatography** has been developed for strongly ionic species and again utilizes reverse-phase chromatography. If the pH of the solvent is such that the solute molecules are in the ionized state and if an ion (the counter-ion) with an opposite charge to the test ion is incorporated in the solvent, the two ions will associate on the basis of their opposite charges. If the counter-ion has a non-polar chain or tail, the ion-pair so produced will show significant affinity for the non-polar stationary phase.

Counter-ions which are frequently used include tetrabutylammonium phosphate for the separation of anions and hexane sulphonic acid for cations. The appropriate counter-ions are incorporated in the solvent, usually at a concentration of about 5 mmol l^{-1}, and the separation performed on the usual reverse phase media. This ability to separate ionic species as well as non-polar molecules considerably enhances the value of reverse-phase chromatography.

Derivatives

The process of chemically modifying the test molecules before the separation procedure is known as preparing derivatives or pre-column derivatization. In liquid chromatography this is done to permit the test molecules to be more easily detected after separation and to increase the sensitivity of the detection system and less frequently to alter the separation process.

Most derivatives are formed by the introduction into the test molecule of a group that confers either some absorption characteristic, usually in the ultraviolet, or a fluorescent property, thus aiding detection (Table 3.7).

Table 3.7   Derivatives used in liquid chromatography (HPLC)

Derivative	Suitable for analytes	Detection	
		Method	**Wavelength (nm)**
p-Nitrobenzyl	Carboxylic acids	Absorbance	254
p-Nitrobenzoyl	Alcohols	Absorbance	254
p-Bromophenacyl	Carboxylic acids	Absorbance	260
3.5-Dinitrobenzoyl	Alcohols	Absorbance	254
	Amines		
Dansyl	Amines	Fluorescence	360/510
	Peptides		
	Phenols		

One application in liquid chromatography which does alter the separation process is the use of a specific series of derivatives to enable the separation of chiral (optical isomers) forms of alcohols, amines and amino acids using reverse-phase separation. FLEC is available in the two chiral forms (+)-1-(9-fluorenyl) ethyl chloroformate and (−)-1-(9-fluorenyl) ethyl chloroformate (Figure 3.12). Reaction of two stereoisomers of a test compound (e.g. T+ and T−) with a single isomer of the derivatizing reagent (e.g. R+) will result in the formation of two types of product, T+R+ and T−R+. It is possible to separate these two compounds by reverse-phase chromatography.

➤ Racemization – see Section 9.1.1.

Derivatives of FLEC are formed without any racemization of the sample and the derivatives of the D and L forms of the test molecules are eluted from reverse-phase columns in sequence. FLEC derivatives are fluorescent, having an excitation maximum at 260 nm and an emission maximum at 315 nm and are particularly useful in the analysis of amino acids.

3.2.3   Gas–liquid chromatography

GLC (Figure 3.13) depends upon the partition of a solute between two physical states or phases, i.e. gaseous and liquid (solution). Hence in GLC there is

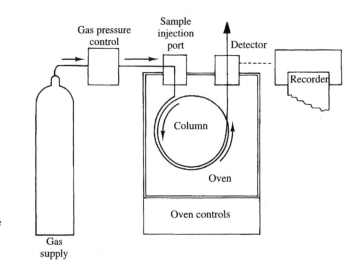

Figure 3.13
A representation of the components of a gas chromatograph.

only one solvent, the stationary phase, which is held immobile in a narrow coiled tube or column.

A solution of the test compounds is introduced into the heated column and is blown through the column by the carrier gas. Upon initial contact of the solute with the liquid stationary phase, an equilibrium is rapidly established between the amount of solute which dissolves in the liquid phase and the amount of solute remaining as a vapour. The precise equilibrium position is a characteristic of the solute and solvent involved but the equilibrium will always be displaced towards the vapour phase if the temperature of the column is raised.

The vapour fraction of the solute is moved down the column by the carrier gas and the equilibrium between the two phases is destroyed. However, in an attempt to re-establish an equilibrium, solute molecules leave the liquid phase restoring the partial pressure above the solution. The solute vapour which has been moved down the column encounters fresh solvent and a new equilibrium is established.

The sample vapours are detected as they leave the column and the compounds which emerge first have either the lowest solubility in the stationary phase or the highest volatility. The detector response is displayed on a recorder and a trace or chromatogram is produced.

> Carrier gases are non-reactive gases used to transport the gaseous samples through the column in GLC.

The impelling force in GLC is the flow of the carrier gas through the column. The gases most frequently used are nitrogen, argon and helium. Gases vary in their viscosity and the more viscous gases, e.g. carbon dioxide, require higher pressures to maintain the flow rate but compared with the less viscous gases, e.g. nitrogen, restrict the extent of diffusion occurring and so tend to give sharper peaks. It is essential that all gases used are pure and dry and that they are chemically inert towards the solutes and liquid phases being used.

The columns used for GLC are several metres in length and are usually coiled in order to save on oven space. Originally the columns were packed with particles of an inert supporting medium (diatomaceous earth derivatives, porous polymers and occasionally glass beads) which had been coated with a film of the stationary phase. This gives a very large surface area and allows tight packing so that there is very little dead space where solute molecules will not be in direct contact with the liquid.

> Capillary columns are designed to give rapid and effective separation by GLC.

These have now been superseded by capillary columns, which offer greatly improved separation efficiency. Fused silica capillary tubes are used which have internal diameters ranging from 0.1 mm (small bore) to 0.53 mm (large bore) with typical lengths in excess of 20 m. The wall-coated open tubular (WCOT) columns have the internal surface of the tube coated with the liquid (stationary) phase and no particulate supporting medium is required. An alternative form of column is the porous-layer open tubular (PLOT) column, which has an internal coating of an adsorbent such as alumina (aluminium oxide) and various coatings. Microlitre sample volumes are used with these capillary columns and the injection port usually incorporates a stream splitter.

Analytical procedure

The chromatogram produced by gas chromatography (Procedure 3.2) is comparable to that produced by HPLC and all the considerations regarding peak height and area measurements, and internal and external standards are relevant. The need to use internal standardization and response factors for quantitative work is more important because of the instability of some detectors and the imprecision associated with injection of very small volumes of relatively volatile samples.

Individual components are identified by their retention time, usually measured along the time axis of the chart paper, but the reproducibility of retention times is significantly affected by alterations in the gas flow and column temperature. It is not adequate to rely on quoted values for retention times but it is necessary to determine the values at frequent intervals using identical experimental conditions for tests and reference compounds.

> Retention time is the qualitative parameter in GLC.

Detectors

The **thermal conductivity detector** or katharometer is the simplest detector but is not commonly used nowadays. It measures the variations in resistance of a solid electrical conductor (a length of platinum wire) which are induced by changes in temperature. The wire is suspended in the column and heated electrically, and under stable conditions will attain a steady temperature depending upon the current flowing and the loss of heat by radiation and thermal conduction (Figure 3.14). Any variation in the composition of the gas surrounding the wire will result in a change in its temperature and therefore a change in its resistance.

> A katharometer measures the changes in the resistance of a wire as the samples emerge from the GLC column.

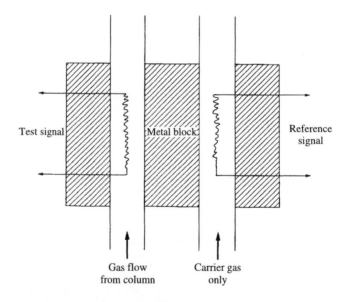

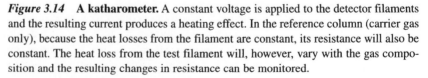

Figure 3.14 **A katharometer.** A constant voltage is applied to the detector filaments and the resulting current produces a heating effect. In the reference column (carrier gas only), because the heat losses from the filament are constant, its resistance will also be constant. The heat loss from the test filament will, however, vary with the gas composition and the resulting changes in resistance can be monitored.

The resistance of the wire is balanced against the resistance of a similar wire in a reference column (containing only the carrier gas) by means of a Wheatstone bridge arrangement so that variations in resistance can be monitored.

A thermal conductivity detector will respond to all compounds and is capable of detecting about 1×10^{-7} mol of substance in the carrier gas but does not show a good linear response to increasing amounts of test substances.

The **flame ionization detector** is commonly used and depends upon the thermal energy of a flame causing some ionization of the molecules in the flame. The ions are collected by a pair of polarized electrodes and the current produced is amplified and recorded (Figure 3.15). For a particular gas composition, there will be a constant degree of ionization but as the composition of the gas mixture changes due to the presence of the test components so the degree of ionization will change. A supply of hydrogen and oxygen is fed to the detector to produce the flame.

> ➤ A flame ionization detector measures changes in ionic composition as the samples emerging from a GLC column are burnt in a flame.

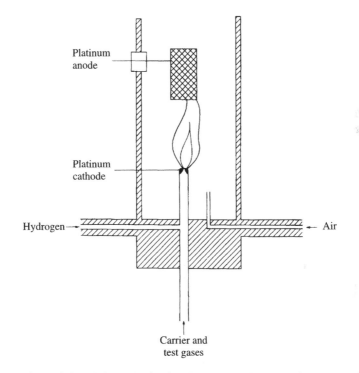

Figure 3.15   **A flame ionization detector.** Hydrogen and oxygen are introduced into the gas mixture as it emerges from the column to allow it to be burnt in the detector. Some molecules are ionized in the flame and cause a current to flow between the two polarized electrodes. The degree of ionization varies with the composition of the gas mixture and the resulting changes in current can be monitored.

Flame ionization detectors are capable of detecting virtually all organic compounds and show a lower limit of detection of approximately 1×10^{-9} mol. They also show good linearity of response and the fact that they do not respond to oxides of carbon or nitrogen or to water makes them particularly convenient for aqueous samples. They have the disadvantage, however, that samples are destroyed unless a stream-splitting device is incorporated.

A modification of flame ionization detection involves the introduction into the flame of atoms of one of the alkali metals, e.g. potassium, rubidium or caesium. Such a device, known as an **alkali flame ionization detector (AFID)**, gives an enhanced response to nitrogen- or phosphorus-containing compounds, detecting levels as low as 1×10^{-10} mol.

The **electron capture detector** also depends upon the ionization of the carrier gases but uses a beta-emitting isotope as a means of ionization (Figure 3.16). The isotope, usually ^{63}Ni or ^{3}H, is held on a foil in the ionization chamber through which the emerging gases pass. The ionization of the carrier gas results in the release of electrons and hence a current will flow between two polarized electrodes. If, however, electrophilic compounds are present (i.e. those containing oxygen, nitrogen, sulphur or a halogen) they will capture free electrons, reducing the current to some extent.

> ➤ An electron capture detector detects electrophilic compounds.

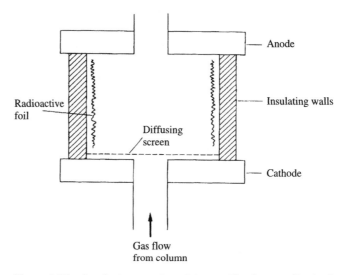

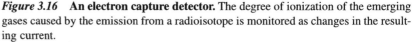

Figure 3.16   **An electron capture detector.** The degree of ionization of the emerging gases caused by the emission from a radioisotope is monitored as changes in the resulting current.

Electron capture detectors are extremely sensitive (1×10^{-12} mol) but are specific for electrophilic compounds. However, they can be used in parallel with flame ionization detectors to identify specific peaks in a chromatogram.

The stationary phase

The major consideration in selecting a stationary (liquid) phase is its polarity relative to the test compounds. A non-polar stationary phase will tend to retain non-polar solutes while a polar liquid will show greater affinity for polar solutes. In non-polar systems, because hydrogen bonding does not occur, elution of the solutes from the column is usually in order of their boiling points. Hydrogen bonding will occur to varying degrees with polar systems and will appreciably modify the elution order because the molecules involved in hydrogen bonds will be retarded.

It is difficult to classify solvents in terms of their polarity and the concept of the selectivity of a solvent was developed initially by Kovats and subsequently by McReynolds, whose classification system is currently the most popular. It is based on Kovats's concept of attributing a retention index (I) to the n-paraffin series of hydrocarbons, selected as reference compounds because of their stability and ease of purification. For a particular stationary phase, a plot of the logarithm of the retention time against the carbon number results in a straight line graph. This can be used to attribute a retention index value to any compound on the basis of an apparent carbon number determined from the retention time of that compound. For convenience the carbon number determined from the graph is multiplied by 100 to eliminate decimals. This method involves isothermal chromatography but it is not uncommon for specific applications for retention index data to be determined from programmed temperature gradient separations.

McReynolds used the retention index of certain solutes to compare different stationary phases and to assess their selectivity compared with a reference liquid phase, squalane. Squalane is considered to be non-polar and any increase in the retention index of the selected solute on the test column compared to squalane may be considered to be due to the greater polarity of that solvent. McReynolds constants have been determined for all stationary phases using a range of solutes of varying polarity (Table 3.8) and may be used to assist in selecting an appropriate stationary phase.

➤ McReynolds constants constitute a classification system for GLC stationary phases based on their polarity.

$$\text{McReynolds constant} = I_{\text{test liquid}} - I_{\text{squalane}}$$

An examination of a table of these constants for different stationary phases and a solute that is most similar to the test substance will indicate the most suitable stationary phase. If it is desired to separate two solutes of differing polarity, a solvent should be selected that shows a significant difference in the constants for the two most appropriate reference solutes.

Table 3.8   Stationary phases for gas–liquid chromatography

Liquid phase	Example	McReynolds constants for selected test compounds				
		Benzene	Butanol	Pentanone	Nitropropane	Pyridine
Squalane	Reference phase	0	0	0	0	0
Paraffin oils	Nujol	9	5	2	6	11
Apiezon grease	APL	32	22	15	32	42
Fluorinated hydrocarbons	Fluorolube	51	68	114	144	118
Diethyleneglycol succinate	DEGS	492	733	581	833	791
Polyethyleneglycol	PEG 600	350	631	428	632	605
	Carbowax 1000	347	607	418	626	589
Silicones, methyl	OV1	16	55	44	65	42
Silicones, phenyl	OV17	119	158	162	243	202
Silicones, fluoro	OV210	146	238	358	468	310
Silicones, cyano	OV275	629	872	763	1106	849

Benzene represents aromatics and olefins. Pentanone represents keto compounds and esters.
Butanol represents alcohols and weak acids. Nitropropane represents nitro- and nitrile compounds.
Pyridine represents N-heterocyles.

Column conditions

The efficiency of a column can be assessed in a similar manner to that described for HPLC and values for the resolution index of two solutes, the number of theoretical plates and the height equivalent to a theoretical plate may also be calculated. Although it is easier to measure gas pressure, it is the actual gas flow, which is affected by the particle size and compression of the packing, that should be used in column assessment investigations.

In many instances, samples contain components with a wide range of volatilities and it may be difficult to separate them quickly and effectively at a fixed temperature; a temperature gradient may be used. The separation is initiated at the lower temperature for a specified period of time depending upon the retention times of the more volatile components. Subsequently the temperature of the column is raised at a specified rate to speed up the elution of the less volatile components. Such temperature programming can cause some problems but these are usually solved in the design of the instrument. As the temperature rises so the gas flow will fall owing to an increase in the back-pressure within the column, and it is usual to have a system that regulates the gas pressure in order to maintain the flow rate. In addition, the increase in temperature may also cause loss of the solvent (bleeding) from the column, resulting in a baseline increase which must be taken into account in peak height measurements. Temperatures above the specified maximum for the stationary phase must not be used as the column will rapidly deteriorate.

► Temperature programming is a form of gradient elution – see Section 12.7.4.

Derivatives

Many substances are not initially appropriate for gas chromatography because of their relatively high boiling points or insolubility. In such cases it is often possible to modify the compound chemically and render it more amenable to separation. In some instances, the chemical modification is used to enable easier detection of the compound, e.g. the introduction of a halogen for use with electron capture detectors.

There are a wide variety of derivatization reactions but the most frequently used are indicated in Table 3.9.

Table 3.9 Derivatives for gas chromatography

Reaction	Derivatives formed	Reagents used	Compounds treated
Silylation	Trimethylsilyl R—OH R—OSi(CH$_3$)$_3$	BSTFA TMS	Widely used acids, phenols, alcohols
Acylation	Trifluoroacetyl Heptafluorobutyryl	TFAA	Amines, phenols, alcohols
Alkylation	Methyl, ethyl or butyl	TMAH	Fatty acids, various drugs

BSTFA, bis(trimethylsilyl)-trifluoroacetamide; TMS, trimethylchlorosilane; TFAA trifluoroacetic anhydride; TMAH, trimethylanililium hydroxide.

3.2.4   Mass spectrometry

Mass spectrometry is a technique that is widely employed as a detector for GLC and HPLC. It is a powerful tool for the rapid identification and quantification of even femtogram quantities of analyte. Since the early 1980s there has been considerable development of the technique and associated instrumentation and improvements are constantly being introduced. Particularly significant developments have been associated with the computing aspects of mass spectrometry, the means of producing ions from neutral species and the overall design specification, including the production of 'bench top' instruments. The use of mass spectrometry has now been extended to the analysis of large molecules, e.g. proteins, and this has considerably enlarged its application in the biochemical field. This is largely due to the widespread use of electrospray and 'time of flight' (TOF) mass spectrometry. In addition, the introduction of mass spectrometry/mass spectrometry (MS/MS) or tandem MS, which are independent of a chromatographic separation process, has opened up a whole new area of analytical applications.

> ➤ In tandem MS, ions separated in the first analyser are analysed in the second.

There are many types of mass spectrometer, each having special design features, some offering very sophisticated modes of analysis and these details will not be described here. Rather, the fundamental principles and instrumental aspects associated with the technique in general are covered. More detailed information is available in specialist books or instrument manufacturers' publications.

The mass spectrum

The production of ions from neutral compounds and the examination of how these ions subsequently fragment is fundamental to mass spectrometry. Neutral sample molecules can be ionized by a variety of processes. The most important of these for the production of positively charged species is the removal of an electron or the addition of one or more protons to give either 'molecular ions' ($M^{+\bullet}$) or 'protonated molecular species' $(M+nH)^{n+}$. This initial stage of ionization is often followed by fragmentation to produce ionized fragments, 'fragment ions'.

> ➤ A molecular ion ($M^{+\bullet}$) is formed when a molecule is ionized by the removal of an electron.

The extent to which further fragmentation pathways now proceed, and hence the fragmentation pattern that is produced, is characteristic of the compound. The mass spectrometer is designed to separate and measure the mass of the ions using their mass-to-charge ratio (m/z) ratios. Ions are usually formed with a single, positive charge ($z = 1$) and in this situation m/z gives the mass of the ion. In the mass spectrum that is produced the relative amounts of the ions is displayed as their relative abundance on the y-axis and their m/z values on the x-axis (Figure 3.17).

> ➤ Mass-to-charge ratio is represented as m/z.

This information is plotted by the computer in alternative ways and it is important for the analyst to be aware of the ways in which the data system is operating. In the normalized or percentage relative abundance (%RA) method, which is commonly used, the height of each peak is shown as a percentage of the biggest peak in the spectrum. The total ion current (TIC) is the sum of all the detector responses for each scan plotted against time and this is equivalent to a GLC trace. This information is particularly useful in quantitative analysis.

> ➤ A mass spectrum is a plot of ion abundance against m/z ratio.
> ➤ The highest peak in the spectrum is called the base peak.

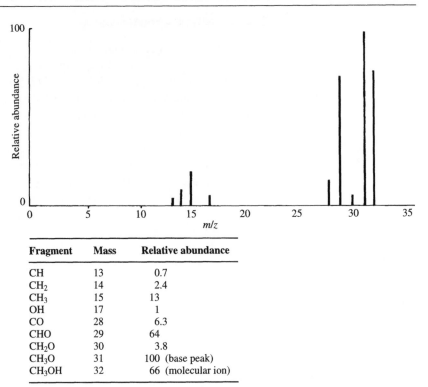

Fragment	Mass	Relative abundance
CH	13	0.7
CH_2	14	2.4
CH_3	15	13
OH	17	1
CO	28	6.3
CHO	29	64
CH_2O	30	3.8
CH_3O	31	100 (base peak)
CH_3OH	32	66 (molecular ion)

Figure 3.17 **Mass spectrum of methanol.**

The strongest peak is called the base peak and is set to read 100. The signal from the non-fragmented ionized molecule is not usually very strong and is sometimes missing completely.

Instrumentation

There is a variety of differently designed instruments available but they all have the same essential features (Figure 3.18).

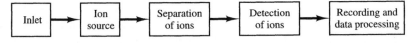

Figure 3.18 **Diagrammatic representation of the components of a mass spectrometer.** Often the sample is introduced through the inlet as a vapour and subjected to bombardment with electrons. The products of this process are separated by a variety of means and detected as they exit the system.

Inlets and ion sources

Mass spectrometry is traditionally a gas phase technique for the analysis of relatively volatile samples. Effluents from gas chromatographs are already in a suitable form and other readily vaporized samples could be fairly easily accommodated. However the coupling of mass spectrometry to liquid streams, e.g. HPLC and capillary electrophoresis, posed a new problem and several different methods are now in use. These include the spray methods mentioned below and bombarding with atoms (fast atom bombardment, FAB) or ions (secondary-ion mass spectrometry, SIMS). The part of the instrument in which ionization of the neutral molecules occurs is called the ion source. The commonest method of

► The ion source is where ionization of the analyte occurs.

ionization is by electron ionization or electron impact (EI) of a volatile sample in an ion chamber maintained under a vacuum. This produces positively charged ions. The energy of the electrons causes an electron to be ejected from the molecule, forming a molecular ion and fragment ions. This technique is commonly used for GC-MS.

Several methods of ionization are based on the use of electric fields. In electrospray (ES) a liquid is pumped through a stainless steel capillary held at a potential of several thousand volts. This results in a spray of highly charged droplets. The droplets are dried using heat and/or a gas flow and after evaporation the charges are retained by the sample molecules. This makes electrospray a suitable ionization method for liquid chromatography/mass spectrometry (LC/MS). Its applicability to large RMM proteins arises from the fact that in electrospray ions formed are often multiply charged and therefore their m/z values fall within the capabilities of conventional instruments.

> ➤ LC/MS refers to the use of a mass spectrometry as a detector for liquid chromatography, e.g. HPLC.

Separation of ions

Sector instruments
The separation of ions according to their m/z ratios is achieved using electric and/or magnetic fields in a number of ways. The trajectories of ions moving in such fields are determined by their m/z values and these can be monitored to ascertain their mass. Double-focusing instruments use the combined effects of electric and magnetic fields to effect separation (Figure 3.19). In a typical instrument, after the ions have been accelerated away from the ion source

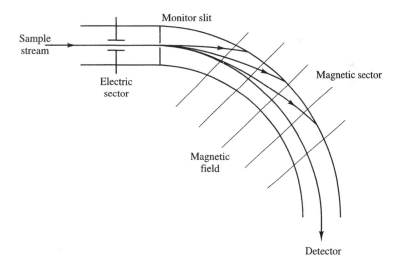

Figure 3.19
Schematic diagram of a sector mass spectrometer.

through a potential of several kilovolts, they pass through slits to reduce the spread. The ion beam then passes through an electric sector, formed by applying a potential across two plates. This has the effect of deflecting the beam into an arc, focusing ions of like energies, irrespective of their mass-to-charge ratios, at the monitor slit. Mass separation is effected in the magnetic sector as the mass-to-charge ratios will determine the curvature of the path taken by the ions. Some instruments use an array of multiple detection devices to record the ions of several different m/z ratios simultaneously while, more commonly, others scan the magnetic field to record one m/z at a time on a point detector.

Quadrupole instruments
The ions are propelled from the ion source into the quadrupole analyser (Figure 3.20) by a low accelerating voltage of only a few volts. They enter the space

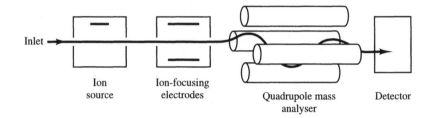

Inlet →

Ion
source

Ion-focusing
electrodes

Quadrupole mass
analyser

Detector

Figure 3.20
Diagrammatic representation of a quadrupole mass spectrometer system.

between four or more parallel rods which are precisely positioned and have a DC voltage and a radio-frequency potential associated with each opposite pair. Under the influence of the combined electric fields the ions oscillate and follow complex trajectories through the rods. The separation of ions according to mass can be effected by changing the DC and radio-frequency voltages while keeping the ratio of the two constant. Although this type of instrument has lower resolution than magnetic sector instruments, it is robust and available in bench top designs, and the rod voltages can be readily changed in order to focus on selected ions (SIM). This makes it very useful for quantitative work.

An instrument that stores ions and then ejects those of selected masses is called an ion trap. Like quadrupole analysers such instruments employ electric fields to effect separation and are based on a similar technology. Such systems can be considered as alternatives to tandem MS.

Another approach to the separation of the ions is based on differences in their velocity after acceleration through a potential. The mass is related to the time taken to reach the detector. This is known as TOF-MS.

➤ Selected ion monitoring (SIM) is when the mass spectrometer is tuned to record spectra for specific m/z values.

Tandem mass spectrometry (MS/MS)
This basically means that two instruments have been linked together. The first analyser can replace the traditional chromatographic separation step and is used to produce ions of chosen m/z values. Each of the selected ions is then fragmented by collision with a gas, and mass analysis of these product ions effected in the second analyser. The resulting mass spectrum is used for their identification. The potential combinations of the various magnetic sector and quadrupole instruments to form such coupled systems is considerable. Ion traps may also be operated in a tandem MS mode.

Computer
Modern mass spectrometers are fully computer controlled for all aspects of operation including automatic scanning, signal processing, data collection,

quantitation and, for identification, library searching. This latter function involves comparing the spectrum obtained with that of known compounds held in the computer database.

Quantitation

Internal standards are used in quantitative work in a similar manner to GLC. These are usually homologues or isotopically labelled analogues of the analyte, e.g. deuterium (^2H).

Self test questions

> **Section 3.2**
> 1. Which of the following solvents is the most polar?
>   (a) Chloroform.
>   (b) Benzene.
>   (c) Methanol.
>   (d) Acetone.
> 2. In GLC, if a stationary phase with a low value McReynolds constant is used which of the following will be eluted quickly from the column?
>   (a) A polar molecule.
>   (b) A non-polar molecule.
>   (c) A large RMM molecule.
>   (d) A molecule with a low boiling point.
> 3. A column with a high number of theoretical plates is likely to give good resolution of samples
>   BECAUSE
>   the theoretical plate number is a measure of the band broadening that occurs during a separation.
> 4. It is only necessary to use a response factor when working with external standards
>   BECAUSE
>   a detector may respond differently to the test substance and the standard.

3.3 Methods based on ionic nature

Ionizable groups in a molecule give it an ionic character and, as a result, under certain conditions it will carry a charge. However, the intensity of this charge and, in the case of many molecules, the sign (positive or negative) depend upon the pH and composition of the solution. These properties form the basis of several methods of separation, namely ion-exchange chromatography and electrophoresis (Figure 3.21).

3.3.1 Ion-exchange chromatography

Although the phenomenon of ion-exchange has been appreciated for many years, it was the development, by D'Alelio in 1942, of synthetic ion-exchange media based on the polystyrene resins that extended the use of ion-exchange as an analytical tool.

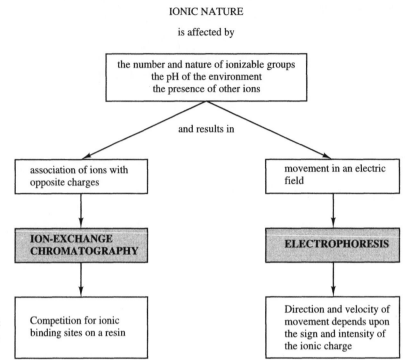

IONIC NATURE

is affected by

the number and nature of ionizable groups
the pH of the environment
the presence of other ions

and results in

association of ions with opposite charges

movement in an electric field

ION-EXCHANGE CHROMATOGRAPHY

ELECTROPHORESIS

Competition for ionic binding sites on a resin

Direction and velocity of movement depends upon the sign and intensity of the ionic charge

Figure 3.21 **Separation methods in which the ionic nature of the test molecule plays an important part.**

An ion-exchange resin consists of an insoluble, porous matrix containing large numbers of a particular ionic group which are capable of binding ions of an opposite charge from the surrounding solution. Hence a cation-exchange resin contains fixed anions and the mobile cations can be any cations from the solution. Ion-exchange chromatography is essentially a displacement technique in which the ions in either the sample or the buffer displace the existing mobile ions associated with the fixed resin ions (Figure 3.22).

Ion-exchange resins are cross-linked polymers which are typically polystyrene, cellulose or agarose based. Polystyrene is hydrophobic in nature and useful for inorganic ions and small molecules while cellulose and agarose are hydrophilic and more useful for the larger, biologically important molecules, e.g. proteins and nucleic acids, which either would be adversely affected by a hydrophobic environment or could not gain access to the small pore structure.

The degree of cross-linking of the resin is significant in that it enhances the rigidity and insolubility of the resin but it also reduces the pore size, and this is undesirable when working with macromolecules. Polystyrene cross-linking is expressed as the percentage of divinyl benzene (the cross-linking group) present in the original preparation and is commonly in the region of 8–10%.

Although, theoretically, any ionic group can be used in an ion-exchange resin, in practice the number is limited (Table 3.10). For ion-exchange to take place the resin-bound group must be ionized, and its capacity to do this is related to its strength (or ability to dissociate). The dissociation of weak acids is suppressed by the presence of only slightly increased concentrations of hydrogen ions, while the ionic form of a weak base (its

➤ Hydrophobic – water hating; hydrophilic – water loving.

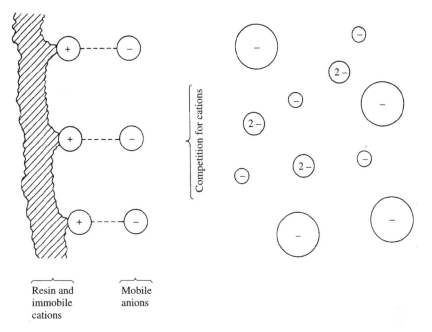

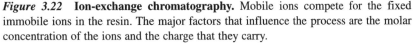

Figure 3.22 **Ion-exchange chromatography.** Mobile ions compete for the fixed immobile ions in the resin. The major factors that influence the process are the molar concentration of the ions and the charge that they carry.

conjugate acid) dissociates as the pH increases:

$$NH_2H^+ \rightleftharpoons NH_2 + H^+$$

Hence an ion-exchange resin cannot be used at a pH that suppresses the ionization of the group, and weak anion exchange resins, for instance, are only effective over the pH range 2–8. The stronger acids and bases, however, are capable of being used over almost the whole pH range.

The major consideration in selecting a resin is the charge carried by the test ions. In the case of uncomplexed inorganic ions, it is relatively constant but for many molecules, it can be altered considerably by variations in pH.

There are three major factors that determine the binding of ions in such a competitive system. The size of the charge carried will result in divalent ions showing greater affinity for the resin than monovalent ions. The intensity of the charge is also significant and small monovalent ions, e.g. hydrogen, will show greater affinity than large monovalent ions, e.g. potassium. Superimposed on both of these considerations is the effect of the concentration of the ions and this is demonstrated by the fact that a high concentration of a low affinity ion is capable of displacing a low concentration of an ion with a higher affinity for the resin. It is the careful control of these three factors that provides the selectivity in ion-exchange separations (Figure 3.22).

Methods in ion-exchange chromatography

Ion-exchange chromatography may be used for either purification of an individual component or fractionation of a mixture. In order to isolate a particular component

Table 3.10   Ion-exchange media

Medium	Nature	Effective pH range	Applications
Anion exchangers			
Quaternary ammonium	Strong	2–11	Strong acids, e.g. nucleotides
Tertiary ammonium	Intermediate	2–7	Weak acids, e.g. organic acids
Diethylaminoethyl	Weak	3–6	Weakly polyanionic, e.g. proteins
Cation exchangers			
Sulphonated	Strong	2–11	Strong bases, e.g. amino acids
Carboxylate	Intermediate	6–10	Weak cations, e.g. peptides
Carboxymethyl	Weak	7–10	Weak polycationic, e.g. proteins

from a mixture, it may be selectively retained on a resin while the unwanted constituents, which under the conditions of the analysis should be uncharged or carry the same charge as the resin, will be eluted. Subsequently the required component can itself be eluted in a small volume of an appropriate buffer. Conversely, the resin may be selected so that the required compound does not bind but unwanted ions do and the sample can be eluted quickly from the column without the contaminating ions. Neutral compounds lend themselves to this technique in which a bed of mixed anion and cation resins can remove unwanted ions and replace them with water (H^+ and OH^-). The technique of fractionation is used for samples that contain mixtures of similar ions and permits the separation and quantitation of each component, the various components being eluted sequentially as the solvent composition is changed. It is in this form that ion-exchange chromatography is used as an HPLC technique.

The resin should be prepared for use by washing it with a solution containing a high concentration of an ion which has a high affinity for the resin, e.g. hydrochloric acid or sodium hydroxide solution (1.0 mol 1^{-1}). The hydrogen and hydroxyl ions will displace all other ions present from cation- and anion-exchange media respectively. After washing in water or buffer to remove the excess acid or alkali, the resin is said to be in the hydrogen or hydroxyl form respectively. For some applications it may be necessary to convert it to a different ionic form, e.g. the sodium form, that has an affinity for the resin comparable to that of the test ion. The resin is finally equilibrated with the chosen buffer, the pH of which must be carefully selected to ensure the correct ionization of both the test ions and the resin.

> ➤ The form of a resin refers to the mobile ion associated with it prior to use.

A small volume of sample is applied to the column and the components of the mixture eluted using the buffer. A single buffer (isocratic separation) or a gradient technique may be used depending on the complexity of the sample. Gradient elution normally involves either changes in pH (causing alterations in ion affinity for the resin) or changes in the concentration of the buffer (causing displacement of the test ions).

> ➤ Gradient separation of amino acids – see Section 10.6.2.

3.3.2   Electrophoresis

Electrophoresis is the term given to the migration of charged particles under the influence of a direct electric current. It is a single-phase system and depends upon the relative mobilities of ions under identical electrical conditions.

Moving-boundary electrophoretic techniques, originally demonstrated by Tiselius in 1937, employ a U-tube with the sample occupying the lower part of the U and the two limbs being carefully filled with a buffered electrolyte so as to maintain sharp boundaries with the sample. Electrodes are immersed in the electrolyte and direct current passed between them. The rate of migration of the sample in the electric field is measured by observing the movement of the boundary as a function of time. For colourless samples, differences in refractive index may be used to detect the boundary. Such moving-boundary techniques are used mainly in either studies of the physical characteristics of molecules or bulk preparative processes.

Zonal techniques are the most frequently used form of electrophoresis and involve the application of a sample as a small zone to a relatively large area of inert supporting medium which enables the subsequent detection of the separated sample zones. A wide range of supporting media have been developed either to eliminate difficulties caused by some media (e.g. the adsorptive effects of paper) or to offer additional features (e.g. the molecular sieving effects of polyacrylamide gel).

Although molecular frictional effects impede the movement of molecules through a liquid, an effect that increases with the size of the molecule, the major factors in electrophoresis are the charge carried by the molecule and the voltage applied. The nature of the charge (positive or negative) will determine the direction of migration while the magnitude of the charge will determine the relative velocity.

➤ Ionization of amino acids – see Section 10.1.3.

This charge, although initially due to the ionization effects mediated by the pH of the buffer, will be appreciably modified by other features of the buffer composition. For a colloidal particle in suspension, e.g. a protein, its charged surface is surrounded by a layer of oppositely charged ions derived from the solution (Figure 3.23). There is an immobile, fixed layer of ions adsorbed on to the surface and a diffuse, mobile layer which becomes more mobile with increasing distance from the surface of the colloid. The charge developed by the molecule is due to its chemical nature and the pH of the buffer and is known as the **electrochemical potential**. For colloids, this is reduced as the salt concentration in the buffer increases, due to the partial neutralizing effect of the adsorbed layer of ions. The net resulting charge is known as the **zeta potential** and is the determining factor in electrophoretic mobility. The combined effect of the fixed and mobile layers of ions also increases the size of the particle and reduces its mobility even further.

➤ The electrochemical potential is the theoretical charge developed by a molecule at a specific pH.
➤ The zeta potential is the actual charge carried by a molecule as a result of interactions with other ions in the solution.
➤ The ionic strength is a measure of either the total anionic or cationic charge of a solution.

The electrophoretic mobility of an ion is inversely related to the ionic strength of the buffer rather than to its molar concentration. The ionic strength (μ) of a buffer is half the sum of the product of the molar concentration and the valency squared for all the ions present in the solution. The factor of a half is necessary because only half of the total ions present in the buffer carry an opposite charge to the colloid and are capable of modifying its charge:

$$\mu = \frac{1}{2}\Sigma\ ct^2$$

The rate of movement of a charged particle is also related to the voltage gradient applied across the supporting medium, which is quoted in volts per centimetre (i.e. the distance between electrodes). The current generates heat,

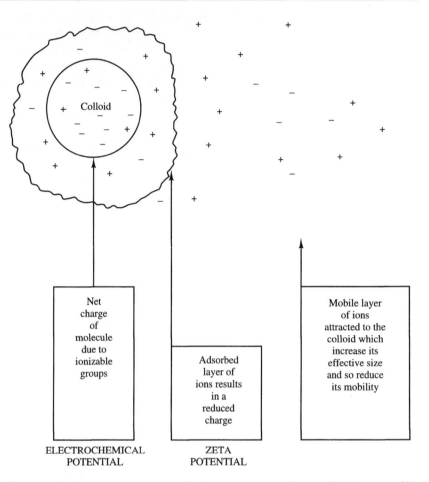

Figure 3.23 **The charge on a colloid.** The charge carried by a colloid because of its chemical composition and the pH of the solution (the electrochemical potential) is reduced by the adsorption of ions from the solution and the resulting charge is known as the zeta potential.

which causes an increase in the conductivity of the electrolyte solution and this further increases the current. This rise in temperature will cause some evaporation of the buffer, which in the small amount held in the supporting medium will result in an increase in its concentration. This will affect both the zeta potential of the colloid and the conductivity of the buffer. Because of these effects it is necessary to control the temperature of the system in some way and to choose between the use of a fixed voltage and a fixed current. Some power sources have the facility to select a fixed wattage output, which is effectively a compromise position.

An additional problem with some supporting media is the phenomenon of **electroendosmosis**, in which the buffer itself moves due to an electrophoretic effect and hence masks the movement of the solute to some extent. However, this feature is exploited in some situations to aid separation. Electroendosmosis is caused by the presence of negatively charged groups on

➤ Electroendosmosis is the movement of the buffer during electrophoresis.

the surfaces of some supporting media, e.g. paper, agar, and these induce an opposite charge in the buffer solution. As a result, under alkaline conditions, the buffer moves towards the cathode, the rate being greater for lower concentrations of buffer.

A wide range of electrophoretic equipment is available and while it varies considerably in appearance, there is a common basic design (Figure 3.24). A well-designed tank should include a cover to prevent any significant evaporation from the supporting medium and an effective cooling system is necessary when gels or high voltage techniques are used. This usually is in the

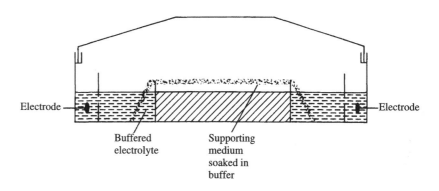

Figure 3.24
An electrophoresis tank.

Electrode

Electrode

Buffered electrolyte

Supporting medium soaked in buffer

form of an electrically insulated plastic or metal base for the supporting medium, through which cold water is circulated. A stabilized voltage supply is necessary and safety devices on the whole equipment are essential to prevent accidents, particularly if high voltages are used.

Supporting media

If the supporting medium is a sheet or membrane, it is soaked in the buffered electrolyte, the excess being removed before the sample is applied. Alternatively a gel, such as agar or polyacrylamide, can be prepared in the buffer and a slab or column used for the separation process.

Filter paper has been an extremely useful and convenient medium for many years and compared with some of the more recent membrane media is capable of handling a relatively large volume of sample, e.g. 10 μl. The random structure of the paper, however, tends to result in irregularities in the separation patterns and the relatively polar nature of the cellulose shows some adsorptive effects giving a significant degree of 'tailing' of the zones. Additionally the large buffer volume in the paper, although an advantage from conductivity and evaporation points of view, because of the relatively long separation times (12–18 hours), results in a significant amount of diffusion and hence broadening of the zones.

Cellulose acetate membrane (CAM) was developed with the aim of reducing the polar nature of paper by acetylating the hydroxyl groups and producing a more regularly defined pore structure. For these reasons, it has considerable advantages over paper and shows minimal adsorptive effects with increased separation rates (1–2 hours) but it still shows electroendosmotic effects. Particular care must be taken over handling the membrane, which is very fragile when dry but is much more resilient when wet. Less sample can be applied to cellulose acetate membrane than to paper, volumes of 1–2 μl being common. After locating the separated bands, the membrane can be made transparent for densitometry by soaking in either an oil or an organic solvent mixture specified by the manufacturer.

➤ Densitometry – see Section 11.3.2.

A variety of gels has been used for preparative techniques and to eliminate some of the disadvantages of the solid support media. Purified agar is used at a concentration of 10 g l^{-1} in a buffered electrolyte and the samples are introduced in a small well or trough cut out of the gel. It shows an appreciable degree of electroendosmosis and, owing to its polar nature, causes the precipitation of some substances, notably the lipoproteins. In such situations the use of **agarose** (5–8 g l^{-1}), a less polar fraction of agar, eliminates virtually any electroendosmotic effect and gives a molecular sieving effect with large molecules such as proteins and nucleic acids.

➤ Electrophoresis of nucleic acids – see Section 13.2.3.

The molecular sieving effect can be increased considerably by using gels with a smaller pore structure. **Starch gel** is prepared by the partial hydrolysis of potato starch; the greater the degree of hydrolysis, the smaller the fragments and, hence, the smaller the pore size of the resulting gel. In such a medium the larger molecules are impeded by the pore structure and therefore do not move as rapidly as the smaller molecules with the same charge (Figure 3.25).

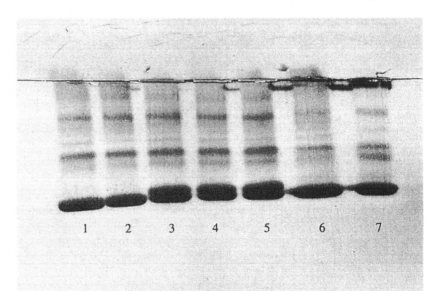

Figure 3.25 **Starch gel electrophoresis of human serum proteins.** Samples 1 and 2 are normal while samples 3, 4 and 5 show an extra band adjacent to the normal albumin which is due to the presence of bis-albumin. Sample 7 contains a myeloma protein that has remained near the origin. (Photograph by permission of Dr D. Brocklehurst, Department of Clinical Chemistry, Doncaster Royal Infirmary, UK.)

The gel is prepared by heating a slurry of the starch in the appropriate buffer (150 g l^{-1}) in a boiling water bath until the mixture becomes translucent. A vacuum is applied to the flask to remove air bubbles and the solution quickly poured into a heated tray and allowed to cool to form a gel about 5 mm thick. During electrophoresis, the gel must be adequately cooled to prevent it melting and separations may take up to 6 hours at about 150 V. These gels are extremely fragile and the method of hydrolysis is entirely arbitrary; it is difficult to reproduce the pore size in successive preparations.

Polyacrylamide shows many advantages over starch gel as a medium for high resolution electrophoresis and because of its synthetic nature its pore size can be more easily controlled. The gel is formed by the polymerization of the two monomers, acrylamide and a cross-linking agent, N, N-methylene-bis-acrylamide (Figure 3.26). The proportion of the two monomers and not their total concentration is the major factor in determining the pore size, the latter having more effect on the elasticity and

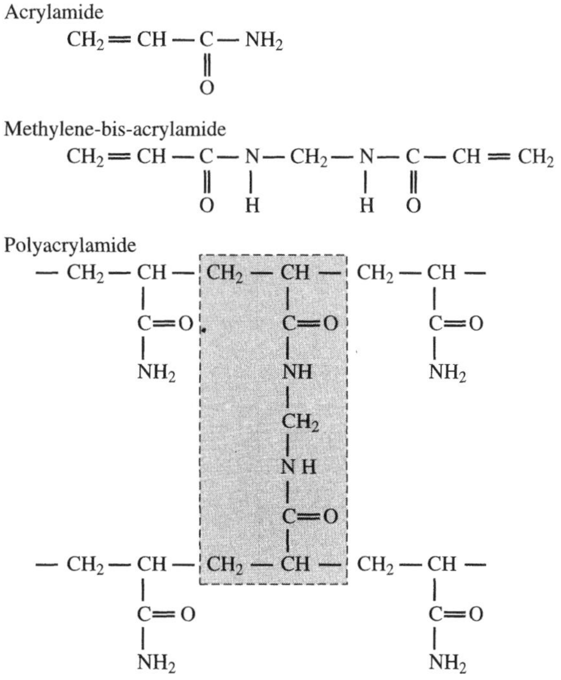

Acrylamide

$$CH_2 = CH - C - NH_2$$
$$\overset{\|}{O}$$

Methylene-bis-acrylamide

$$CH_2 = CH - \overset{\|}{\underset{O}{C}} - \overset{}{\underset{H}{N}} - CH_2 - \overset{}{\underset{H}{N}} - \overset{\|}{\underset{O}{C}} - CH = CH_2$$

Polyacrylamide

Figure 3.26 **Polyacrylamide.**

Preparation of gel

A mixture is prepared containing:

1 volume of the selected buffer,

2 volumes of the selected mixture of monomers,

1 volume of TEMED (3.0 g 1^{-1}).

Dissolved oxygen (which tends to inhibit the reaction) is removed under a vacuum.

Four volumes of ammonium persulphate (3.0 g 1^{-1}) are added and gently mixed before pouring into the mould. The mixture is covered with a thin-layer of water to exclude air and allowed to polymerize (approximately 30 min). Protective gloves should be worn and the preparation of the gel carried out in a fume cupboard because the monomers are toxic.

transparency of the gel. A minimum total concentration of approximately 20 g l^{-1} is necessary for gel formation although concentrations in the region of 70 g l^{-1} are frequently used. The pore size of the gel decreases with the increasing proportion of the bis-acrylamide, with a limiting value of approximately 5% of the total giving minimum pore size. For protein electrophoresis, gels containing 2.5% of the cross-linking agent are often used.

Polymerization of the gel may be achieved either by ultraviolet photoactivation with riboflavin, or, preferably, using ammonium persulphate as a catalyst. It is necessary to include an initiator for the reaction, TEMED (tetramethylethylene diamine) being commonly used.

Polyacrylamide gel (PAG) electrophoresis is performed either in cylindrical glass tubes or in flat beds (Figure 3.27). The method is comparable with other electrophoretic techniques but care must be taken to keep the current low to prevent any significant heating effect.

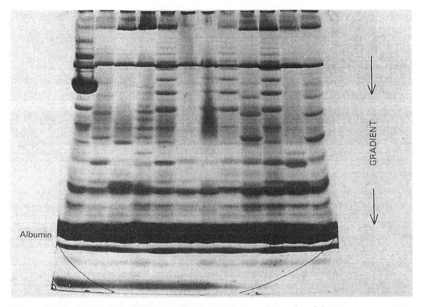

Figure 3.27 **Polyacrylamide gradient gel electrophoresis of human serum proteins**. The proteins are separated in a gel which has an increasing concentration gradient with a parallel decrease in pore size, which restricts the movement of the larger molecules. Note the large number of different protein bands that can be demonstrated. (Photograph by permission of Dr D. Brocklehurst, Department of Clinical Chemistry, Doncaster Royal Infirmary, UK.)

➤ DNA sequencing – see Section 13.5.
➤ Protein SDS electrophoresis – see Section 11.3.2.

PAG electrophoresis is an important technique in many areas. In DNA sequencing it provides a fundamental method of separation, and in protein studies it is important both in the form described here and also in conjunction with the detergent, sodium dodecyl sulphate (SDS), giving a valuable method for assessing the relative molecular mass of a protein.

3.3.3 Iso-electric focusing

➤ Iso-electric pH – see Section 10.1.3.

Electrophoresis depends primarily upon the charge carried by a molecule, but at its iso-electric pH (pI) a molecule will not migrate in an electric field. If electrophoresis is performed in an electrolyte that has a pH gradient, molecules will migrate to the point where the pH of the solution is the same as their iso-electric pH and remain there as long as the gradient and potential differences are maintained. This is the basis of iso-electric focusing techniques (Figure 3.28).

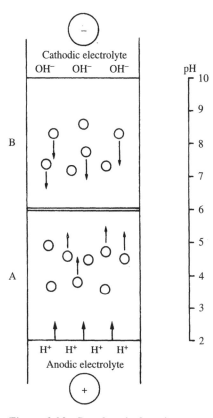

***Figure 3.28* Iso-electric focusing.** A protein which has an iso-electric pH of 6, in position A in the pH gradient is on the acid side of its iso-electric pH and will carry a positive (H^+) charge. It will migrate electrophoretically towards the cathode. In position B, the charge on the protein will be negative and it will migrate towards the anode. In both cases, movement is towards its iso-electric pH position, where it will remain.

Artificial pH gradients cannot be produced in the same way as concentration gradients but can only be formed electrophoretically. If an acid is used as the anodic electrolyte and an alkali as the cathodic electrolyte, when a voltage is applied the hydrogen and hydroxyl ions will move towards each other. A pH gradient formed in this way would be extremely fragile and has to be stabilized with suitable buffering agents. These will also migrate electrophoretically to their iso-electric pH points and because they will have no charge at that point may also show no electrical conductivity, so disrupting the electrophoretic process. The use of polyamino-polycarboxylic acids, which show a significant electrical conductivity at their pI values, enables the formation of a stable pH gradient. These compounds are commercially available often under the trade name of Ampholine carrier ampholytes.

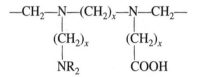

where R $=$ H or $-(CH_2)_x-COOH$
and $x = 2$ or 3

Ampholines are available with a wide range of pI values as a result of their precise chemical nature and this allows the formation of pH gradients varying from 2.5 to 11 or over narrower pH ranges. They have relative molecular masses below 1000 and hence may be easily separated from the sample macromolecules after electrophoresis by either dialysis or gel permeation techniques. They show all the qualities necessary for iso-electric focusing techniques, i.e. high buffering capacity, high solubility and good, although minimal, conductivity at their pI values. They have generally no toxic effects on cells or enzymic systems nor do they show any antigenic effects and these features are often important when dealing with biological samples.

A pH gradient is produced by incorporating a mixture of Ampholines with appropriate pI values in either a polyacrylamide slab or column. An acid is used at the anode and an alkali at the cathode, the pH of these being approximately the same as the pI values of the two extremes of the Ampholine range. Phosphoric acid and sodium hydroxide are suitable for wide range separations while various amino or carboxylic electrolytes are used for intermediate pH ranges.

The mixture of Ampholines will initially show a pH that is approximately the mean of all their pI values but the application of a voltage will cause each one to migrate towards one of the electrodes but, owing to the pH gradient which also develops, each Ampholine will stop moving electrophoretically at its iso-electric pH. The gradient will be maintained as long as the voltage is applied but it is necessary to stabilize it using either a gel or a sucrose concentration gradient in liquid column techniques.

Most flat-bed electrophoresis tanks are suitable for iso-electric focusing, provided that adequate support and cooling of the gel is possible. More recent techniques tend to use relatively high voltages (1–2 kV) but satisfactory separations can be achieved using lower voltages of 400–500 V although the separation time will be longer (3–4 hours).

The prepared Ampholine gel is set up in the tank and thick filter paper strips are soaked with either the anodic or cathodic electrolyte and placed along the appropriate edge of the gel. The samples may be applied either to small filter paper squares laid on the surface of the gel or, for bulk preparative work, incorporated in the gel. The appropriate voltage is applied through terminals attached to the electrode wicks and after about 30 min can be switched off to permit the removal of the sample filter papers before continuing the separation.

It is difficult to know when separation is complete but if two samples of coloured protein, e.g. haemoglobin, are placed adjacent to each other but in different positions on the gel, they will eventually form sharp bands at the

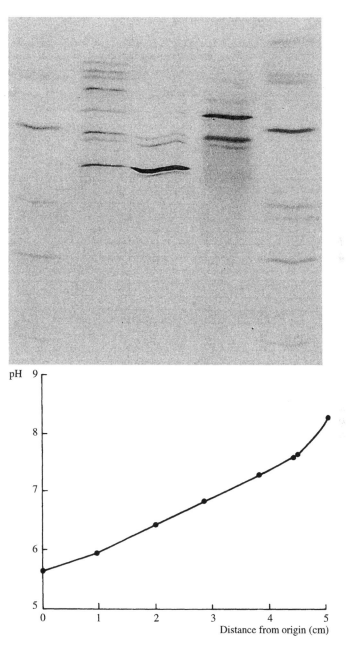

Figure 3.29 **Iso-electric focusing of haemoglobin variants and derivatives.** The two outside separations are of a mixture of coloured protein markers with a range of known p*I* values. These are used to calibrate the pH gradient formed in iso-electric focusing media and so enable the p*I* value of test proteins to be determined. The three inner separation patterns are of different haemoglobin derivatives and illustrate the analytical value of iso-electric focusing techniques.

same position, indicating completion of the process. When separation is complete, it is essential to fix the protein in the gel as soon as possible usually by immersion in 5% trichloroacetic acid solution or another suitable precipitating agent followed by an appropriate staining method (Figure 3.29).

3.3.4 Isotachophoresis

Isotachophoresis or displacement electrophoresis, as it is sometimes called, is a recent development of a principle described in 1920. In this technique the electrolyte between the electrodes is not a uniform single solution but a discontinuous sequence of different electrolyte solutions (Figure 3.30), the ions of which all migrate at the same velocity. Under identical conditions of concentration and voltage, different ions have different electrophoretic mobilities and to cause all the ions to move at the same speed a different voltage gradient is required for each group of ions. These different gradients develop spontaneously during the electrophoretic process.

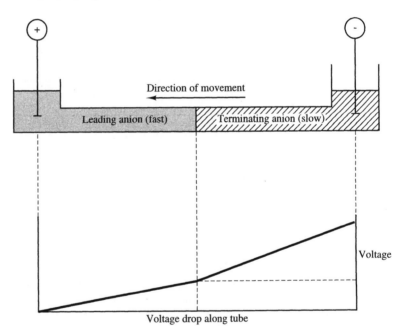

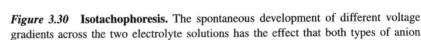

Figure 3.30 **Isotachophoresis.** The spontaneous development of different voltage gradients across the two electrolyte solutions has the effect that both types of anion migrate at the same velocity.

In practice isotachophoresis is usually performed in narrow tubes with electrodes at either end and is one form of capillary electrophoresis. For the separation of a particular type of ion, e.g. an anion, two buffered electrolyte solutions are selected that have different anions but a common cation with a buffering capacity. One of the anions (termed the leading electrolyte) should show a greater mobility than the other anion and occupies the anodic end of the tube. The slower anion (the terminating electrolyte) occupies the cathode end of the tube, with a sharp boundary separating the two solutions. The application of a potential difference across the tube will result in the common buffering cation migrating towards the cathode and the two anions towards the anode. The impossible situation of an ionic gap resulting from the different mobilities of the two anions is prevented by the spontaneous development of a discontinuous voltage drop along the tube (Figure 3.30), with a larger gradient developing across the terminating electrolyte than across the leading electrolyte, so maintaining the same velocity for all the ions.

If a sample containing various anions is introduced in between the leading and the terminating electrolyte, the different ions will move at a speed that depends upon their ionic mobility and the voltage gradient. A test ion which is in the terminating electrolyte will move quickly because of the higher voltage but when it moves into the leading electrolyte it will slow down because of the lower voltage gradient. As a consequence the ions will move into zones in decreasing order of their ionic mobilities and a voltage pattern will develop along the tube to maintain a constant velocity of all ions. This forms the basis of the qualitative or preparative form of isotachophoresis.

In addition to this qualitative aspect, isotachophoresis may also be used quantitatively. Once the sample zones have developed, the conductivity of each zone and hence the voltage drop across it will be related to the concentration of the ions and in order to maintain uniform conductivity the concentration (i.e. the volume occupied by each component) will alter. Hence at equilibrium, not only will the components of a mixture be separated from each

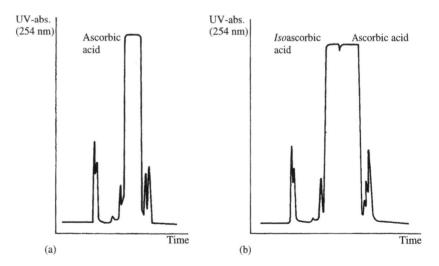

Figure 3.31 **Analytical isotachophoresis.** Ascorbic acid (vitamin C) is naturally present in many foods and is often added to others. Occasionally the cheaper isomer, *iso*ascorbic acid, which has no vitamin action, is used and is distinguishable from the natural isomer by many analytical methods. (a) shows the analysis of a sample of commercial fruit juice while (b) shows the same fruit juice to which a known amount (4 nmol) of *iso*ascorbic acid has been added. (Reproduced by permission of LKB, Stockholm, Sweden.)

other but the volume occupied by each component will be proportional to its initial concentration in the sample (Figure 3.31). Detectors similar to those described for HPLC may be suitable but isotachophoresis is a specialized technique and special instrumentation is necessary (Figure 3.32). This now resembles that used for other types of capillary electrophoresis.

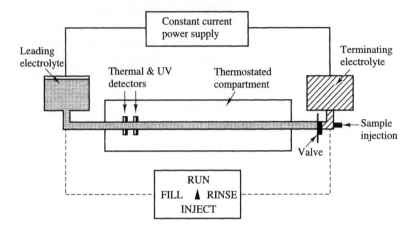

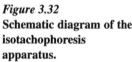

Figure 3.32
Schematic diagram of the isotachophoresis apparatus.

3.3.5 Capillary electrophoresis

Capillary electrophoresis (CE) is a family of related techniques that began to gain popularity in the late 1980s, when commercial instrumentation became available. It employs narrow bore (20–100 μm) fused silica capillaries, externally coated with plastic (polyimide) for increased durability in lengths of 10–100 cm. (Figure 3.33). Although high voltages (up to 30 000 V) are used, the heat is dissipated quickly because of the relatively large surface area and cooling is not normally required.

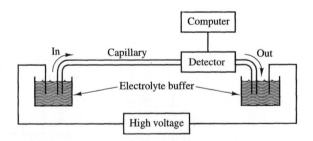

Figure 3.33 **Capillary electrophoresis.** The sample is introduced through the inlet, washed in with a wash buffer and then separated under the influence of a potential difference between the two electrodes. The zones are monitored as they pass through the detector and the data captured and computed.

The sample is usually introduced into the capillary by dipping it into the sample vial and applying a slight positive pressure. This results in a small volume of the sample entering the capillary. Depending upon the pressure and the time, the amount can be carefully controlled. The sample is then moved into the capillary as a zone and the entry to the capillary cleaned at the same time by drawing a set volume of a wash buffer. Alternative techniques include suction, siphoning and an electrokinetic technique, but all achieve the same result. The bands that separate are detected while they are still in the capillary, usually by absorbance, either in the ultraviolet or visible region of the spectrum, sometimes involving a diode array. Alternatively fluorescence or refractive index detectors may be employed and some instruments link the capillary to a mass spectrometer.

Capillary electrophoresis has been shown to have several advantages over other separation techniques and since its introduction the number of applications described for it has risen rapidly. It lends itself to the analysis of a wide range of ionic and neutral molecules of varying size and shape. Analysis times are similar to those for HPLC, typically being less than 30 minutes, but separation efficiency is often superior and theoretical plate numbers of greater than 1×10^6 are typical. Nanolitre sample volumes are used, a factor of great importance when limited amounts are available. Reagent costs are low because only microlitre volumes are required for each analysis. The capillaries must be appropriately conditioned or regenerated before use and should be flushed with water and air for storage. Modern instruments can be programmed to perform all the steps in the analysis from the introduction of the sample through to the interpretation of the results, which are presented as a series of peaks on a chart recorder, and preparation of the capillary for the next sample.

There are several modes of operation of CE which rely on variations of the principles of electrophoretic separation.

Free solution electrophoresis (capillary zone electrophoresis).

> ➤ Capillary zone electrophoresis is based on conventional electrophoresis of the analyte.

This was the first type of capillary electrophoresis to be developed. Migration of the analytes through a buffer solution, which sometimes contains chemical additives to enhance the separation, is affected by their net charge and the electroendosmotic flow. The magnitude of the electroendosmotic effect is due to negatively charged silanol groups on the capillary wall, which attract positively charged ions from the buffer, and the overall effect is the movement of the buffered solution towards the cathode. Endosmosis is particularly evident with buffers of high pH and low concentration. A variety of internal polymeric coatings can be applied to the capillaries to modify the electroendosmotic effect and thus extend the applications of the method.

Micellar electrokinetic capillary chromatography (MEKC or MECC)

> ➤ Micellar electrokinetic chromatography is based on the effects of the interaction of the analyte with surfactant micelles on the electrophoretic process.

This is used to measure small neutral or charged molecules and is applicable to a wide range of analytes including drugs, nucleosides, peptides and vitamins. Surfactants such as sodium dodecyl sulphate (SDS) are added to the buffer solution to produce anionic aggregates called micelles. The surfactants have long-chain hydrophobic tails (between 10 and 50 carbon units) and hydrophilic heads which, when bound in the micelles, point outwards into the solution. Separation depends on the interaction of the analytes with these charged micelles, which modifies their electrophoretic mobility. When negatively charged surfactants such as SDS are used, cationic analytes are strongly attracted to the micelles in a manner that can be likened to the situation in ion-pair chromatography, and the overall negative charge of the micelle is reduced.

Anions and uncharged analytes tend to spend more time in the buffered solution and as a result their movement relates to this. While these are useful generalizations, various factors contribute to the migration order of the analytes. These include the anionic or cationic nature of the surfactant, the influence of electroendosmosis, the properties of the buffer, the contributions of electrostatic versus hydrophobic interactions and the electrophoretic mobility of the native analyte. In addition, organic modifiers, e.g. methanol, acetonitrile and tetrahydrofuran are used to enhance separations and these increase the affinity of the more hydrophobic analytes for the liquid rather than the micellar phase. The effect of chirality of the analyte on its interaction with the micelles is utilized to separate enantiomers that either are already present in a sample or have been chemically produced. Such pre-capillary derivatization has been used to produce chiral amino acids for capillary electrophoresis. An alternative approach to chiral separations is the incorporation of additives such as cyclodextrins in the buffer solution.

Capillary gel electrophoresis

➤ Capillary gel electrophoresis is based on differences in analyte size.

This technique is used for the analysis of large molecules such as proteins and nucleic acids. The molecular sieving property of the gels is important in the separation process. The terms 'chemical gel' and 'physical gel' are often used to describe the gel matrices. Chemical gels are cross-linked with relatively high viscosity and have a defined pore structure. Polyacrylamide comes into this category. Physical gels are solutions of entangled polymers, e.g. alkylcelluloses. They have relatively low viscosity and the contents of the capillary can be easily flushed out after use. The concentration and chain length of the linear polymers determines the separation characteristics of the gel solution.

Capillary procedures offer several advantages, including speed, resolution, sensitivity and technical simplicity, compared with the traditional methods on which they were based. An added advantage for iso-electric focusing in capillaries is the fact that it can be performed without a gel, but a coating on the internal surface of the capillary is usually required to reduce electroendosmosis. Similarly, isotachophoresis can be conveniently performed in capillary electrophoresis apparatus.

Capillary electrochromatography (CEC)

➤ Capillary electrochromatography is based on the effect of interaction of the analyte with the stationary phase on its electrophoretic mobility.

This is a more recently developed technique which is a hybrid between HPLC and capillary electrophoresis. The capillary is packed with HPLC media and the mobile phases are aqueous buffers. A voltage is applied to generate an electroendosmotic flow and the analytes separate by interaction with the stationary phase and electrophoretic forces; no pump being required as for HPLC. Improved separation efficiencies have been reported.

Section 3.3

1. Which of the following ion-exchange media could be used effectively at pH 5?
   (a) Strong cation exchanger.
   (b) Weak cation exchanger.
   (c) Strong anion exchanger.
   (d) Weak anion exchanger.
2. Which of the following electrophoretic media show molecular sieving effects?
   (a) Cellulose acetate membrane.
   (b) Agar gel.
   (c) Polyacrylamide gel.
   (d) Agarose gel.
3. In iso-electric focusing, an analyte will not move from a position where the buffer pH is the same as its iso-electric pH
   BECAUSE
   a molecule will have a positive charge when it is at a pH greater than its iso-electric pH.
4. A molecule will be displaced from an ion-exchange medium by another molecule carrying an opposite charge
   BECAUSE
   the binding of a molecule to an ion-exchange medium is dependent upon the charge that it carries.

3.4   Methods based on size

The separation of macromolecules presents major problems and techniques such as dialysis permit only fairly gross separations to be achieved. However, the development of gel permeation chromatography introduced an extremely valuable method by which molecules may be separated from each other on the basis of their size (Figure 3.34). A wide range of molecular sizes can be separated by this method but for any particular gel the range is relatively narrow and there are a large number of different gels available commercially.

Ultracentrifugal techniques strictly speaking separate on the basis of the mass and not the size of particle. A variety of ultracentrifugal techniques are available and appropriate to different analytical problems. Preparative techniques generally involve iso-density or maximum velocity ultracentrifugation, whereas it is possible to investigate the relative molecular mass, diffusion characteristics and shape of a molecule using an analytical ultracentrifuge and equilibrium or maximum velocity techniques.

3.4.1 Dialysis and ultrafiltration

Dialysis is a technique for separating macromolecules from smaller solute molecules. The word refers to the diffusion of the solute molecules through a

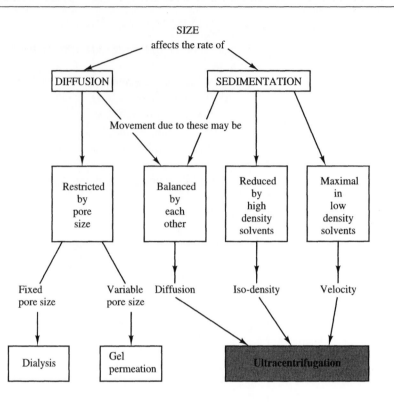

Figure 3.34
Separation methods in which the size of the molecule plays an important part.

membrane which, owing to its pore size, restricts the movement of large molecules. The passage of the small molecules is due solely to a concentration gradient across the membrane. In ultrafiltration the solvent and solutes are forced through the membrane under pressure and the movement of the large molecules is again restricted by the pore size.

Cellophane is frequently used for dialysis and it has a pore size of approximately 4–8 μm, which makes it impermeable to molecules with a relative molecular mass in excess of about 10 000. The development of a variety of membrane materials in which the pore size is much more rigorously controlled, has led to wider applications of ultrafiltration (Table 3.11). Various cellulose and polycarbonate membranes are available with pore sizes down to 5 nm which are capable of excluding molecules with a relative molecular mass of about 50. The internal structure of such membranes, as well as the pore size, determines their exclusion range and as a result precise specifications of membranes vary from one manufacturer to another.

Table 3.11 Applications of ultrafiltration

Particle size (nm)	Particles	Application
0.1–1.0	Salts, monomers	Water purification
1–100	Proteins, viruses	Fractionation, desalting
100–1000	Bacteria	Sterilization

3.4.2 Gel permeation chromatography

Gel permeation chromatography is also known as gel filtration and molecular size exclusion chromatography. The gel structure contains pores of varying diameter up to a maximum size. The test molecules are washed through a column of the gel and molecules larger than the largest pores in the gel are excluded from the gel structure. Smaller molecules, however, penetrate the gel to a varying extent depending upon their size and this retards their progress through the column. Elution, therefore, is in order of decreasing size (Figure 3.35).

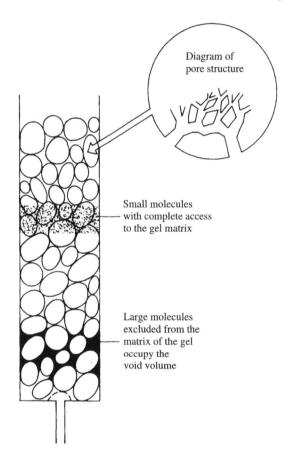

Diagram of pore structure

Small molecules with complete access to the gel matrix

Large molecules excluded from the matrix of the gel occupy the void volume

Figure 3.35
Separation by gel permeation chromatography.

➤ The exclusion limit of a gel is the RMM above which molecules cannot enter the gel pores.

There are a variety of products available for gel permeation chromatography (Table 3.12) and they are usually classified according to their exclusion limit, i.e. the relative molecular mass above which all molecules are excluded from the gel structure. Sephadex, the original medium, is based on dextran (a linear glucose polymer) that is modified to give varying degrees of cross-linking, which determines the pore size of the material. It is a strongly hydrophilic polymer which swells considerably in water and before a column is prepared, the gel must be fully hydrated with a large volume of liquid.

Polyacrylamide beads are also available with a wide range of pore sizes and are prepared commercially in a manner similar to that described for poly-

.acrylamide electrophoresis. They are also hydrophilic and swell significantly in aqueous solutions but are chemically more stable than the dextran gels. Agarose gels are particularly useful when gels with a very large pore size are required. They are also hydrophilic but are usually sold in the swollen form. A major problem is the fact that they soften at temperatures above 30 °C. Mixed gels of polyacrylamide and agarose are available in which the polyacrylamide provides a three-dimensional structure which supports the interstitial agarose gel.

Polystyrene gels are hydrophobic, and as a result are used primarily with non-aqueous solvents and for organic chemical applications rather than with biological samples.

Table 3.12   Range of gel permeation media

Trade name and manufacturer	Chemical nature	Fractionation range (RMM)
Pharmacia		
Sephadex G 10	Cross-linked dextrans	Up to 700
to		
G 200		$5000-800\ 000$
Sepharose 6B	Agarose	Up to 4×10^6
to		
2B		Up to 40×10^6
Sepharose C1−6B		Up to 4×10^6
to		
C1−2B		Up to 40×10^6
Biorad		
Biogel P2	Cross-linked polyacrylamide	$200-2500$
to		
P300		$100\ 000-400\ 000$
Biogel A 0.5	Agarose	$10\ 000-500\ 000$
to		
A 150		$1 \times 10^6 - 150 \times 10^6$
Biobeads S $\times$ 1	Polystyrene	$600-14\ 000$
to		
S $\times$ 12		Up to 400
LKB		
Ultragel AcA22	Agarose and polyacrylamide	$100\ 000-1.2 \times 10^6$
to		
AcA54		$5000-70\ 000$

Column conditions

Columns for gel permeation chromatography must have both a minimal dead space volume at the outlet, to prevent broadening of the zones developed during the separation, and a suitable bed support, e.g. nylon mesh, which will retain the gel without becoming blocked. The length of the column affects the resolution, while the diameter influences the amount of sample that can be applied without reducing the resolution. Gels with a small bead size result in columns with lower void volumes (Figure 3.35) and hence improved resolution. However, columns of small beads give a high resistance to liquid flow resulting in slower separations. For preparative work, therefore, larger beads are preferable.

The major technical difficulty in gel permeation chromatography involves the careful packing of the column to ensure a uniform structure and it is absolutely essential that the level of the solvent is never allowed to fall below the level of the gel otherwise disruption of the gel bed occurs. Additionally, because most gels are compressible, only an absolute minimum of pressure should be applied to prevent alteration of the pore characteristics. In practice hydrostatic pressures greater that 30–40 cm should not be used and pumps should be used only with non-compressible gels.

Elution parameters

> ➤ The effluent volume is the total volume that has come off the column by the time the analyte is eluted.

The effluent volume is the most convenient parameter to measure in gel permeation chromatography because flow rates are often variable, making the use of retention times unsuitable. The sequence of different solutes emerging from a column will therefore be reported as the total volume of solvent that has emerged from the column when the substance appears in the effluent (V_e).

Molecules larger than the exclusion limit of the gel will be eluted when a volume of solvent equal to the void volume (V_o) leaves the column and

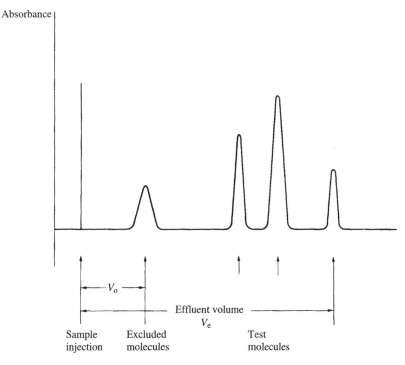

Figure 3.36 **Gel permeation chromatogram.** All molecules larger than the exclusion limit of the gel appear at V_o (the void volume). Molecules which can gain access to the gel structure to varying degrees are eluted in order of decreasing size.

> ➤ The void volume is the amount of liquid that is outside the gel structure.

similarly those molecules that can gain access to all parts of the gel matrix will have an effluent volume equal to the total bed volume (V_t). Molecules of an intermediate molecular size will be eluted in volumes between V_t and V_o (Figure 3.36).

The measurement of effluent volume is not very reliable because of the effect of the geometry and packing characteristics of any column. It is often more useful to use a reduced parameter, such as V_e/V_o, which is not so dependent upon column characteristics and is comparable with the calculation for R_F values in thin-layer chromatography.

It is possible to calculate a partition constant (K_d) for a solute in terms of the access gained to the gel matrix:

$$V_e = V_o + K_d \times V_i$$

where V_i is the gel inner volume. Hence

$$K_d = \frac{V_e - V_o}{V_i}$$

However, because it is difficult to measure V_i precisely, the equation may be modified to determine the 'available' part of the resin, K_{av}:

$$K_{av} = \frac{V_e - V_o}{V_t - V_o}$$

In order to separate two solutes satisfactorily, their values for K_{av} must be significantly different from each other.

Applications of gel permeation chromatography

Gel permeation chromatography is commonly used in the fractionation of mixtures of substances that vary in their relative molecular mass and it is extremely useful for labile molecules, such as enzymes.

The elution volume of a solute is determined mainly by its relative molecular mass and it has been shown that the elution volume is approximately a linear function of the logarithm of the relative molecular mass. It is possible to determine the relative molecular mass of a test molecule using a calibration curve prepared from the elution volumes of several reference substances of known relative molecular mass. This should be done using the same column and conditions (Figure 3.37) and in practice it may be possible to calibrate the column and separate the test substance at the same time by incorporating the reference compounds in the sample. Such a method is rapid and inexpensive and does not demand a highly purified sample, provided that there is a specific method for detecting the molecule in the eluate.

> ➤ Desalting is the name given to the technique of removing unwanted small molecules from preparations of macromolecules.

Solutes of low relative molecular mass may be removed from preparations of macromolecules, often called desalting, using a small column of a gel which excludes the macromolecules but retains the small solutes. The macromolecules will be eluted in the void volume, which for a small column will be only 2–3 ml.

A dry gel may also be used to concentrate a dilute aqueous solution of a macromolecule. When it is added to the solution, it will swell as it takes up water but will exclude the larger molecules. Hence, after sedimentation the supernatant fluid will contain the macromolecule, leaving the majority of the water in the pores of the gel. Such a technique is quicker and more effective than dialysis.

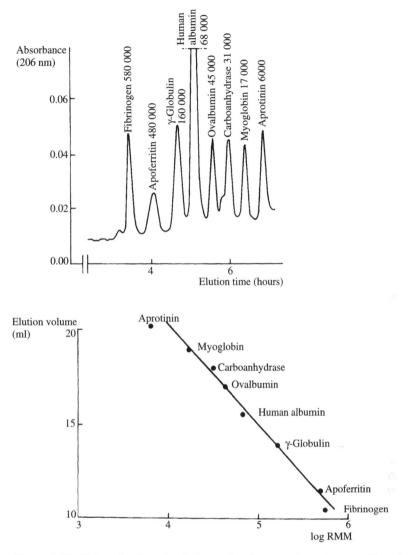

Figure 3.37 **Determination of relative molecular mass by gel permeation chromatography.** A mixture of proteins (approximately 10 μg of each) was separated on an UltroPak TSK SW column and the elution volume for each protein plotted against the logarithm of its relative molecular mass (RMM).
(Reproduced by permission of LKB, Stockholm, Sweden.)

3.4.3   Ultracentrifugation

Although the use of high rotor speeds is not always necessary, most ultracentrifuges are designed to operate safely at rotor speeds of up to 60 000 or 100 000 rpm. Despite the use of high speed, the sedimenting forces involved are only small and the technique is extremely gentle, providing a valuable method for the separation of large, labile molecules or particles. Preparative ultracentrifugation uses rotors with relatively large capacities to separate various components of a mixture, while analytical ultracentrifugation involves the measurement of sedimentation velocities in small volumes under carefully controlled conditions.

Because particles sediment in a tube during centrifugation, it may appear that a force is acting radially outwards, whereas in fact there is no such force. Any particle that is rotating in a circle will always tend to move away at a tangent to the circle and to prevent this a force has to be applied towards the centre of rotation. This is known as the centripetal force and is provided in a centrifuge by the rotor arms and spindle. Particles suspended in solution, when rotated in a centrifuge tube, are not connected to the axis of rotation and therefore tend to move at a tangent to the circle. Their movement, however, relative to the centrifuge tube is radially down the tube and appears to be determined by a force equal and opposite to the centripetal force (Figure 3.38).

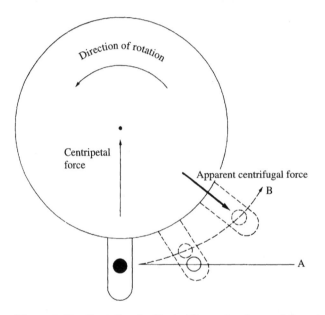

Figure 3.38 **Centrifugal effects.** The path of a particle unimpeded by frictional effects is at a tangent (A) to the circle described by the rotation of the centrifuge tube. When impeded by friction, its path (B) more closely resembles circular movement with a reduced rate of sedimentation in the tube.

It is convenient, therefore, in ultracentrifugation to assume that this movement in an apparently stationary tube is due to centrifugal force and to compare this force with that due to gravity by quoting the intensity of the centrifugal force with respect to that of gravity. This is known as the relative centrifugal force (RCF) and enables quick and simple comparisons of the forces applied by different centrifuges and different rotors. The effective centrifugal force may be calculated as:

▶ Relative centrifugal force is a measure of the intensity of centrifugal force compared with that due to gravity.

$$F = M\omega^2 r$$

where M is the mass of the particle, ω is the angular velocity in radians per second and r is the radius of rotation in centimetres.

The effect of centrifugation is therefore dependent upon the mass of the particle and the convenience of using the relative centrifugal force is that the value for the mass of the particle is eliminated from the equation, permitting more general comparisons of centrifuge performance:

$$RCF = \frac{f_c}{f_g} = \frac{M\omega^2 r}{Mg} = \omega^2 r \times g^{-1}$$

It is more usual to monitor the speed of a centrifuge in terms of revolutions per minute (rpm) rather than in radians per second. Hence:

$$RCF = \left(\frac{2\pi \ \text{rpm}}{60}\right)^2 r \times g^{-1}$$

In comparing the relative centrifugal force developed by different centrifuges it is important to appreciate that the value varies considerably from the top to the bottom of a centrifuge tube (Figure 3.39).

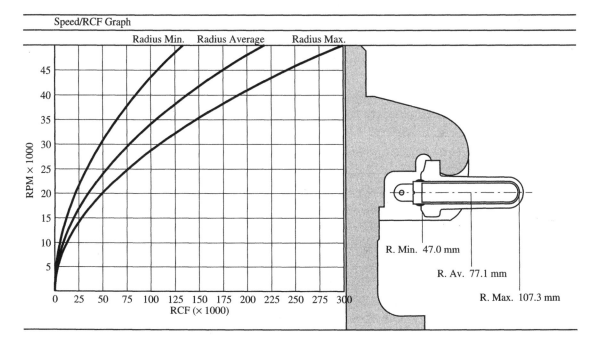

Figure 3.39 **Relative centrifugal force.** The relative centrifugal force developed by any rotor can be calculated but will vary depending upon the distance from the centre of rotation.

For a particle to sediment it must displace an equal volume of the solvent from beneath it. This can only be achieved by centrifugation if the mass of the particle is greater than the mass of the solvent displaced. Hence, density as well as mass is an important consideration and explains the phenomenon of low density particles floating rather than sedimenting during centrifugation.

When a particle is moving it will be necessary to pass solvent molecules regardless of the direction of movement and the resulting frictional effects will always oppose the movement. This effect is proportional to the velocity of the molecule and is affected by its size and shape as well as the viscosity of the medium:

Friction $= fv$

where f is the frictional coefficient of the particle in the solvent, and v is the velocity of the particle.

Because of the centrifugal force a particle will accelerate, its velocity increasing until the frictional force equals the centrifugal force. Under these conditions the net force acting on the particle is zero and the particle will not accelerate any further but will have achieved a maximum velocity:

Net force $= (M_p - M_s)\omega^2 r - fv$

where M_p is the mass of the particle and M_s is the mass of equal volume of solvent.

It is convenient to classify particles by their rate of sedimentation per unit centrifugal field, a parameter known as the sedimentation coefficient (s), which has units of seconds. A Svedberg unit (S) is defined as a sedimentation coefficient of 1×10^{-13} seconds.

> ➤ A Svedberg unit is a measure of the rate of sedimentation of a molecule due to centrifugation.

Maximum velocity methods
Rotor speeds that are high enough for all the test particles to achieve their maximum velocity are used in preparative and analytical separations.

Moving-boundary ultracentrifugation
A centrifuge tube is filled with the sample and centrifuged at a speed high enough for all the components of the sample to achieve their maximum velocity. The components begin to separate out, leaving clear solvent (supernatant)

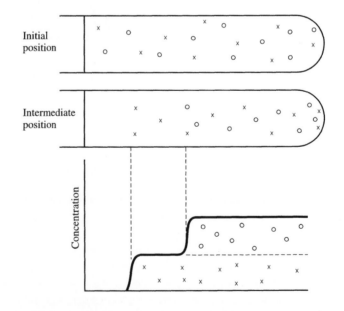

Figure 3.40 **Moving-boundary ultracentrifugation.** The heavier particles (o) sediment more rapidly than the lighter particles (✕) and as a result concentration boundaries develop as the supernatant fluid is cleared of particles. Monitoring the movement of a boundary will permit the determination of sedimentation coefficients.

at the top of the tube, and a series of concentration boundaries develop depending upon the number of components in the sample (Figure 3.40). The movement of a boundary is characteristic of the particle in the bulk of the solution ahead of the boundary and although it may not be possible to purify a particular component by this technique it is possible to measure its rate of sedimentation by following the movement of the boundary.

Separation into components can only be achieved by stopping the process when sedimentation of the desired component has occurred. The sediment is then resuspended in fresh solvent and centrifuged at a lower speed, when the heavier particles will sediment leaving the component in suspension. Such a method is known as differential sedimentation and is particularly useful for the fractionation of cellular components. The method outlined in Procedure 3.3 is simple and is designed to separate four main cellular fractions, namely, nuclear, mitochondrial, microsomal and soluble.

> ➤ Differential centrifugation is a preparative technique involving sequential centrifugation stages each at a slower speed.
> ➤ Sub-cellular fractions – see Section 8.4.3.

Procedure 3.3: Preparation of sub-cellular fractions using ultracentrifugation

Equipment
A suitable homogenizer (e.g. a Potter PTFE pestle type)
A preparative ultracentrifuge

Reagents
Buffer

Tris	0.05 mol l^{-1}
Sucrose	0.25 mol l^{-1}
Magnesium chloride	0.003 mol l^{-1}
Adjust to pH 7.4	

Method
1. Cut the tissue into pieces approximately 0.5 cm thick and weigh.
2. Add approximately 8 ml of cold buffer solution for each gram of tissue.
3. Cut the tissue into small pieces and homogenize in a Potter homogenizer at 1000 rpm for 2 min.
4. Transfer quantitatively to a cooled measuring cylinder and make up the volume with buffer solution to give a 10% suspension by weight.
5. Centrifuge a convenient aliquot (X ml) at 800 g for 10 min. The precise centrifuge speed obviously depends upon the design of the centrifuge.
6. Remove the supernatant fluid (I) and resuspend the deposit and re-homogenize using the Potter homogenizer as before but this time for only 1 min.
7. Centrifuge the re-homogenized deposit at 800 g for 10 min as before. Remove the supernatant fluid (II) and resuspend the deposit in the same volume of buffer as was originally used (X ml). This is the **nuclear fraction**.
8. Centrifuge both supernatant fluids at 6000 g for 15 min and discard the supernatant fluid II but retain that for sample I. Pool the deposits and resuspend in the same volume of buffer as before (X ml). This is the **mitochondrial fraction**.
9. Centrifuge the supernatant fluid from sample I at 60 000 g for 1 hour.
10. Remove the supernatant fluid. This is the **soluble fraction**. Resuspend the deposit in the same volume of buffer as before (X ml). This is the **microsomal fraction**.

Notes

The reason for the second homogenization and centrifugation of the original deposit is that the disruptive process is only partial and rather than risk excessive damage to the mitochondria by homogenization, two shorter periods are used. In addition to this, a significant proportion of mitochondria may be carried down in the initial deposition and the second centrifugation considerably improves the mitochondrial yield. The resuspension of all deposits in a standard volume permits the content of each fraction to be related to the content of the original tissue by virtue of the constant dilution factor.

Moving-zone ultracentrifugation

A layer or zone of the sample is superimposed on the solvent in the centrifuge tube and again, under the influence of a centrifugal force, the particles sediment and discrete zones or bands develop depending upon the relative mobility of each particle (Figure 3.41). As centrifugation continues, so each zone will pellet at the bottom of the tube in sequence, resulting finally in a mixed pellet. Centrifugation is stopped before this occurs and the different zones can be removed. Zonal techniques are almost solely preparative and the resolution of the zones will be maintained more effectively if a slight density gradient is used.

Zonal techniques may be used for the separation of a wide range of particles and macromolecules, e.g. mitochondria, nuclei, ribosomes and proteins. The technique may be used for bulk preparative work using a zonal rotor which is filled with a solvent gradient while running at a slow speed. The sample is similarly introduced and the rotor speed is then increased to the desired value. After centrifugation is complete, the contents are drawn off while the rotor is running slowly by displacing them with a more dense solution.

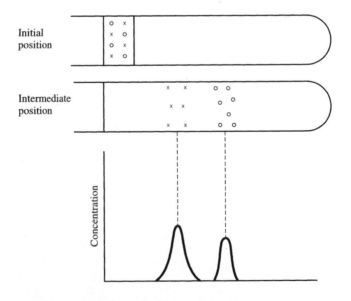

Figure 3.41 **Moving-zone ultracentrifugation.** Zones develop during centrifugation, the lower ones being heavier particles.

Iso-density methods

Regardless of the rotor speed and maximum velocity, sedimentation (or flotation) will not occur in a solution of equal density to the sample. Iso-density methods use this lack of movement in a manner comparable to a pH gradient in iso-electric focusing techniques. The methods are a combination of sedimentation and flotation, achieved by using a density gradient that straddles the density of the particles concerned. On centrifugation, the particles sediment until they reach a solvent zone with the same density. This results in the development of a zone for each type of particle present in the sample.

Pre-formed density gradients

The centrifuge tube is filled with a solvent in a graded range of concentrations prepared from mixtures of two different stock solutions which determine the limits of the gradient. While it is possible to produce a stepwise gradient by layering decreasing concentrations of the solution carefully on top of each other, it is preferable to produce a smooth gradient using a gradient former. Solutions of either sucrose or caesium chloride are usually used and the sample is layered on the surface and centrifuged. Under the influence of the centrifugal force, each component will move to an area of iso-density, the fastest particle not necessarily being the most dense.

Density gradient techniques are particularly useful in the separation of cellular components and macromolecules but are not generally as efficient as differential sedimentation. Table 3.13 indicates the order of separation of cellular organelles after ultracentrifugation to equilibrium on a density gradient of sucrose ranging from a high density of 1.25ρ at the bottom of the tube to a minimum value of 1.10ρ at the top. The cell extract is prepared in a sucrose solution with a density of 1.05ρ and layered on top of the gradient. Some organelles, such as mitochondria, show fairly consistent density values but others do vary considerably depending upon the cellular source and the manner of preparation of the extract. Density values for rough endoplasmic reticulum vary considerably depending upon the proportion of dense ribosomes present.

Table 3.13   Density gradient ultracentrifugation of sub-cellular organelles

Organelle	Density (ρ)
Plasma membrane	1.14
Rough endoplasmic reticulum	1.16
Mitochondria	1.17
Lysosomes	1.20

A cell extract was separated on a sucrose gradient ranging from 1.10ρ at the top to 1.25ρ at the bottom and centrifuged to equilibrium.

Equilibrium density gradients

This is a method for the separation of molecules with very similar densities and depends upon the formation of a density gradient by the effect of centrifugal force on the supporting solute molecules but because of their low mass the technique demands a long period of centrifugation, e.g. 2–3 days. The test compounds are suspended initially in a solution of selected density determined usually by preliminary pre-formed gradient technique. The gradient forms and the molecules separate into zones simultaneously during the long period of centrifugation.

The materials usually used to produce such gradients are salts of heavy alkali metals, caesium chloride being the most frequently used. This salt has a high solubility and its low relative molecular mass permits rapid diffusion enabling the gradient to be formed reasonably quickly. Concentrations of salt up to about 2 g ml^{-1} can be used and are chosen on the basis of information about the density of the test particles or macromolecules. The technique is used frequently in the separation of viral particles and nucleic acids.

➤ Centrifugation of nucleic acids – see section 13.2.

Equilibrium methods

Centrifugation of a suspension of macromolecules at relatively low speeds will not result in complete sedimentation, because of the effect of diffusion, but a concentration gradient will be produced owing to the equilibrium between sedimentation and diffusion. Such a gradient may be studied in the analytical centrifuge and the resulting measurements can be used to determine the relative molecular mass of the molecule. It is not a preparative technique and requires very long periods of centrifugation for the equilibrium to be established.

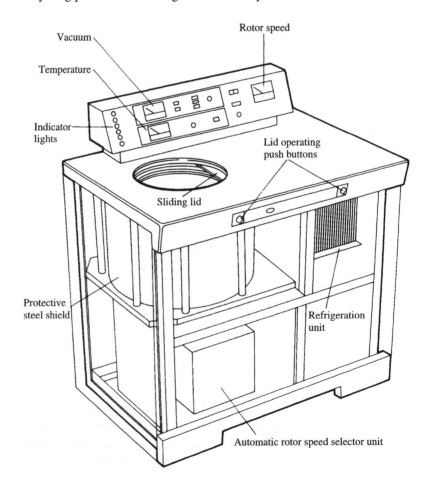

Figure 3.42
An ultracentrifuge.

Instrumentation

The major difference between a preparative and an analytical ultracentrifuge lies in the design of the rotors and the need for an optical monitoring system in the latter. The basic features of an ultracentrifuge are illustrated in Figure 3.42 and the safety aspects are a major consideration in their design. A vacuum is necessary to minimize the frictional effects of high speed rotation and a cooling system is essential to protect the high speed bearings, etc., and any labile materials being studied.

> ➤ Preparative rotors are used for small bulk preparation of fractions of a sample.

Rotors are generally made of aluminium or titanium, the latter being more durable and safer at high speed. They are protected by a hard anodized film to prevent corrosion but have a limited life owing to metal fatigue and careful records of their use must be kept. Preparative rotors may be either the swing out or fixed angle type (Figure 3.43) of varying sample capacity. The large capacity rotors have a lower maximum safe speed, with swing out heads having lower limits than the corresponding fixed angle rotors. Centrifuge tubes are cylindrical and usually made of polypropylene or polycarbonate with a capping or sealing device. It is essential that tubes should be filled to capacity to prevent collapse and that they are carefully balanced before centrifugation.

Figure 3.43 **Types of preparative rotors.** (Photograph by courtesy of MSE Ltd, Crawley, Sussex, UK.)

> ➤ Analytical rotors are used to study the sedimentation rate of a sample.

Analytical rotors are usually solid with holes where the cells containing the samples are inserted, the simplest type of rotor incorporating the cell together with a reference. These cells are usually made of quartz or sapphire and are sector shaped to prevent distortion of the boundaries that develop during sedimentation.

The optical system required to monitor the movement of the boundary during centrifugation presents design problems in maintaining a vacuum in the centrifuge while also monitoring the rapidly rotating sample cell. Absorption of ultraviolet radiation may be used to detect zones of protein, nucleic acid, etc., the zones being detected either photographically or absorptiometrically. A reference cell is necessary to compensate for absorbance by the solvent.

Variations in concentration that are produced as a boundary develops result in changes in the refractive index along the length of the cell. In Rayleigh interference optics, a fringe pattern produced by the superimposition of two parallel rays of light is distorted by a change in refractive index. The degree of distortion is proportional to the change in the refractive index and hence to the change in concentration. The Schlieren optical system utilizes the same phenomenon but results in a trace that represents the change in the refractive index gradient and shows the boundary as a peak.

While fractions or samples can be removed from the centrifuge tubes using a pipette it is possible with appropriate equipment to pierce the base of a plastic tube to allow the contents to drip slowly out. Alternatively the contents of the tube may be displaced by slowly injecting a dense solution to the bottom of the tube (Figure 3.44) or by puncturing the tube in the appropriate position and withdrawing the desired band with a syringe.

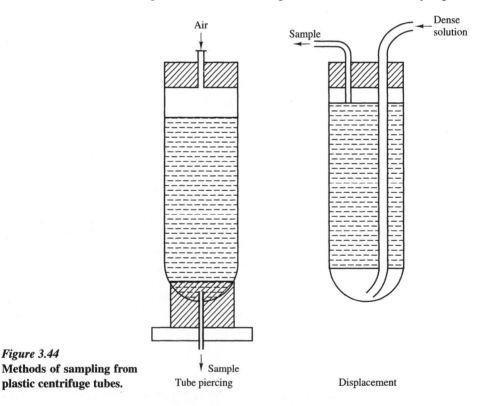

Figure 3.44
Methods of sampling from plastic centrifuge tubes.

Determination of relative molecular mass

During ultracentrifugation, when the frictional force equals the centrifugal force and the particle achieves maximum velocity, the relative molecular mass of the particle may be represented by the Svedberg equation:

$$M = \frac{RT}{(1 - \overline{V}\rho)} \times \frac{s}{D}$$

where s is the sedimentation coefficient, R is the gas constant, T is the absolute temperature, D is the diffusion coefficient, ρ is the density of the solvent and $\overline{V}$ is the partial specific volume of the solute.

The partial specific volume is the increase in the volume of a solution caused by the presence of the solute and is defined as the volume occupied by 1 kg of the solute in 100 kg of the solution. The value may be calculated using the densities of both the solvent (ρ) and the solution (ρ_s):

$$\overline{V} = \frac{\dfrac{100}{\rho_s} - \dfrac{(100 - n)}{\rho}}{n}$$

where n is the percentage concentration of the solute in the solvent (w/w).

The value of the partial specific volume for proteins is approximately 0.7 while for nucleic acids the value varies around 0.5.

The sedimentation coefficient may be determined by measuring the velocity of the particle at a fixed centrifuge speed (ω, in radians per second) and, from a series of observations, plotting the logarithm of the distance moved (x) against the time taken in seconds (t). The relationship is expressed by the equation:

$$s = \frac{\ln x}{\omega^2 t} = \frac{2.303 \log_{10} x}{\omega^2 t}$$

and the value for s can be calculated from the slope of the resulting graph. Although relative molecular mass may be calculated using the Svedberg equation, the calculations are complicated by difficulties in determining the diffusion coefficient and correcting for differences in viscosity and temperature. For these reasons the sedimentation coefficient of a particle is often used instead of its mass, the two being proportional to one another.

Meniscus depletion is an alternative method for the determination of the relative molecular mass. If a homogenous sample is centrifuged at a speed great enough to cause the particles to move away from the meniscus, resulting in a solute concentration of zero at the meniscus and producing a concentration gradient due to the solute, the relative molecular mass is proportional to the slope of a plot of the logarithm of concentration against the distance squared. This relationship is not absolute and depends upon measurements being taken early in the process and assumes that the density of the solution is constant throughout the solution despite the very slight concentration gradient that does develop. The concentration of solute may be determined from the steepness of the interference fringes formed in a Rayleigh pattern.

The relative molecular mass (RMM) may then be calculated using the equation:

$$RMM = \frac{4.606RT}{(1 - \overline{V}\rho)\omega^2} \times \frac{\log \text{concentration}}{\text{distance}^2}$$

➤ Immunoassays – see
Section 7.4.

3.5   Methods based on shape

The basis of many biochemical processes within a cell lies in the shape relationships that exist between the reacting molecules, e.g. an enzyme active site and its substrate. The affinity and specificity that such molecules show for each other form the basis of methods such as immunoassays, and they can also be exploited in affinity chromatography.

3.5.1 Affinity chromatography

A binding molecule (or ligand) is linked to an insoluble polymer in such a way that the affinity of the ligand for its complementary molecule is not significantly altered. The polymer is subsequently used in a manner comparable to adsorption or ion-exchange media but has the additional advantage of specificity. The process may also be used in a reverse manner by insolubilizing the complementary molecule and using it to separate a specific binding molecule, e.g. the isolation of specific antibodies from an antiserum. Antibodies probably provide the largest source of ligand molecules but enzymes, transport proteins and membrane receptors are all used.

Methods of insolubilization
A variety of techniques has been used varying from simple adsorption to entrapment within a polymer matrix but the commonest method is the covalent linking of the ligand to a suitable polymer, such as dextran, agarose or polyacrylamide.

The polymer should be hydrophilic and of a pore size large enough to allow the test molecules to gain access. If the ligand group is linked closely to the polymer backbone it is possible that steric interference will impair its binding ability

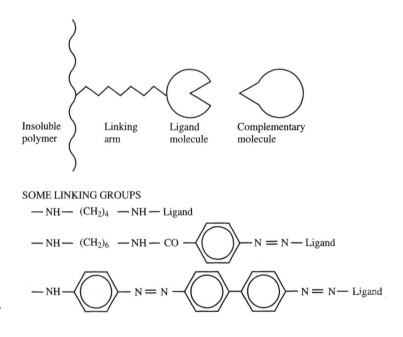

Figure 3.45
Affinity chromatography.

and it is desirable that a relatively long spacer arm (6–8 atoms) is used in the linking reaction (Figure 3.45).

Several linking reactions have been described but the two common examples both involve aromatic amino groups in the polymer. These can be converted to reactive diazo groups by the action of nitrous acid and will subsequently react with free amino groups in the ligand proteins. Alternatively, they can be converted to isothiocyanate groups by thiophosgene and will then react with tyrosine and histidine residues in the proteins (Figure 3.46). Other functional groups in the polymer can be used in a similar manner. A range of prepared gels is available commercially and the best gel for a particular situation must be determined by trial and error.

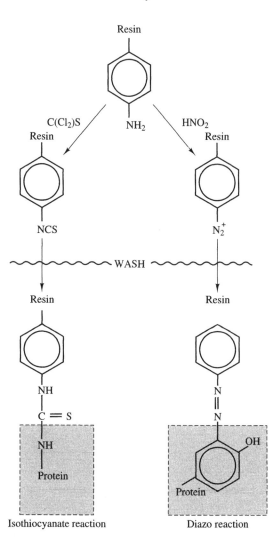

Figure 3.46 **Examples of reactions used for insolubilizing proteins using aromatic amino resins.**

Isothiocyanate reaction   Diazo reaction

Methods of affinity chromatography
Affinity chromatography techniques are not generally appropriate for the fractionation of mixtures because of the very positive and specific affinity between

the ligand and solute molecules. Separation may be achieved in a column where the solute molecules bind to the matrix, and, after washing to remove unwanted substances, the test substance can be eluted. Alternatively, a batch procedure may be used in which the affinity medium is mixed with the sample solution and allowed to react. The solid-phase complex is then removed from the bulk solution by filtration or centrifugation and after washing, the purified solute is separated from the complex.

In order to elute molecules which are strongly bound to the ligand group it is necessary either to reduce their affinity for each other or to introduce molecules that are more strongly bound or in greater concentration and will therefore be able to displace the test molecules. Generally, non-specific methods are preferred and involve altering either the ionic strength or the pH of the buffer. These will result in conformational changes in the proteins and hence their binding characteristics. Changes in the dielectric constant of the solvent caused by the introduction of organic solvents will result in altered hydrophobic bonding and again aid the dissociation of the complex. It is possible to use specific agents which compete for the binding sites, such as alternative substrates and inhibitors for enzymes.

Applications

The nature of the binding between the ligand and its complementary molecule restricts affinity chromatography to a particular type of biological compound and some examples are given below.

1. *Enzymes.* The specificity of an enzyme for its substrate, coenzyme or competitive inhibitor provides the basis for many affinity chromatographic separations. Enzymes may be extracted and purified using insolubilized substrates, coenzyme or inhibitors. Less frequently, enzymes are used as the ligands.

> Immobilized enzymes – see Section 8.7.

2. *Antibodies.* The reaction between an antibody and its antigen does not result in the chemical modification of the antigen compared with the action of an enzyme and provides the basis for producing chromatographic media capable of selecting the complementary molecules. Either the antigen is insolubilized and used to isolate and purify the appropriate antibodies or with the increased availability of monoclonal antibodies, the reverse procedure is used.

3. *Lectins.* Some proteins extracted from certain seeds are capable of binding compounds containing carbohydrate groups. These proteins are known as phytohaemagglutinins or lectins. Affinity chromatographic media using such lectins have been used to investigate cell membrane structures and aid in the study of cell interactions. They are also used in conjunction with quantitative column chromatographic methods and in some electrophoretic separations of carbohydrate-rich proteins.

> A phytohaemagglutinin is a plant extract which is capable of agglutinating blood cells.

4. *Receptor proteins.* Hormones act on cells via specific membrane receptors and can be used to purify these receptors from cell homogenates. Receptors for compounds such as insulin, oestrogens and acetylcholine among others have been purified in this way.

5. *Nucleic acids.* Immobilized polynucleotides can be used to extract nucleic acid binding proteins as well as complementary strands of nucleic acids.

Self test questions

Sections 3.4/5
1. Which of the following techniques can be used to determine relative molecular mass?
   (a) Affinity chromatography.
   (b) Gel permeation chromatography.
   (c) Zonal centrifugation.
   (d) Maximum velocity centrifugation.
2. Which of the following affect the speed of sedimentation of a particle in an ultracentrifuge?
   (a) Radius of the rotor arm.
   (b) Gravity.
   (c) Mass of the particle.
   (d) Temperature.
3. In gel permeation chromatography, the larger molecules are eluted from the column last
   BECAUSE
   the pore size restricts the access of molecules larger than the exclusion limit of the medium in gel permeation chromatography.
4. Two particles with the same mass but different densities cannot be separated from each other by ultracentrifugation without using a density gradient
   BECAUSE
   if the density of a particle is less than the density of the solvent, the particle will float rather than sediment.

3.6   Further reading

Dean, P.D.G., Johnson, W.S. and Middles, F.A. (1985) *Affinity chromatography – a practical approach*, IRL Press, UK.

Rickwood, D. (ed.) (1993) *Preparative centrifugation – a practical approach*, IRL Press, UK.

Rickwood, D., Ford, T. and Steensgaard, S. (1994) *Centrifugation: Essential data*, Wiley-Liss Inc., USA.

Stock, R., Rice, C.B.F., Braithwaite, A. and Smith, F.J. (1995) *Chromatographic methods*, 5th edition, Blackie Academic and Professional, UK.

Johnstone, R.A. and Rose, M.E. (1996) *Mass spectrometry for chemists and biochemists*, 3rd edition, Cambridge University Press, UK.

Katz, E. (ed.) (1995) *High performance liquid chromatography*, John Wiley, UK.

Grant, D.W. (1995) *Capillary gas chromatography*, John Wiley, UK.

Baker, D.R. (1995) *Capillary electrophoresis*, John Wiley, UK.

4 Electroanalytical methods

The subject of electrochemistry deals with the study of the chemical interaction of electricity and matter generally, but it is the interaction with solutions that is of particular value in analytical biochemistry. The electrical properties of a solution depend upon both the nature of the components and their concentration and permit qualitative and quantitative methods of analysis to be

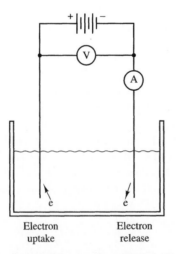

Electron   Electron
uptake   release

Figure 4.1 **An electrolytic cell.** The current which flows between the electrodes depends not only on the voltage that is applied but also on the electrical properties of the solution.

developed. These electrical properties of solutions are measured using electrodes in arrangements known as electrochemical cells. There are two main types of electrochemical cell:

1. *Electrolytic cells.* When a potential difference is applied across two electrodes that dip into a solution, a current will flow between them (Figure 4.1). The amount of current that flows depends upon the voltage applied and the electrochemical properties of the solutions. This provides the basis for conductimetric and polarographic methods of analysis. In a similar manner, the total amount of chemical change which takes place at an electrode is related to the total amount of current. This forms the basis of coulometric methods of analysis.

2. *Voltaic or galvanic cells.* In some cells, the chemical nature of the electrodes and solutions results in a chemical reaction taking place at the electrodes with the production of electrical energy (batteries) but without the need for an external voltage (Figure 4.2). This type of cell is used in potentiometric methods in which no voltage is applied to the cell and although no current actually flows between the electrodes (which are said to be unpolarized), they develop a potential relative to the solution due to the nature of the electrodes and the solution. This potential can be measured and related to the concentration of the ions in the solution.

> ➤ Electrodes are said to be polarized when a voltage is applied to them and as a result there is a potential difference between them.

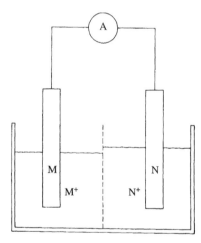

Figure 4.2 A galvanic cell. A reaction taking place at the electrodes results in a potential difference developing between them and, in some cases, a measurable current will flow between them.

4.1   Potentiometry

The potential developed by a single electrode in a solution is caused by the tendency of the solution either to donate or accept electrons and can be

➤ The Nernst equation expresses the potential developed by an electrode in terms of the activity and number of ions involved in the reaction.

calculated using the Nernst equation:

$$E = E^0 - \frac{2.3026RT}{nF} \times \log a$$

where E is the electrode potential at the specified concentration,
E^0 is the standard electrode potential,
R is the gas constant,
T is the absolute temperature,
n is the number of electrons involved,
F is the Faraday constant,
a is the activity of the ion.

For measurements made at 25°C the equation simplifies to:

$$E = E^0 - \frac{0.059}{n} \log a$$

➤ A half-cell describes a single electrode and its associated chemical reaction which forms part of a voltaic cell.

➤ A half-reaction is the reaction that occurs at a single electrode but that requires another reaction to accept or donate the electrons involved in the overall process.

➤ An active electrode is composed of an element that is in an equilibrium with its ions in the surrounding solution.

There can be no chemical reaction in such a system without a complementary electron donor or acceptor to complete the process. Each of these electrode systems is known as a half-cell and the potential developed by a half-cell cannot be measured in absolute terms but only compared with that of another half-cell. The chemical reaction occurring at each half-cell is known as a half-reaction.

An active electrode consists of an element (M) in its uncombined state which is capable of establishing an equilibrium with a solution that contains its ions:

$$M = M^+ + e^-$$

The ionization of atoms or molecules results in a potential being developed by such an electrode, the intensity of the potential being related to the concentration of the ions. The effective concentration of the ions (known as the activity of the ions) is more significant than the molar concentration. The values for activity and concentration are only the same in very dilute solutions.

➤ The activity of an ion reflects the proportion of the molecules that are actually ionized at the time.

➤ Inert electrodes are used to make electrical contact with solution and are not part of the chemical reaction.

Inert electrodes, such as silver, platinum, and carbon, are used solely to make electrical contact with the solution and only reflect the potential of the solution. They are used to measure the potential of solutions containing mixtures of ions which have a tendency to transfer electrons between them, e.g. ferric and ferrous ions:

$$Fe^{3+} + e^- = Fe^{2+}$$

Such reactions are know as redox reactions and, in this case, the potential developed by the electrode system depends on the tendency of ferrous ions to donate an electron compared with the tendency of the ferric ions to accept one.

➤ Standard electrode potential is the potential developed by an active electrode when in equilibrium with a molar solution of its ions.

The standard electrode potential of an element is defined as its electrical potential when it is in contact with a molar solution of its ions. For redox systems, the standard redox potential is that developed by a solution containing molar concentrations of both ionic forms. Any half-cell will be able to oxidize (i.e. accept electrons from) any other half-cell which has a lower electrode potential (Table 4.1).

Table 4.1   Standard electrode potentials at pH 7.0

Half-reaction	E'_0
Oxygen/water	0.81
Ferric/ferrous	0.77
Ferricyanide/ferrocyanide	0.36
Oxygen/hydrogen peroxide	0.30
Cytochrome c ferric/ferrous	0.22
Dehydroascorbic acid/ascorbic acid	0.08
Pyruvate/lactate	-0.19
FAD/FADH$_2$	-0.22
NAD$^+$/NADH	-0.32
2H$^+$/H$_2$	-0.42
Ferrous/iron	-0.44

It is impossible to measure the potential of a half-cell directly and a reference half-cell must be used to complete the circuit. The hydrogen electrode (Figure 4.3) is the standard reference electrode against which all other half-cells are measured and is arbitrarily attributed a standard electrode potential of zero at pH 0. Because it is difficult to prepare and inconvenient to use, the

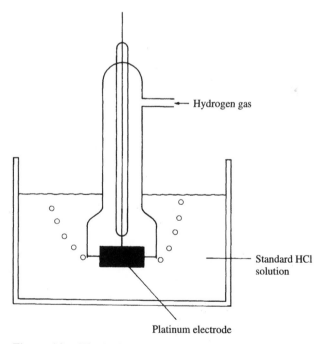

Hydrogen gas

Standard HCl solution

Platinum electrode

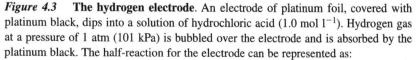

Figure 4.3   **The hydrogen electrode**. An electrode of platinum foil, covered with platinum black, dips into a solution of hydrochloric acid (1.0 mol l^{-1}). Hydrogen gas at a pressure of 1 atm (101 kPa) is bubbled over the electrode and is absorbed by the platinum black. The half-reaction for the electrode can be represented as:

$$2H^+ + 2e^- = H_2$$

> ➤ A saturated calomel electrode is conveniently used as a reference electrode in potentiometric measurements.

saturated calomel electrode is frequently used instead. This has a standard electrode potential of $+0.242$ V relative to the hydrogen electrode. It consists of a mercury electrode in equilibrium with mercuric ions in the slightly soluble salt, mercuric chloride (Figure 4.4). A high concentration of chloride ions is maintained by using a saturated solution of potassium chloride, which also provides a means of ensuring electrical contact with any other half-cell used.

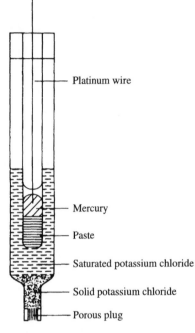

Platinum wire

Mercury

Paste

Saturated potassium chloride

Solid potassium chloride

Porous plug

Figure 4.4 **The saturated calomel electrode**. A platinum wire makes electrical contact with an electrode which is composed of a paste of metallic mercury, mercuric chloride (calomel) and potassium chloride. A saturated solution of potassium chloride completes the half-cell and provides electrical contact through a porous plug.

4.1.1 Potentiometric measurements

The sensitivity of instruments using low resistance circuits is determined primarily by the sensitivity of the galvanometer (Figure 4.5). Electrode systems that have a high resistance, e.g. glass electrodes, require a high impedance voltmeter, which converts the potential generated into current which can be amplified and measured. Such instruments are commonly known as pH meters but may be used for many potentiometric measurements other than pH.

Titrations can often be conveniently followed potentiometrically and in many cases it is not the actual value of the electrode potential that is important but the pattern of changing potential as the composition of the solution varies – pH and redox measurements are particularly well suited to such methods. In many instances the equivalence point will be indicated by a significant change in potential (Figure 4.6) but sometimes the change at the equivalence point is difficult to detect and it might be more convenient to use a derivative plot. Instead of plotting potential (E) against volume of titrant (V), a graph is plotted of the change in potential for the unit change in volume $\Delta E/\Delta V$ against volume. Such a first-derivative plot (Figure 4.7) indicates the equivalence point as a peak or spike.

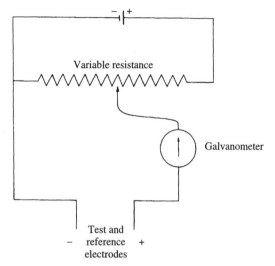

Figure 4.5   **A potentiometer circuit**. The voltage from the test circuit is balanced against a known voltage by means of a variable resistance using a galvanometer to indicate the position at which no current flows in either direction.

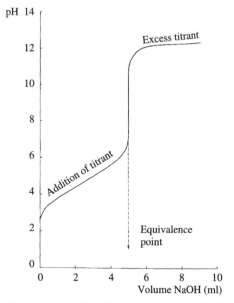

Figure 4.6   **A titration curve**. Acetic acid (10 ml of a 0.1 mol l^{-1} solution) was titrated with a sodium hydroxide solution (0.2 mol l^{-1}) and the pH of the resulting solution plotted against the amount of alkali added.

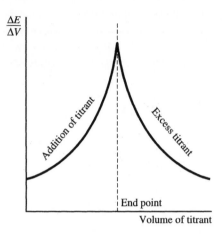

$\dfrac{\Delta E}{\Delta V}$

Addition of titrant

Excess titrant

End point

Volume of titrant

Figure 4.7
A first-derivative plot of a potentiometric titration curve.

If two identical electrodes are placed in separate solutions that are similar in every way except for the concentration of the test ions, the potential developed between the electrodes will be related to the ratio of the two concentrations, e.g. Ag−AgCl electrodes in solutions containing chloride ions. A calibration curve of potential developed in a series of known concentrations of test ions can be used in the analysis of unknown samples. Often it is advisable to plot the graph as potential versus logarithm of concentration to give the straight line relationship as indicated by the Nernst equation. It is necessary to add a constant amount of a high concentration of a non-reacting electrolyte to give all solutions tested the same ionic strength. This is because the potential depends upon the activity of the ions rather than concentration.

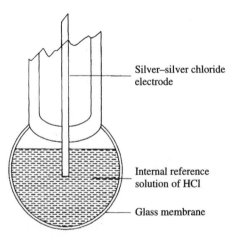

Silver–silver chloride electrode

Internal reference solution of HCl

Glass membrane

Figure 4.8
The glass electrode.

4.1.2 Measurement of pH

The glass electrode is most commonly used for routine measurement of pH because the use of the hydrogen electrode is impracticable (Figure 4.8). It consists of a silver–silver chloride electrode in a reference solution of hydrochloric

> ➤ A glass electrode is selectively permeable to hydrogen ions and is used for measuring the pH of solutions.

acid (usually 0.1 mol l^{-1}) contained in a glass membrane. The membrane is made of a special glass, usually a hydrated aluminosilicate containing sodium or calcium ions. It is selectively permeable to hydrogen ions and the potential that develops across the membrane depends upon the hydrogen ion concentration of the test solution compared with the reference acid solution inside the electrode. This potential can be measured against a reference calomel electrode using a high impedance voltmeter. For ease of operation the glass electrode is often combined with a reference calomel electrode in a single probe (Figure 4.9). Glass electrodes must be stored in water between use in order to keep the glass membrane fully hydrated.

> ➤ A combined electrode is a double electrode system consisting of a glass electrode and a calomel electrode combined in one probe.

There is an almost linear relationship between potential and pH over the pH range of 2–10 but at extreme pH values special glass electrodes must be used. Calibration of the instruments is essential and may be achieved with either buffers whose pH has been previously measured using a hydrogen electrode or solutions of very pure chemicals prepared in specific concentrations, e.g. a solution of potassium hydrogen phthalate (0.05 mol l^{-1}) has a pH of 4.00 at 15 °C. For measurements over a range of pH values it is necessary to standardize the instrument on at least two standard buffer solutions which cover the required range, making sure that the correct temperature setting is used.

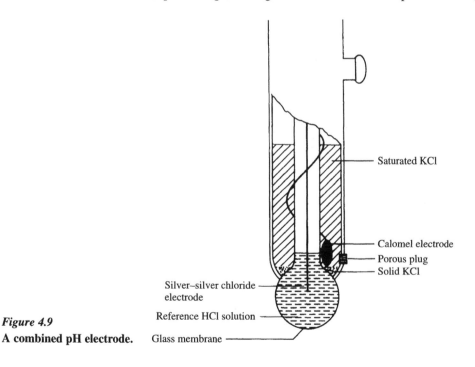

Figure 4.9
A combined pH electrode.

Saturated KCl

Calomel electrode
Porous plug
Solid KCl

Silver–silver chloride electrode

Reference HCl solution

Glass membrane

4.1.3   Ion-selective electrodes

The membrane of the glass electrode used for pH measurements is selectively permeable to hydrogen ions and from this basic concept a whole range of ion-selective electrodes have been developed. Varying the composition of the glass membrane can change the permeability of the glass and several cation-sensitive electrodes have been developed in this way (Table 4.2). Although such electrodes may be described as specific for ions such as Na^+, K^+, Ca^{2+} and NH_4^+, they are in fact only selective and not specific and often show significant interference from other ions, particularly hydrogen ions. In all cases the basic design of the electrode is the same (Figure 4.10) but the nature of the ion-selective membrane varies.

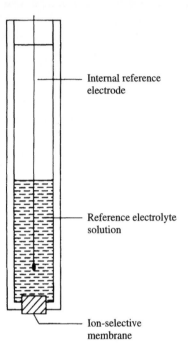

— Internal reference electrode

— Reference electrolyte solution

Figure 4.10
The basic design of an ion-selective electrode.

— Ion-selective membrane

Table 4.2   Ion-selective electrodes

Test ion	Membrane material	Major interfering ions
Solid-state electrodes		
Fluoride	LaF	I^- Br^- Cl^-
Chloride	$AgCl/Ag_2S$	S^- I^-
Bromide	$AgBr/Ag_2S$	S^- I^-
Iodide	AgI/Ag_2S	S^-
Sulphide	AgI/Ag_2S	
Cupric	Ag_2S/CuS	Hg^+ Ag^+
Lead	Ag_2S/PbS	Hg^+ Ag^+
Cadmium	Ag_2S/CdS	Hg^+Ag^+ Cu^{2+}
Silver	Ag_2S	
Liquid-membrane electrodes		
Potassium	Valinomycin in diphenyl ether	
Ammonium	Macrotetrolides in tris(2-ethylhexyl)phosphate	
Calcium	Calcium dialkylphosphate in dioctylphenylphosphonate	
Calcium/magnesium	Dialkylphosphoric acid in aliphatic alcohol	

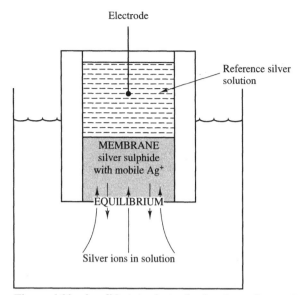

Figure 4.11   A solid-state electrode showing a first-order response. An electrode designed to measure the activity of silver ions uses a crystalline membrane of silver sulphide. An equilibrium between the mobile silver ions of the membrane and the silver ions in the solutions results in the development of a potential difference across the membrane.

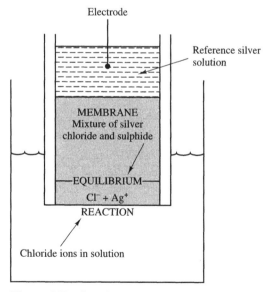

Figure 4.12   A solid-state electrode showing a second-order response. The electrode shown in Figure 4.11 can be modified by the incorporation of silver chloride into the membrane to enable the activity of chloride ions in a sample to be measured. A surface reaction between the test chloride ions and the membrane silver ions alters the activity of the latter, resulting in a change in the potential difference across the membrane.

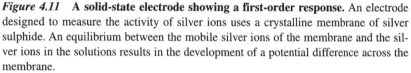

Test anion ⇌ Membrane cation

In solid-state electrodes the membrane is a solid disc of a relatively insoluble, crystalline material which shows a high specificity for a particular ion. The membrane permits movement of ions within the lattice structure of the crystal and those ions which disrupt the lattice structure the least are the most mobile. These usually have the smallest charge and diameter. Hence, only those ions that are very similar to the internal mobile ions can gain access to the membrane from the outside, a feature that gives crystal membranes their high specificity. When the electrode is immersed in the sample solution, an equilibrium is established between the mobile ions in the crystal and similar ions in the solution and the resulting potential created across the membrane can be measured in the usual manner.

The simplest solid-state membranes are designed to measure test ions, which are also the mobile ions of the crystal (first-order response) and are usually single-substance crystals (Figure 4.11). Alternatively, the test substance may be involved in one or two chemical reactions on the surface of the electrode which alter the activity of the mobile ion in the membrane (Figures 4.12 and 4.13). Such membranes, which are often mixtures of substances, are said to show second- and third-order responses. While only a limited number of ions can gain access to a particular membrane, a greater number of substances will be able to react at the surface of the membrane. As a result, the selectivity of electrodes showing second- and third-order responses is reduced.

Liquid-membrane electrodes consist of an ion-selective material dissolved

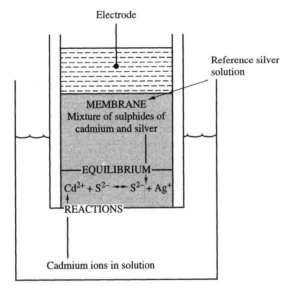

Figure 4.13 **A solid-state electrode showing a third-order response.** An alternative modification to the electrode described in Figure 4.11 will permit the measurement of cadmium ions in solution. The membrane is composed of a mixture of silver and cadmium sulphides. The surface reaction between the cadmium ions in the test solution and the sulphide ions in the membrane will affect the equilibrium between the sulphide ions and the silver ions in the membrane.

Test cation ⇌ Membrane anion ⇌ Membrane cation

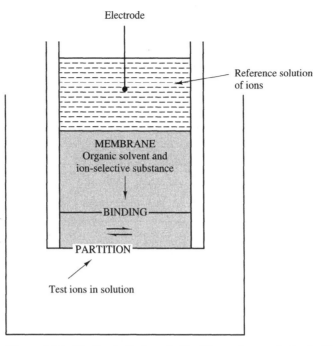

Electrode

Reference solution
of ions

MEMBRANE
Organic solvent and
ion-selective substance

BINDING

PARTITION

Test ions in solution

Figure 4.14   **The basic design of a liquid-membrane electrode.** The partition of the
ions from the sample in the immiscible solvent of the membrane will cause a change in
the potential across the membrane. The effect can be enhanced by the binding of the
test ions by an ion-selective substance which can be incorporated in the membrane.

in a solvent that is not miscible with water. The liquid is held in a porous, inert
membrane (often plastic) which allows contact between the test solution on
one side and the reference electrolyte on the other. The ions from the reference
solution will partition themselves between the two immiscible solvents (the
aqueous and the organic phases), giving the electrode a particular potential.
The presence of the test ions in the sample affects the activity of the reference
ions in the membrane resulting in a change in the potential difference across
the membrane. While the solvent is chosen to give the best partition effects for
a particular ion, the specificity and sensitivity of the process (Figure 4.14) can
be greatly improved by the incorporation of a specific chelating or ion-selective substance. These are often large counter-ions to the test ions and have a
low solubility in water, tending to form undissociated complexes with the test
ions.

Ion-selective electrodes are often incorporated in automated analysers,
particularly for the measurement of sodium and potassium. The electrodes
may be sited within the instrument or be available in a miniature disposable
form (Figure 4.15)

Use of ion-selective electrodes

Because potentiometric measurements reflect the activity of an ion rather than
its concentration, it is usually necessary to calibrate the system with standard
solutions of known activity. It is possible to calculate the concentration of an

ion in a sample if the value of the activity coefficient for the ion is known:

$$\log \frac{a}{c} = \log \gamma$$

where a is the activity of the ion,
  c is the concentration of the ion,
  γ is the activity coefficient.

The equation is probably more useful in the form

$$\log c = \log a - \log \gamma$$

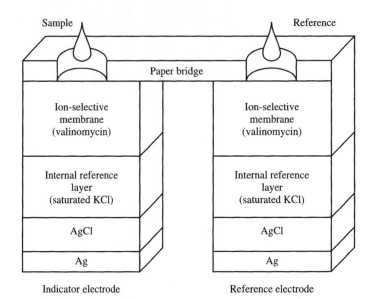

Figure 4.15   **Representation of the disposable slide for the measurement of potassium by the Vitros Chemistry System.** The difference in potential between the two half-cells, one receiving the sample and the other a reference solution of known potassium ion concentration, is mathematically converted to give the concentration of potassium ions in the sample.

It is often more convenient to relate the potentiometer reading directly to concentration by adjusting the ionic strength and hence the activity of both the standards and samples to the same value with a large excess of an electrolyte solution which is inert as far as the electrode in use is concerned. Under these conditions the electrode potential is proportional to the concentration of the test ions. The use of such solutions, which are known as TISABs (total ionic strength adjustment buffers), also allows the control of pH and their composition has to be designed for each particular assay and the proportion of buffer to sample must be constant.

> ➤ Total ionic strength adjustment buffers (TISABs) are used to equalize ionic activity in different solutions.

Ion-selective electrodes are particularly useful for monitoring the disappearance of an ion during a titration. In many cases it is not necessary to calibrate the instrument because there is often a significant change in the potential at the end-point of a titration. However, some electrodes have a slow response time and care must be taken to ensure that titration is not performed too quickly.

Apart from interference the greatest problem in the use of ion-selective electrodes is that of contamination. Any insoluble material deposited on the surface of the electrode will significantly reduce its sensitivity and oil films or protein deposits must be removed by frequent and thorough washing. It is possible to wipe membranes with soft tissue but they can be easily damaged. Solid-state membranes are more robust but they must not be used in any solution which might react with the membrane material.

Self test questions	**Section 4.1**

Section 4.1

1. If half-cell A has an electrode potential of 0.2 V and half-cell B has an electrode potential of -0.1 V, which of the following statements are true?
   (a) A can reduce B.
   (b) B can reduce A.
   (c) A can oxidize B.
   (d) B can oxidize A.
2. Which of the following factors will affect the measurement of pH using a glass electrode?
   (a) Temperature.
   (b) Valency of ions present.
   (c) Presence of salts.
   (d) Presence of protein.
3. Potentiometry involves the use of electrode systems known as voltaic cells
   BECAUSE
   in potentiometric methods a voltage is applied across two electrodes.
4. The glass electrode is an example of an ion-selective electrode
   BECAUSE
   an ion-selective electrode incorporates a membrane which preferentially restricts the flow of all ions except the analyte ion.

4.2   Conductimetry

The simplest application of an electrolytic cell is the measurement of conductance. If a fixed voltage is applied to two electrodes which dip into a test solution, depending upon the conductivity of the solution, a current will flow between them. Although conductivity measurements do not give any information about the nature of the ions in the solution they can be used quantitatively. They are, however, more frequently used to monitor the changing composition of a solution during a titration.

The current passing between the two electrodes is carried by the ions in the solution (Figure 4.16) and is related to the number present, which is a result of both the molar concentration of the compound and its extent of ionization. The anions donate electrons to the anode while the cations accept

➤ The ionic mobility of
an ion is its speed of
migration under the
influence of a fixed
potential difference.

electrons from the cathode. It is this transfer of electrons that determines the amount of current flowing and the contribution of each ionic species is determined by its ionic mobility (velocity).

Because of the nature of the reaction by which molecules dissociate into ions there is not always a simple relationship between the increase in conductivity with increasing molar concentration and often a limiting value is reached. For instance, increasing concentrations of sulphuric acid result in increased conductivity up to a concentration of about 35% (v/v), above which the conductivity decreases.

The current flowing through a conductor is defined by Ohm's law as:

➤ Ohm's law expresses
the current flowing in
terms of the voltage
applied and the
resistance of the system.

$$\text{current } (I) = \frac{\text{voltage applied } (V)}{\text{resistance } (R)}$$

or in terms of conductance (C)

$$I = C \times V$$

The specific conductance (κ) of a solution is defined as the conductance (S) per centimetre of a solution that has a cross-sectional area of 1 cm^2, and is measured in S cm^{-1} (or in non-SI units as Ω^{-1} cm^{-1}). The molar conductance (Λ) is the specific conductance of a solution corrected for the concentration of ions in the solution. $\Lambda = \kappa \times$ volume of solution which contains 1 gram mole.

The value for κ will normally decrease as the concentration of the solution decreases but the value for Λ will increase because of the increased dissociation of molecules in dilute solutions. A value for the molar conductance at infinite dilution (Λ_0) can be determined by plotting the calculated values for Λ against the molar concentration of the solution used and determining the plateau value for Λ. From such investigations it is possible to determine the ionic mobilities of ions (Table 4.3) and calculate the molar conductance of an

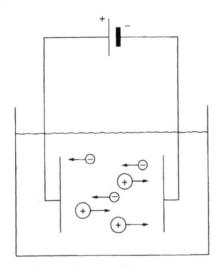

Figure 4.16 **The conductivity of a solution.** The total transfer of electricity is not necessarily shared equally because different ions move at different speeds. The Kohlrausch law says that the total conductivity of an electrolyte is the sum of the conductivities of the anions and the cations.

electrolyte at infinite dilution, e.g. the value for potassium sulphate (234) is the sum of the ionic mobilities of potassium (74) and sulphate (160).

Table 4.3   Molar ionic conductivity

Cation	Molar conductance ($\Omega^{-1}\,cm^2\,mol^{-1}$)	Anion	Molar conductance ($\Omega^{-1}\,cm^2\,mol^{-1}$)
Na^+	50	HCO^-	45
Ag^+	62	HSO_4^-	52
K^+	74	MnO_4^-	61
NH_4^+	74	NO_3^-	71
Zn^{2+}	106	Cl^-	76
Mg^{2+}	106	I^-	77
Cu^{2+}	107	Br^-	78
Fe^{2+}	108	CO_3^{2-}	119
Ca^{2+}	119	SO_4^{2-}	160
Fe^{3+}	205	OH^-	197
H^+	350	PO_4^{3-}	240

4.2.1   Conductimetric measurements

The basic instrumentation for conductimetric measurements is a Wheatstone bridge arrangement (Figure 4.17). An alternating rather than direct current is used in order to prevent polarization of the solution (i.e. anions moving to the anode and cations moving to the cathode) and any electrolysis that might result from this polarization.

With fixed resistances, R_1 and R_2, the variable resistance R_v is adjusted until no current flows through the galvanometer. Under these conditions:

$$R_1 \times R_v = R_{cell} \times R_2$$

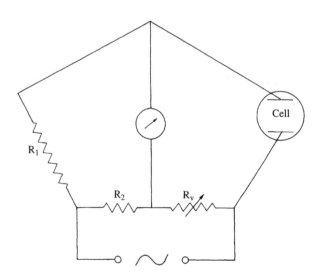

Figure 4.17
A Wheatstone bridge

This gives a value for resistance of the cell which can be converted to conductance by calculating the reciprocal.

The electrodes of conductivity cells are usually made of platinum coated with platinum black with a known area. Although in many cells the distance between the electrodes is adjustable, for any series of experiments it must be held constant and for many calculations the precise value is required. The cells must be thermostatically controlled because any changes in temperature will cause significant alteration of conductivity values.

4.2.2 Applications of conductimetric measurements

Although conductimetry is useful in determining various physical constants, e.g. dissociation and solubility constants, its major analytical application is for monitoring titrations.

When two electrolyte solutions that do not react with each other are mixed, provided that there is no appreciable change in volume, the conductance of the solution will rise because of the increase in the numbers of ions. However, if there is a reaction between the ions, one ion being replaced by another, the conductance of the solution will alter depending upon the relative

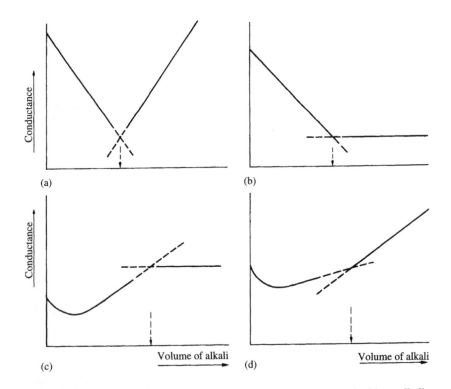

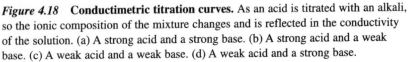

Figure 4.18 **Conductimetric titration curves.** As an acid is titrated with an alkali, so the ionic composition of the mixture changes and is reflected in the conductivity of the solution. (a) A strong acid and a strong base. (b) A strong acid and a weak base. (c) A weak acid and a weak base. (d) A weak acid and a strong base.

mobilities of the ions involved. In the reaction

$$A^+B^- + C^+D^- = AD + C^+B^-$$

in which CD is the titrant, the effect of the reaction is to replace A^+ by C^+ in the solution. If these two ions have different ionic mobilities, the conductance of the solution will change accordingly. This is illustrated in the titration of a strong acid by a strong base: the hydrogen ions are replaced by the cation from the base, which will have a smaller ionic mobility and hence a lower conductivity (Table 4.3) and will result in a gradual fall in the conductance of the solution. However, at the end-point of the reaction, the increasing concentration of hydroxyl ions will result in an increasing conductance (Figure 4.18). An extrapolation of the lines before and after the end-point will give an intercept at the end-point.

Conductimetric titrations, particularly when they are automated, are useful if either of the reactants is deeply coloured, so preventing visual monitoring of the titration using indicators, or if both reactants are very dilute.

Flow through conductance cells are useful detectors in ion-exchange chromatographic separations where the analytes are ionic when they enter the detector cell. At the low concentrations encountered, conductivity is proportional to the mobility of the ions involved as well as their concentration, and this, together with the background conductivity of the solvent, demands that standards are used. Conductivity increases with an increase in temperature and it is important in such measurements that temperature is monitored and appropriate corrections made when calculating the results.

4.3   Coulometry

Coulometry is an electrolytic method of analysis. In general, electrolytic methods have limited applications in analytical biochemistry but they are useful in the analysis of substances which, while not strictly biochemical, are often important in biological and physiological chemistry.

Faraday's laws of electrolysis form the basis of quantitative coulometric analysis. They are:

1. The weight of substance liberated during electrolysis is directly proportional to the quantity of electricity passed.
2. The weights of substances liberated by the same quantity of electricity are in direct proportion to their equivalent weights.

> ➤ Coulometric methods are quantitative methods based on constant voltage coulometry.

There are two types of coulometric analysis using either constant current or constant voltage. In the latter (known as coulometric methods), a fixed voltage is applied which causes the test substance to react at an electrode, and the current that flows decreases as the number of test ions remaining falls. The total amount of current that flows during the reaction can be measured and related to the amount of test substance originally present. This type of measurement has only limited biochemical applications. In the past it was used for the measurement of oxygen, but devices known as oxygen electrodes are more commonly used nowadays. The alternative technique of constant current

➤ Coulometric titrations
are quantitative methods
based on constant
current coulometry.

coulometry (known as coulometric titration) is more useful in analytical bio-chemistry because it permits the generation of specific, often very labile, reagents at electrodes.

Table 4.4 Coulometric titrations

Analyte	Titrant generated	Generating electrolyte	Electrode
Calcium	EDTA	Mercuric EDTA in ammonia buffer	Mercury cathode
Chloride	Silver ions	—	Silver anode
Iron^{2+}	Ce^{4+}	Cerous sulphate in H$_2$SO$_4$	Anode
Oxygen	Cr^{2+}	Chromium sulphate in H$_2$SO$_4$	Anode
(Acids)	Hydroxyl ions	Sodium sulphate solution	Cathode
(Bases)	Hydrogen ions	Sodium sulphate solution	Anode

4.3.1 Coulometric titrations

When a constant current is passed through the cell, the time taken for the reaction to go to completion can be used as a measure of the amount of test substance originally present. The electrode reaction may directly involve the test ions (as in constant voltage techniques) or alternatively may gener-ate another substance which will react with the test substance. Reagents can be generated at an electrode which cannot be produced in any other way because they are so unstable. The end-point can be detected using a visual system, e.g. indicator dyes, but more usually by means of an additional electrode system, e.g. potentiometry. Coulometric titrations offer methods of analysis (Table 4.4) that are very precise and readily lend themselves to automation and the analysis of very small samples. While such methods do not strictly require standardization, the quantity of electricity being the standard, it is usual to use control samples to check or adjust the analysis conditions.

The electrolytic system consists of a generating or working electrode and an auxiliary electrode (Figure 4.19). The auxiliary electrode is usually held in a separate glass tube with a fine porous glass disc to prevent the titrating reagent produced at the working electrode from reaching it and so reducing the efficiency of the reaction. In addition a pair of indicator elec-trodes is necessary in order to detect the end-point of the reaction. The elec-trodes vary depending upon the nature of the test reaction taking place in the cell. In order for these electrodes to function effectively, there must be ions present at all stages of the reaction and it is therefore usual to incor-porate a non-reacting carrier electrolyte in the reaction mixture.

The amount of substance titrated can be calculated from the following equation:

$$W = \frac{ItM}{Fn}$$

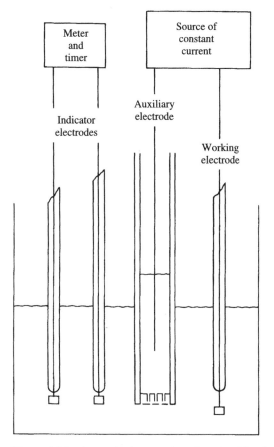

Figure 4.19 **A constant current coulometry titration cell.** The reagent is produced at the working electrode and reacts with the sample. The indicator electrodes detect the changing potential or conductivity of the solution and the amount of change that takes place is measured and related to the concentration of the reactant in the sample.

where W is the weight of substance,
           M is the relative molecular mass,
           F is the Faraday constant (96 493),
           n is the number of equivalents,
           t is the time in seconds,
           I is the current in amperes.

However, most instruments designed for coulometric titration are at least semi-automated and directly convert the time into a measure of the sample concentration.

The determination of chloride using an instrument known as a **chloride meter** is probably the most common application of coulometry in biochemistry. The instrument is designed to generate silver ions electrolytically from a silver anode. These ions are removed from the solution as undissociated silver chloride, which is either deposited on the anode or precipitated in the solution. A low concentration of carrier electrolyte (nitrate ions) permits a small current to flow between the indicating electrodes during the titration but this rises rapidly at the end-point, when silver ions appear in the solution. This increase in current can be used to trigger a relay circuit which switches the timer off and records the total time for the reaction.

4.4   Voltammetry

> ➤ Polarographic methods involve the measurement of current as the voltage applied is gradually increased.
> ➤ Amperometric methods involve the measurement of current that results from the application of a fixed voltage.

Voltammetry is the term given to electrochemical techniques which monitor the relationship between the voltage applied to an electrode system and the current that flows as a result of the reaction. It covers a wide range of different electrode techniques, many of which are specifically designed to monitor a particular chemical reaction. Voltammetry is generally divided into two main subdivisions of polarography and amperometry.

During electrolysis, ions are discharged at the appropriate electrodes and a continuous supply of the ions is necessary for the current to continue flowing. An ion is said to be discharged when it either accepts or donates an electron at an electrode and as a result loses its charge. If there is only a low concentration of reactive ions they will be quickly discharged around the electrode and the current will fall to a level that depends upon the rate at which more ions can defuse from the bulk of the solution to the electrode. As the voltage is increased, so the rate of diffusion will increase up to a maximum level, which is determined not only by the nature of the ion but also by its concentration. This results in a maximum current, which is known as the diffusion current. The level of the diffusion current is used in quantitative polarographic techniques but the voltage at which it occurs forms the basis for qualitative polarography. Variations of voltammetry in which the voltage is held constant and the variations in current measured are known as amperometric techniques and are frequently used to monitor the end-point of titration or the presence of a particular analyte in the effluent from a chromatographic column.

4.4.1   Polarography

In polarographic methods the voltage that is applied to an electrode system is gradually increased and the resulting current is continuously measured. The information is usually presented in a graphical form known as a polarogram, which is a plot of current against voltage. If there is a very high concentration of non-reacting ions present, i.e. ions that are not discharged at the voltage applied, they will be responsible for carrying most of the current. The amount of current carried by these ions will be almost independent of the voltage applied and is known as the residual current. However, as the voltage is increased and reaches the discharge potential of other ions, the current will rise rapidly owing to their discharge but it will stabilize at a new maximum level (the diffusion current) dependent upon their subsequent rate of diffusion to the electrode. These changes are reflected in the shape of the polarogram (Figure 4.20).

The potential or voltage at which a particular ion will be discharged is

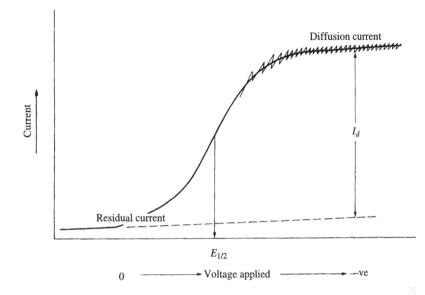

Figure 4.20 **A polarographic titration curve.** A polarogram has oscillations about a mean current–voltage curve owing to the regularly changing area of the mercury drop.

> ➤ Half-wave potential is a qualitative measurement in polarography.

characteristic of the ion although it is slightly affected by its concentration. However, the half-wave potential ($E_{1/2}$) is independent of concentration and can be used to identify a particular ion, giving polarography a qualitative application. The difference between the residual current and the diffusion current is dependent on concentrations and is used for quantitive measurements (Figure 4.20). It is necessary in quantitative analysis to standardize the instrument using solutions of known concentration and to prepare a calibration graph using the difference between residual current and the diffusion current as the variable parameter.

In classical polarographic techniques, a dropping mercury electrode is used. This is a complex device in which continuously produced small droplets of mercury are used as the active electrode in order to prevent poisoning of the electrode and to provide constant conditions throughout the analysis. For many applications, specifically designed electrodes are available which are simpler to use.

One specific variant of the technique is known as direct current cyclic voltammetry (DCCV), in which the voltage sweep is over a limited range and a short time and is immediately reversed. The cycle is repeated many times and the pattern of current change is monitored. The technique uses relatively simple electrodes and is used to study redox reactions and there are a range of sophisticated variants of the technique.

4.4.2   Electrochemical detectors for HPLC

Amperometric devices are gaining popularity as HPLC detectors. A fixed voltage is applied and the resulting current is measured. The system consists of a flow cell containing a working and a reference electrode. There is a range of electrodes available for use in the oxidation mode including glassy carbon, which is hard and impervious to organic solvents, platinum and gold, which tend to become poisoned by absorbed solutes, and others with improved properties which are being commercially developed.

In designing a method it is necessary to know the operating voltage of the electrode for a particular analyte. The classical polarogram (Figure 4.20) shows a half-wave potential ($E_{1/2}$) and the diffusion current. As the potential applied is increased beyond the diffusion current level, eventually the mobile phase will react at the electrode and there will be a further significant increase in the current flowing. The potential used should be between the half-wave potential for the analyte and the discharge potential of the mobile phase. Most frequently the voltage at the beginning of the diffusion current is selected. By using more than one electrode it is possible to measure different compounds simultaneously by selecting a different operating voltage for each electrode.

4.4.3   Oxygen electrode

The dissolved oxygen content of a solution can be determined by measuring the diffusion current that results at a selected voltage. The Clark electrode was developed for this purpose and various modifications have subsequently been introduced. It consists basically of a platinum electrode separated from the sample by a membrane which is permeable to oxygen, e.g. Teflon or polyethylene. A reference electrode of silver/silver chloride in potassium chloride is used to complete the system (Figure 4.21). When a voltage that is sufficient to give the

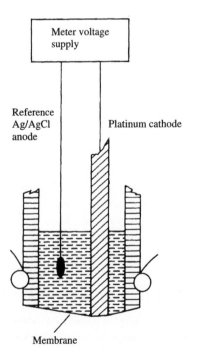

Figure 4.21
An oxygen electrode.

diffusion current for oxygen (often -0.6 to -0.8 V with respect to the reference electrode) is applied, the amount of current that flows is related to the partial pressure of oxygen (PO_2) in the sample. It is necessary, however, to calibrate the electrode using solutions of known oxygen content. Other gases as well as oxygen can be reduced at the operating voltage of the electrode, e.g. halogens, halogenated hydrocarbons and sulphur dioxide, and will interfere with quantitative measurements, while hydrogen sulphide will poison the electrode.

4.4.4 Anodic stripping voltammetry

This is a technique for trace metal analysis and is a reversal of the usual polarographic method. Using a mercury electrode, either as a hanging drop or a graphite rod coated with mercury, a negative (cathodic) potential is applied relative to a reference silver/silver chloride electrode. Any metal ions, with discharge potentials less than that applied, will be deposited on the electrode and will form an amalgam with the mercury. It is important to control the time of electrolysis and the method of stirring the sample because the amount of metal deposited will be taken as a measure of its concentration in the sample.

The analysis stage involves scanning the voltage applied to the electrode towards a more positive (anodic) potential. As the potential becomes more positive, so the metal atoms are re-oxidized, giving an anodic wave of current. The current will show a maximum value as all the atoms of a particular element are stripped off the electrode as ions. The height of the peak that results is proportional to the amount of metal and the voltage at which stripping occurs will be characteristic of the element.

The technique is useful for the quantitation of many metals including lead, copper, mercury, cadmium and zinc with detection limits as low as 10 pg. Its sensitivity makes it a very suitable method for trace metal analysis in biological samples.

4.5   Biosensors

Biosensors are analytical devices that incorporate a biological component and a transducer. These must be in close proximity with one another and preferably in intimate contact, i.e. the biological component immobilized on to the transducer. Such devices are available in disposable forms, e.g. for measurement of blood glucose in diabetic patients, evaluation of the freshness of uncooked meat. Other designs are suitable for continuous use, e.g. on-line monitoring of fermentation processes, the detection of toxic substances.

4.5.1 Biosensor components

Biological components
The possible biological components fall into two main categories (Table 4.5) and their function is to recognize and bind a specific analyte. It is the properties of the biological component that give specificity to biosensors and make them suitable for the analysis of samples without pre-treatment.

Table 4.5 Biological components of biosensors

Component	Examples
Biocatalysts	Enzymes
	Microbial, plant, animal cells
	Sub-cellular organelles
Bioreceptors	Antibodies
	Lectins
	Cell membrane receptors
	Nucleic acids
	Synthetic molecules

For **biocatalysts**, binding is followed by a chemical reaction and release of products. Enzymes were the first catalysts used in biosensors and a large number of these natural proteins are available. Although they remain the most commonly employed, the use of purified enzymes is not always satisfactory and in some situations cell preparations containing the required enzymes in their natural environment may be preferable. This approach reduces specificity but can be used to advantage when the analysis of the range of related substances is required as in the case of pollution monitoring.

In devices incorporating **bioreceptors** binding is non-catalytic and essentially irreversible. Of major significance to the development of this type of biosensor was the commercial availability of monoclonal antibodies. This, together with the availability of other bioreceptors and improvements in methods for monitoring binding, e.g. optical, piezoelectric, has resulted in the introduction of devices capable of detecting trace amounts of a wide range of substances such as drugs, hormones, viruses and pathogenic bacteria.

➤ Monoclonal antibodies – see Section 7.2.1.

Transducers

The transducer converts the reaction between the analyte and the biological component into an electrical signal which is a measure of the concentration of the analyte. A wide range of transducers are available and are broadly divided into electrochemical, optical, thermal and piezoelectric.

4.5.2   Electrochemical biosensors

Electrochemical biosensors are the most common especially when the biological component is an enzyme. Many enzyme reactions involve electroactive species being either consumed or generated and can be monitored by amperometric, potentiometric or conductimetric techniques, although the latter are the least developed and will not be discussed further.

Amperometric biosensors

These devices measure the current that is produced when a constant potential difference is applied between the transducing and reference electrodes. A current is produced when the species of interest is either oxidized at the anode or reduced at the cathode. The amount of current flowing is a function of the concentration of that species at the electrode surface. This type of transducer is applicable to substances that are easily electrochemically oxidized or reduced at an electrode.

Biosensors for the measurement of glucose which are based on glucose oxidase are examples of devices which use amperometric detection. The overall reaction can be monitored in several ways.

$$\text{glucose} + O_2 = \text{gluconic acid} + H_2O_2$$

Originally, a Clark oxygen electrode was used to measure a reduction in current due to the consumption of oxygen. Anodic detection of the hydrogen peroxide by oxidation at a platinum or carbon electrode was then introduced but, owing to the high electrode potential required, suffered from interference from other electroactive compounds in the sample.

Attempts to reduce interference and minimize the effect of variations in oxygen tension have resulted in the development of biosensors with improved linear ranges which operate at lower electrode potentials. They incorporate artificial electron acceptors, called mediators, to transfer electrons from the flavoenzyme (e.g. glucose oxidase) to the electrode and thus are not dependent on oxygen. Ferrocene (bis(η^5-cyclopentadienyl)iron) and its derivatives are examples of redox mediators for flavoenzymes. The reaction now becomes

$$\underset{\text{(ferricinium)}}{\text{glucose} + 2Fe^+} = \underset{\text{(ferrocene)}}{\text{gluconic acid} + 2Fe} + 2H^+$$

The ferrocene is detected at the anode, which also regenerates the ferricinium ion for further use.

Construction of enzyme-based amperometric biosensors usually involves coating the electrode surface with both the enzymes and the mediator. The method adopted and concentrations used are important factors in the performance of the device. For extended use, covalent immobilization is the most satisfactory method.

Potentiometric biosensors

These measure the potential difference between the transducing electrode and a reference electrode under conditions of zero current. Three types of potentiometric detectors are commonly employed: ion-selective electrodes (ISE), gas-sensing electrodes and field effect transistors (FET).

Gas-sensing electrodes are modified ISEs and consist of either a hydrophobic gas-permeable membrane or an air gap between the sample solution and a pH electrode. Only gases that can permeate through the membrane and produce a change in pH, e.g. carbon dioxide, ammonia, will be detected by the electrode. Gas-sensing electrodes have been used in devices incorporating entrapped microorganisms as the biological component which metabolize the analyte of interest and produce a detectable gas.

Field effect transistors are miniature, solid-state, potentiometric transducers (Figure 4.22) which can be readily mass produced. This makes them ideal for use as components in inexpensive, disposable biosensors and various types are being developed. The function of these semiconductor devices is based on the fact that when an ion is absorbed at the surface of the gate insulator (oxide) a corresponding charge will add at the semiconductor surface and the current flowing between the source and drain electrodes will change. The resulting voltage shift is a measure of the absorbed ionic species.

Variation of the nature of the gate electrode results in the different types of FET. For example, in the metal oxide semiconductor FET (MOSFET) palladium/palladium oxide is used as the gate electrode. This catalytically decomposes gases such as hydrogen sulphide or ammonia with the production of hydrogen ions, which pass into the semiconductor layer. An enzyme may be coated on the palladium, e.g. urease, which catalyses the production of ammonia from urea and thus provides a device for the measurement of this substrate.

In ion-selective FETs (ISFETs), an ion-selective membrane replaces the gate electrode. When an enzyme-loaded gel is combined with the membrane, the device can be used to measure substrates which enzymically generate charged species.

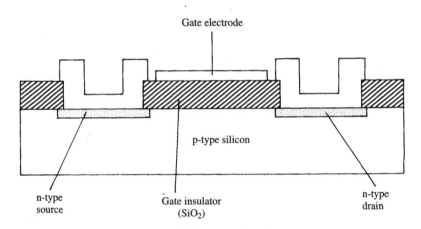

Gate electrode

p-type silicon

n-type source

Gate insulator (SiO₂)

n-type drain

Figure 4.22 **Schematic diagram of a field effect transistor.** The silicon–silicon dioxide system exhibits good semiconductor characteristics for use in FETs. The free charge carrier concentration, and hence the conductivity, of silicon can be increased by doping with impurities such as boron. This results in p-type silicon, the 'p' describing the presence of excess positive mobile charges present. Silicon can also be doped with other impurities to form n-type silicon with an excess of negative mobile charges.

Self test questions

Sections 4.2–5

1. Which of the following can be used for both qualitative and quantitative analysis?
   (a) Conductimetry.
   (b) Coulometry.
   (c) Amperometry.
   (d) Polarography.

Self test questions

2. What is the technique in which a reactive ion is generated at an electrode for analytical purposes?
   (a) Conductimetry.
   (b) Coulometry.
   (c) Amperometry.
   (d) Polarography.
3. Enzymes are unsuitable for use in biosensor systems
   BECAUSE
   biosensors require both a component to bind and convert the analyte to a product and a transducer to detect the product.
4. The increase in conductivity of a solution is directly proportional to an increase in ionic concentration
   BECAUSE
   the conductivity of a solution is the sum of the conductivity of the ions present.

4.6   Further reading

Koryta, J. (1993) *Ions, electrodes and membranes,* 2nd edition, Wiley-Liss Inc., USA

Koryta, J., Dvorak, J. and Kavan, L. (1996) *Principles of electrochemistry,* John Wiley, UK.

Eggins, B.R. (1996) *Biosensors,* John Wiley, UK.

5 Radioisotopes

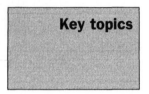

Key topics

- Nature of radioactivity
- Detection and measurement of radioactivity
- Biochemical uses of isotopes

Atoms that have the same atomic number, and hence are the same element, but have different masses are known as isotopes. Radioisotopes, or more correctly, radionuclides spontaneously and continuously emit characteristic types of radiation. They are particularly useful in analytical biochemistry, the unique nature of the radiation providing the basis for many specific and sensitive laboratory methods (Table 5.1).

Table 5.1 Physical data of some frequently used isotopes

Element	Symbol	Half-life	Beta emission	Gamma emission
Calcium	^{45}Ca	165 d	+	−
Carbon	^{14}C	5760 a	+	−
Chlorine	^{36}Cl	3×10^5 a	+	−
Cobalt	^{60}Co	5.26 a	+	+
Hydrogen	^3H	12.2 a	+	−
Iodine	^{125}I	60 d	Electron capture	+
Iodine	^{131}I	8.04 d	+	+
Iron	^{59}Fe	45 d	+	+
Magnesium	^{28}Mg	21.4 h	+	+
Nitrogen	^{13}N	600 s	Positron	+
Phosphorus	^{32}P	14.3 d	+	−
Potassium	^{40}K	10^9 a	Electron capture	+
Potassium	^{42}K	12.4 h	+	+
Sodium	^{22}Na	2.6 a	Positron	+
Sodium	^{24}Na	15 h	+	+
Sulphur	^{35}S	87.2 d	+	−

5.1   Nature of radioactivity

> ➤ A proton is a nuclear particle with a mass of 1 and a positive charge of 1.
> ➤ A neutron is a nuclear particle with a mass of 1 and no charge.

The atomic nucleus is made up of protons and neutrons. The number of protons determines the atomic number and hence the identity of an element and is equal to the number of orbital electrons, a feature necessary to ensure the electrical neutrality of the atom. The atomic mass of the nucleus is made up by the additional neutrons that are present. Hence:

atomic number = the number of protons
atomic mass   = the sum of the number of protons and neutrons

This information is normally shown as a superscript (atomic mass) and a subscript (atomic number) to the symbol for the element. Hence $^{14}_{6}C$ represents the isotope of carbon (atomic number 6) with the atomic mass of 14. In practice the subscript is often omitted because the atomic number is unique to the element that is represented by the appropriate letter (e.g. C for carbon). For simplicity in the spoken form and often in the written form, isotopes are often referred to as carbon-14, phosphorus-32, etc.

The stability of the atomic nucleus depends upon a critical balance between the repulsive and attractive forces involving the protons and neutrons. For the lighter elements, a neutron to proton ratio (N : P) of about 1 : 1 is required for the nucleus to be stable but with increasing atomic mass, the N : P ratio for a stable nucleus rises to a value of approximately 1.5 : 1. A nucleus whose N : P ratio differs significantly from these values will undergo a nuclear reaction in order to restore the ratio and the element is said to be radioactive. There is, however, a maximum size above which any nucleus is unstable and most elements with atomic numbers greater than 82 are radioactive.

5.1.1   Types of radioactivity

> ➤ An alpha particle contains two protons and two neutrons and a double positive charge.

If a nucleus is too heavy and its atomic number exceeds 82, it may revert to a more stable arrangement by releasing both neutrons and protons. This is effected by the emission of an **alpha particle**, which contains two protons and two neutrons and is a helium nucleus, $^{4}_{2}He^{2+}$.

Alpha particles are relatively large particles and are emitted with a limited number of energy levels. They carry a double positive charge and as a result attract electrons from the atoms of the material through which they pass, causing ionization effects. They have an extremely short range, even in air, and as a result present very little hazard as an external source of radiation but their effects within living cells or tissues can be serious.

> ➤ Beta radiation is the emission of a high speed electron (a negatron, β^-) from a neutron.

A nucleus that has an excess of neutrons will undergo neutron to proton transition, a process that may restore the N : P ratio but that requires the loss of an electron to convert the neutron to a positively charged proton, and as a result the atomic number increases by one. The particle emitted is a high speed electron known as a negatron (β^-) and the atom is said to emit **beta radiation,** e.g.

$$^{14}_{6}C \rightarrow {}^{14}_{7}N + \beta^-$$

Conversely, nuclei that contain an excess of protons undergo proton to neutron transition with the emission of a positively charged beta particle known as a positron (β^+) and with the reduction of the atomic number by one. A positron has only a very short existence, combining immediately with an electron of a nearby atom. The two particles disintegrate in the process with the emission of two gamma rays, e.g.

$$^{11}_{6}\text{C} \rightarrow \, ^{11}_{5}\text{B} + \beta^+$$

An alternative mechanism to positron emission, for the conversion of a proton to a neutron, involves a process known as **electron capture** (EC) in which the nucleus captures an orbital electron from an inner shell to restore the N : P ratio. Subsequently, an electron from another orbital falls into the vacancy left in the inner shell and the energy released in the process is emitted as an X-ray, the atomic number again being reduced by one, e.g.

$$^{125}_{53}\text{I} + \text{e}^- \rightarrow \, ^{125}_{52}\text{Te} + \text{radiation}$$

Beta emission from an atom shows a range of energy levels and although the maximum level is characteristic of the isotope, only a small proportion of the emissions may be of this energy. As a result, beta particles show a maximum range in a particular absorbing material with many emissions penetrating considerably less. Penetration distances are about ten times greater than for alpha particles, values of 5–10 mm in aluminium being common. Because they are light, beta particles are easily deflected by other atoms and are less effective as ionizing agents than are alpha particles.

Although a nucleus may have an N : P ratio in the stable range it is still possible for it to be in an unstable energetic state and to emit protons as electromagnetic radiation of extremely short wavelength known as **gamma rays**. These have neither mass nor charge and as a result cannot cause direct ionization effects, but the energy associated with them is absorbed by an atom causing an electron to be ejected, which in turn produces secondary ionization effects.

➤ Gamma radiation is the emission of very shortwave electromagnetic radiation.

The atoms emitting gamma radiation suffer no change in either mass or atomic number although very few elements emit solely gamma radiation. Sodium-24 ($^{24}_{11}\text{Na}$) for example, emits beta radiation (negatron emission) and is converted to magnesium-24 ($^{24}_{12}\text{Mg}$). It also, however, emits gamma radiation in the process.

In order to handle radioisotopes safely it is necessary, among other things, to define fairly carefully the penetrating power of the radiation emitted by any isotope. Alpha particles, having only a relatively limited number of energy levels, are absorbed by contact with other atoms. The absorbing power of a material is referred to in terms of its equivalent thickness. The thickness required can be calculated by dividing the equivalent thickness by the density of the material.

➤ The equivalent thickness of a material is the weight of material per square centimetre necessary to stop alpha particles.

Beta radiation, because of its wide range of energies, shows an approximately linear relationship between the thickness of the material required to absorb radiation and the logarithm of the activity of the radiation. The percentage of beta particles passing through a given thickness of absorbing material has a sigmoidal relationship to the logarithm of the maximum energy of emission. This information permits the selection of a particular thickness of absorbing material which will totally stop the beta radiation from one source and yet will allow the penetration of a relatively constant proportion of radiation from another source with a higher maximum emission energy.

The absorption of gamma radiation is more difficult to assess and the most convenient means of expressing the absorbing power of a particular material is to quote the half-thickness.

➤ The half-thickness of a material is the thickness of material necessary to reduce the intensity of radiation by one-half.

5.1.2 Decay

The emission of radiation by an atom results in a change in the nature of the atom, which is said to have decayed. The rate at which a quantity of an isotope decays is proportional to the number of unstable atoms present and a graph of

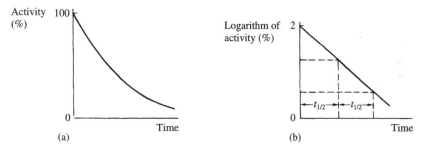

Figure 5.1 **Radioactive decay.** The rate of radioactive decay is proportional to the number of unstable atoms present and although theoretically there should always be some activity left (a), in practice the activity does eventually fall to zero. A plot of the logarithm of the activity against time (b) results in a straight line from which the half-life can be determined.

activity against time results in a typical exponential curve. For this reason the actual life span of a radioactive sample cannot be measured and a more meaningful expression of the rate of decay is provided by the half-life $t_{1/2}$ (Figure 5.1). The half-life can be determined using the equation that describes the rate of radioactive decay:

➤ Half-life is the time taken for the activity of a sample to fall by one-half and is independent of the actual activity.

$$\log_e \frac{N_t}{N_0} = -\lambda t$$

where N_t = the activity at time t,
   N_0 = the activity at zero time,
   λ = the radioactive decay constant,
   t = time.

In its linear form the equation is:

$$\log_e N_t = \log_e N_0 - \lambda t$$

and converting from natural logarithms it becomes:

$$\log_{10} N_t = \log_{10} N_0 - 0.4343\lambda t$$

This is the equation for a straight line with a slope of -0.4343λ.

If N_t is made equal to half N_0, then t will equal the half-life $(t_{1/2})$ and the equation becomes:

$$\log_{10} 1 = \log_{10} 2 - 0.4343\lambda t_{1/2}$$

$$t_{1/2} = \frac{0.3010}{0.4343\lambda} = \frac{0.6931}{\lambda}$$

A plot of the logarithm of the activity of the sample $(\log_{10} N_t)$ against time (t) is linear, and the half-life can be determined from the graph either by measuring the time interval between two readings of activity which vary by a factor of two or by using the value for the gradient of the line and substituting in the equation:

$$\text{Half-life } (t_{1/2}) = \frac{0.3010}{\text{gradient}}$$

5.1.3 Units of radioactivity

➤ A curie is defined as 3.7×10^{10} disintegrations per second.
➤ A becquerel is defined as 1 disintegration per second.
➤ Specific activity is a measure of the relative amount of radioactive atoms in a sample.

The curie unit (Ci) is based on the activity of 1 g of pure radium-226, which undergoes 3.7×10^{10} transformations per second. It is therefore defined as the quantity of a radioactive isotope which gives 3.7×10^{10} disintegrations per second. The SI unit of activity is the becquerel (Bq), which is equal to one nuclear transformation per second. Hence:

$$1 \text{ Ci} = 3.7 \times 10^{10} \text{ Bq}$$

The specific activity of an isotope indicates the activity per unit mass or volume and is quoted as becquerels per gram (Bq g^{-1}) or millicuries per gram (mCi g^{-1}). A sample in which all the atoms of a particular element are radioactive is said to be carrier-free and is very difficult to achieve in practice.

5.1.4 Safety

➤ Hazard assessment – see Section 1.2.4.

Great care must always be exercised in handling radioisotopes. It is not only the powerful emitters that are dangerous but also weak emitters with long half-lives, e.g. tritium, carbon-14, which may be incorporated into the body and over a period of time can constitute a serious hazard.

There are many regulations regarding the handling and disposal of radioisotopes and these must be fully understood and observed when setting up a radioisotope facility. The hazards must be fully assessed and the laboratory must be equipped and approved for the intended applications. All manipulations must be assessed for specific hazards and standard operating procedures (SOPs) fully implemented.

There are, however, certain general principles of good practice which should constantly be kept in mind when working with radioisotopes.

1. All working surfaces and floors must be covered with an approved material and they must be regularly monitored for contamination.
2. Suitable protective clothing, including disposable gloves, must be worn.

3. Containers of a decontamination fluid must be readily accessible in which all contaminated glassware, etc. must be immersed after use.

4. Isotopes must be handled in an appropriate manner using forceps or handling devices if necessary. Adequate shielding must be used and film badges worn at all times.

5. Records of amounts and nature of all isotopes held in the laboratory must be kept, and used isotopes must be disposed of in the approved way.

6. After working, approved washing and decontamination procedures of both laboratory staff and equipment must be completed before the laboratory is vacated.

Self test questions

Section 5.1

1. Which of the following isotopes emit gamma radiation?
   (a) ^{14}C.
   (b) ^{3}H.
   (c) ^{131}I.
   (d) ^{32}P.

2. What is the half-life of an isotope if the activity of the sample falls from 200 disintegration min^{-1} to 150 disintegration min^{-1} in 50 minutes?
   (a) 200 minutes.
   (b) 120 minutes.
   (c) 100 minutes.
   (d) 50 minutes.

3. The element represented as $^{12}_{6}C$ is a stable isotope of carbon
   BECAUSE
   in the element $^{12}_{6}C$ the neutron to proton ratio is $1 : 1$.

4. Alpha emission is a very effective ionizing agent
   BECAUSE
   alpha particles consist of high speed electrons.

5.2   Detection and measurement of radioactivity

Radiation can be detected in several ways, all of which depend upon its direct or indirect ionization effects. The three methods most frequently used are the ionization of gases, the excitation of liquids or solids and the induction of chemical change. Radioactivity is measured as either the number of individual emissions in a unit time (differential measurements) or the total cumulative effect of all emissions in a given time (integral measurements). Generally, differential measurements are used for quantitative analytical work and integrating methods for dosimetry and autoradiography. Regardless of the instrument or method of recording, any critical measurement of radioactivity demands that a background count (blank) must be done and the test value corrected for those emissions that are not due to the sample. Additionally, in order to get consistent results, the mean of replicate counts must be calculated and, where appropriate, statistical assessment must be undertaken because the fundamental basis of radioactive decay is a random process.

➤ Differential measurements are counts of the number of emissions in a unit time.
➤ Integral measurements assess the cumulative emissions over a given time.

5.2.1 Geiger counters

The ionizing effect of radiation on a gas may be detected if a potential difference is applied across two electrodes positioned in the gas. The electrons displaced as a result of ionization will be attracted to the anode and a current will then flow between the electrodes. This type of detector is based on the Geiger–Muller tube, which consists of a hollow metal cylinder (the cathode) within a glass envelope and a wire (the anode) running along the central axis of the cathode (Figure 5.2). The whole tube is filled with the counting gas, often a mixture of neon and argon, under relatively low pressure. Radiation

Anode

Radiation

Cathode

Figure 5.2 **Geiger–Muller tube.** The tube is filled with an ionizable gas mixture, such as neon and argon, and a voltage applied across the electrodes. Ionization of the gas by incident radiation causes a current to flow between the electrodes.

entering the tube causes ionization of the detector gas and, as a result, a pulse of current flows between the electrodes, together with the emission of ultraviolet radiation.

One problem with a Geiger–Muller tube is the fact that once it has fired, the ultraviolet radiation emitted causes further ionization of the gases (self-excitation). It is necessary to quench this self-excitation and this is usually achieved by incorporating a halogen, e.g. bromine, into the gas mixture during manufacture. The bromine molecules absorb this extra energy and are split into their individual atoms. These atoms subsequently recombine when the counter is not being used and so regenerate the system.

The end window of the tube must be thin enough to permit the weaker radiations to enter the tube (aluminium, 6–8 mg cm^{-2}; mica, 2 mg cm^{-2}) but even so alpha particles and very weak beta emissions are either completely or partially absorbed. The emissions from the biologically important isotopes of tritium and carbon-14 fall into this category and alternative detectors should be used for these isotopes.

Gamma rays, being powerful, enter the tube very easily but cause little direct ionization. They do, however, produce emission of photoelectrons from the glass or metal walls of the tube and the internal electrodes. These photoelectrons then produce ionization of the gas with the resulting pulse of current.

If the voltage applied to a Geiger–Muller tube is increased from zero, no response to radiation is detected despite the presence of a radioactive isotope until a minimum or starting voltage is reached. At this point a current is generated which increases rapidly to a plateau and finally at very high voltages (the breakdown potential) the tube goes into continuous discharge. The counter should be operated at a voltage selected from the plateau region and will be severely damaged if voltages above the breakdown potential are applied for any significant period of time. A detector for liquid samples (Figure 5.3) is a common variant of the Geiger–Muller tube and is designed either to be dipped into or to contain a liquid.

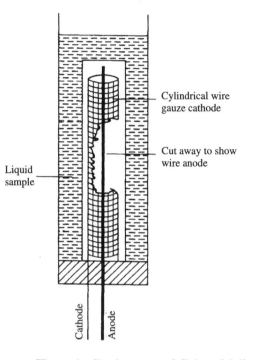

Figure 5.3 A Geiger–Muller tube for measuring the activity of liquid samples.

Cylindrical wire gauze cathode

Cut away to show wire anode

Liquid sample

Cathode

Anode

The main disadvantage of Geiger–Muller counters is the fact that after an ionization has occurred and the electrons have been discharged (a very rapid process) the resulting positive ions are discharged slowly and as a result prevent other electrons from reaching the anode. The period of time during which the detector is insensitive to further ionization is known as the dead-time and is often in the region of 500 μs. Correction factors are used to correct all readings for the counts lost during this time.

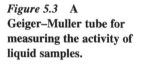

➤ Lost counts correction is necessary because Geiger–Muller tubes have a 'dead-time' in their operation.

5.2.2   Scintillation counters

Some substances, known as fluors or scintillants, respond to the ionizing effects of alpha and beta particles by emitting flashes of light (or scintillations). While they do not respond directly to gamma rays, they do respond to the secondary ionization effects that gamma rays produce and, as a result, provide a valuable detection system for all emissions.

A range of scintillants is available, many designed for maximum efficiency with specific isotopes. Crystals of sodium iodide containing a small amount of thallous iodide are very efficient detectors of gamma radiation and

➤ Gamma counters are a type of scintillation counter and measure gamma radiation by detecting the emission of light induced in a scintillant.
➤ Photomultiplier – see Section 2.3.3.

instruments based on this design are often called gamma counters. The crystal is positioned on top of a photomultiplier which can detect the flashes of light and both are covered with a light-proof material that is thin enough to allow gamma radiation to pass into the crystal. (Figure 5.4). When a sample is placed on the crystal, the pulses of current generated by the photomultiplier are counted electronically.

Weak beta radiation and alpha particles often cannot penetrate the covering material but the use of a scintillant, which, together with the sample, will dissolve in a suitable solvent, enables a similar technique to be used. Liquid scintillation counters usually consist of two light-shielded photomultiplier

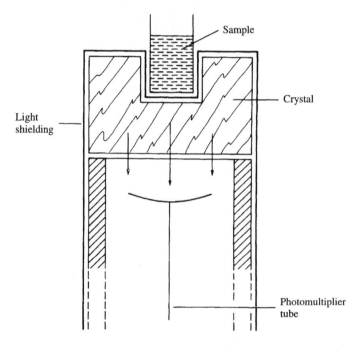

Figure 5.4 **A gamma counter.** The scintillations from a crystal of fluor caused by gamma radiation are detected by a photomultiplier.

tubes with the sample and a scintillant mixture contained in a glass vessel placed between them. Only those pulses that occur simultaneously from both photomultiplier tubes are due to scintillations and these are counted electronically, giving a detection system that can discriminate between genuine signals and random noise from the detectors. This technique is called coincidence counting. A range of scintillants suitable for such work is available (Figure 5.5), each showing particular light emission characteristics. They are dissolved in a suitable solvent, which should be miscible with the sample; toluene or xylene are frequently used for organic samples and dioxane for aqueous samples.

The number of ionizations or scintillations detected by both types of detector is measured electronically and presented either as the total number of counts (a scaler) or as the number of pulses per minute (a count-rate meter)

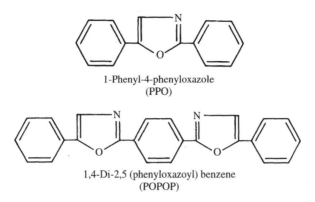

1-Phenyl-4-phenyloxazole
(PPO)

1,4-Di-2,5 (phenyloxazoyl) benzene
(POPOP)

Figure 5.5   **Scintillants**. A range of organic scintillants is available with different solubility and emission characteristics. Scintillant 'cocktails' or mixtures contain a primary scintillant such as PPO and often contain a secondary scintillant which absorbs the radiation produced by the primary scintillant and re-emits it at a longer wavelength, e.g. POPOP.

Figure 5.6   **Autoradiograph of a leaf using calcium-45.** The leaf was stood in a solution containing a small amount of the phosphate salt of calcium-45 for several hours. The leaf was then taped to the outside of an envelope containing a photographic film for a week before developing the film in the normal manner. This is a print from the film and as a consequence the effect of radiation is seen as white on black.

> ➤ Detectors are designed to either measure the total number of emissions (scalers) or the number of pulses per minute (count-rate meters).

Additionally, some counting devices (pulse height discriminators) are designed to record signals at specified voltages and can measure one isotope in the presence of another.

5.2.3 Autoradiography

Ionizing radiation has the same effect as light on photographic film and the extent of blackening of the film is related to the amount of radiation. Autoradiography is particularly useful for demonstrating the location of radioactive isotopes in tissues or chromatograms (Figure 5.6).

The sample is placed on a photographic film which is protected from the light and allowed to remain in contact long enough for an adequate exposure. The exposure time is dependent upon the intensity of the radiation and can usually only be determined by trial and error. It is possible, however, to predict an approximate exposure time from the fact that a total emission of 10^7 beta particles per square centimetre is often required.

> ➤ Dosimetry is an autoradiographic technique to measure the total dose of radiation received by a worker over a period of time.

Autoradiography is also a convenient way of monitoring the amount of radiation to which a worker has been exposed (dosimetry). If a small badge containing photographic film is worn continually and the photographic film developed after a set period of time, an estimate of the radiation received can be made from the degree of darkening of the film. No sophisticated instrumentation apart from photographic facilities are required for these types of application.

5.3	Biochemical uses of isotopes

5.3.1 Tracers

Radioactive isotopes provide a very convenient way of monitoring the fate or metabolism of compounds that contain the isotopes. When used in this way, the isotope is described as a tracer and compounds into which the radioactive atom has been introduced are said to be labelled or tagged. The labelled molecules need only comprise a very small proportion of the total amount of the unlabelled radioactive substance because they act in the same way as the nonradioactive substance but can be detected very much more easily. The varied applications of tracers in biochemistry range from studies of metabolism in whole animals or isolated organs to sensitive quantitative analytical techniques, such as radioimmunoassay. Phosphorus-32 is used in work with nucleic acids, particularly in DNA sequencing and hybridization techniques. In these instances the isotope is used as a means of visualizing DNA separations by autoradiographic techniques.

> ➤ DNA techniques – see Sections 13.3 and 13.5.

While it is generally assumed that radioisotopes are metabolized in exactly the same way as the non-radioactive substances, occasionally some differences in reactivity are seen, due mainly to differences in the atomic weights of the two isotopes. The effect is most obvious for elements with the lowest atomic number, where the difference in mass between isotopes is the greatest. For instance, the difference between hydrogen (mass 1) and tritium (mass 3) is more significant than the difference between iodine-127 and iodine-125. This isotopic effect is extremely slight in most cases and can be ignored unless critical studies of equilibrium constants and rates of reaction are involved.

Ideally, molecules should be labelled by introducing a radioactive isotope in place of a normal atom, e.g. carbon-14 replacing a carbon-12 in a carbohydrate. This method of labelling involves the synthesis of the molecule either *in vivo* or *in vitro* and the use of enzymes often permits the isotope to be introduced in a particular position in the molecule. The position of the labelled atom should be indicated wherever possible as for example in glucose-1-^{14}C.

Alternatively, it may be necessary to label a molecule by introducing an additional atom and in this case it has to be assumed that its presence does not alter the reactivity or metabolism of the molecule. Proteins, for instance, are often labelled with an isotope of iodine. It is very important that any labels used are firmly attached to the molecule otherwise invalid results will be obtained.

➤ Labelling proteins – see Section 7.4.3.

The choice of an isotope for tracer studies requires an appreciation of not only the radiochemical properties of the element, but also the effects that they might have both biochemically and analytically. The isotope should have a half-life that is long enough for the analysis to be completed without any significant fall in its activity. Occasionally this might present a problem in that some elements only have radioactive isotopes with very short half-lives, e.g. fluorine-18 has a half-life of 111 min. Conversely, isotopes with very long half-lives should not be used for *in vivo* studies because accumulation in the tissues of the recipient is unacceptable.

The biochemically important elements of hydrogen, carbon and calcium are weak beta emitters and are difficult to detect using Geiger–Muller tubes. Scintillation counters are necessary for these weak emissions and are preferable for gamma emitters, while the Geiger–Muller detectors are suitable for the more powerful beta emitters. However, for autoradiography, alpha and weak beta emitters give the best resolution, while gamma emitters are usually unsuitable because they cause general fogging of the film.

5.3.2 Isotope dilution analysis

Isotope dilution analysis is the definitive method for the quantitative analysis of many compounds but is usually only carried out in specialist laboratories.

If a known amount of an isotope with a known specific activity is mixed with an unknown amount of the non-radioactive substance, the reduction in the specific activity can be used to determine the degree of dilution of the isotope and, hence, the amount of the non-radioactive isotope present. The isotope and the test substance must be thoroughly mixed before a representative sample of the mixture is purified and its specific activity determined.

Calculation

Isotope: Amount M
                Specific activity S
Mixture: Amount (unknown) $M_x + M$
                Specific activity S_x

The specific activity of the original radioisotope with a total activity (A) is:

$$S = \frac{A}{M}$$

The mixture with the non-radioactive isotope contains the same total activity but the amount of the substance is increased. Hence the specific activity is reduced to:

$$S_x = \frac{A}{M_x + M}$$

Substituting in this second equation an expression for the amount of activity (A) in the sample we have:

$$S_x = \frac{S \times M}{M_x + M}$$

Hence:

$$M_x = M\left(\frac{S}{S_x} - 1\right)$$

5.3.3   Radioactivation analysis

Radioactivation analysis is used for the analysis of trace amounts of suitable elements. It is a technique that is not generally applicable to biological material although it has been used for the measurement of lead in hair and nails.

It involves the simultaneous irradiation of the sample and a standard known mass of the same element to produce a radioactive isotope of the element. The activities of both the sample and the standard are then determined and, because their specific activities will be the same, it is possible to calculate the mass of the unknown sample.

Activation analysis requires the use of a powerful source of neutrons as the activator and is suitable only for elements which form an isotope whose half-life is longer than the isotopes of other elements which may be produced.

Self test questions

Sections 5.2/3

1. Which forms of radiation can a Geiger counter detect?
    (a) Alpha particles.
    (b) Beta radiation.
    (c) Gamma radiation.
2. Which effect of radiation does autoradiography depend upon?
    (a) Ionization.
    (b) Excitation.
    (c) Chemical change.
    (d) None of these.

Self test questions

3. Isotopes used in tracer studies should have a short half-life
   BECAUSE
   isotopes with a short half-life are less dangerous than those with a long half-life.
4. Autoradiography is only suitable for detecting gamma radiation
   BECAUSE
   beta radiation cannot penetrate the gelatin layer of a photographic plate.

5.4   Further reading

Choppin, G.R. and Rydberg, J. (1983) *Nuclear chemistry – its theory and applications,* Pergamon Press, UK.

Ehmann, W.D. and Vance, D.E. (eds) (1991) *Radiochemistry and nuclear methods and analysis,* Blackwell Scientific Publications, UK.

Knoll, G.F. (1989) *Radiation detection and measurement,* 2nd edition, John Wiley, UK.

Slater, R.J. (ed.) (1990) *Radioisotopes in biology – a practical approach,* IRL Press, UK.

6 Automated methods of analysis

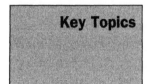

The increased demand for analyses has led to the introduction of machines that will perform all or part of an analytical procedure. Many of the manipulations in laboratory methods are common to a variety of different tests (e.g. pipetting) and lend themselves readily to mechanization.

The introduction of instruments into a laboratory that will perform these tasks will result in a reduction in the analysis time and mean that an increased workload can be met without the need for additional staff. This should reduce the overall cost per test even though the expense associated with the purchase and running of the instruments may be high.

In addition to cost considerations, the fact that automated instruments are designed to give consistently good analytical precision is a factor that has greatly influenced their widespread acceptance. It is recognized, however, that reliability depends upon the quality of their design and manufacture and subsequently their effective and regular servicing in the laboratory situation.

An automated system, by definition, should perform a required act at a predetermined point in the process and should have a self-regulating action. This implies that intervention by the analyst is not required during the procedure and that only those systems that incorporate a microprocessor or computer to control and monitor their performance can be designated as automated. Some systems may not comply strictly with this definition but are a valuable means of mechanizing laboratory activities.

The first so-called automated analysers were developed to carry out analyses by replacing manual pipetting by the mechanical transfer of fixed volumes of sample and reagents. They were designed to mimic the manipulations of a convential manual procedure and sometimes had the capacity to measure the amount of reaction product, usually spectrophotometrically. The precision of the pipetting was often poor and the instruments were cumbersome and prone to breakdown. Their modern counterparts offer a much higher degree of precision and reliability and a wide range of designs are available, each offering different levels of mechanization. The simplest are the auto-pipettors/diluters, which can be adjusted to measure and expel selected volumes of solution using a syringe system. They relieve the analyst of the pipetting but require constant attendance. Instruments which are more highly mechanized and will pipette samples and reagents sequentially into racks of tubes play an important role in the mechanized preparation of samples. These were the forerunners of what are now known as sample processors. An operator still may be required to move the racks of tubes to different positions within the instrument so that each of the steps in a particular procedure may be completed. When a method demands other manipulations which the mechanized analyser cannot perform, e.g. centrifugation, then it is necessary for the operator to transfer the reaction tubes to the appropriate laboratory instrument.

From this approach, a series of automated analysers known as discrete analysers were developed by the incorporation of a microprocessor to control their operation. The design and capabilities of the different analysers vary but they all require only small volumes of sample and reagents and often 'ready-to-use' reagents are supplied by the manufacturer. Many of the newer instruments are designed to use disposable, dry reagent strips or film devices rather than liquid reagents. The modern analysers show high levels of sensitivity and specificity owing to the development of reliable analytical methods which suffer only minimal interference effects. Of particular importance in this respect is the elimination of the centrifugation stage, which was previously needed to remove the protein which is often present in biological samples.

The 1950s saw the introduction of a completely new approach to automation, in the form of continuous flow analysis. This made a significant contribution to the advance of automated analysis and subsequent development has been in the form of flow injection analysis. The original instruments were single channel and capable of measuring only one constituent in each sample. Multichannel instruments were then developed which could simultaneously carry out several different measurements on each sample. These were useful in laboratories where many samples required the same range of tests.

A microprocessor or computer is now incorporated into most discrete and flow analysers and it is an absolute requirement of multichannel instruments. It allows the operator to set the optimum analysis conditions for each different assay and to give instructions regarding the acceptable limits of the results for samples of known concentrations. With this information the computer is capable of continual self-evaluation of the performance of the analyser and the quality of results being produced. All the necessary calculations will be done before the concentration of each sample is printed out, although sometimes the printer will be instructed to produce data in the form of graphs or histograms. The results from single-channel instruments are usually presented as a list of sample numbers with their corresponding concentrations. Multichannel analysers require all the data for each sample to be collected together before they are printed out and in their case each sample must carry its own identification, e.g. barcode, which the computer can recognize.

➤ Quality assurance – see Section 1.2.3.

A natural extension of analytical automation is some means of data processing all the results that are generated. This usually takes the form of a central computer which accepts information from different analysers for presentation in a useful manner. The identification of a sample and the tests performed are typed in using a keyboard and the computer collates all the data on each sample. As well as collecting information, computing and statistically assessing results, an important facility of the computer lies in its ability to store information for future recall via a visual display unit.

➤ LIMS – see Section 1.2.4.

With the introduction of robotics the scope of laboratory automation has been expanded. These robotic instruments are fully computerized and can be programmed to perform either single stages or entire analytical procedures. The more complex systems comprise a series of modules which sequentially perform the various steps in an analytical procedure from sample preparation right through to the production of the final report. Although the introduction of a robotic system into a laboratory may require a considerable period of performance evaluation before it can be routinely used with confidence, once functioning it will continue unattended, thus releasing staff to perform other duties; it may also be used overnight. Other advantages are in the improved precision of complex, labour-intensive procedures and reduced exposure of laboratory workers to hazardous samples and reagents.

There is a wide range of mechanized and automated instruments available from a variety of manufacturers and the choice of which one to purchase should involve several considerations. The nature of the expected workload must be assessed with respect to both the numbers of samples and the variety of tests that will be performed. This will indicate the level of mechanization or automation required. Neither the existing laboratory equipment nor the effects of automation on the current and future staffing policy should be overlooked. The cost is an important aspect and this includes the price of the instrument, its running costs in terms of reagents and personnel and its servicing requirements. The reliability of the instrument and its life expectancy are also important and these facts, together with information regarding the quality of the results that can be obtained, must be sought before a decision can be made. It is advisable to request an instrument on a trial basis so that its performance can be assessed on site.

6.1   Discrete analysers

➤ Photometric detectors – see Section 2.3.3.
➤ Automated flow analysers use a series of narrow tubes through which the samples and reagents are pumped.

Instruments in which each test is performed in its own container or slide are known as discrete analysers, in contrast to flow analysers in which the samples follow each other through the same system of tubing. All discrete analysers have a common basic design incorporating a pipetting system, a photometric detector and a microprocessor. A development of the single test instrument is the parallel fast analyser, which analyses several samples simultaneously but for only one constituent. However, the change-over from one analytical procedure to another is quick and simple.

> Multitest analysers simultaneously measure several analytes in a sample.
> Random access analysers offer the flexibility to select the order of testing and the test menu for different samples.
> Glucose analyser – see Section 9.3.1.
> Enzyme kinetic analyser – see Section 8.6.
> Dedicated analysers are specifically designed to measure one or a very limited range of analytes.

The multitest analysers simultaneously measure more than one constituent in each sample and are often designed to meet the needs of a particular type of laboratory, e.g. clinical chemistry departments. The instruments offer different test combinations (menu) but these are often determined by the manufacturer, leaving little opportunity for modification within the laboratory.

Earlier instruments carried out the same menu of tests on each sample but the introduction of random access analysers made it possible for the operator to select individually a test combination for each sample from the total range available.

Other automated systems may be purchased for a specific purpose and are called dedicated instruments, e.g. glucose analyser. Others have fairly restricted applications, an example being the reaction rate analysers which are specifically designed for the kinetic measurements of enzyme activity. Some of the more recently developed instruments employ individual pre-prepared disposable test packs or strip devices which contain all the reagents for each particular assay in a dry form.

6.1.1   Centrifugal analysers

A new concept in automated analysis was developed in the late 1960s, resulting in the introduction of centrifugal fast analysers. The development project was sponsored in the United States of America by the Institute of General Medical Sciences and the Atomic Energy Commission, and the prototype was named the GeMSAEC system. Several instruments are now available and they are all based on the original idea in which batches of samples are treated in parallel and centrifugal force is used to transfer and mix the solutions, the reactions being continuously monitored photometrically. These instruments have considerable flexibility and are capable of being used for a variety of different types of analysis simply by changing the reagents in the dispensers and altering the spin programme. In addition, they use only small volumes of reagents, not only reducing the running costs but also permitting the analysis of very small volumes of sample.

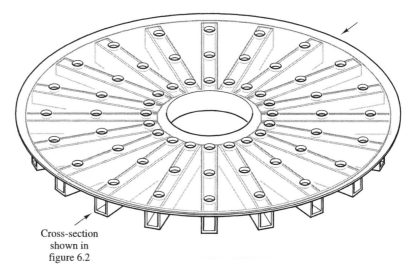

Figure 6.1
Diagrammatic representation of a typical transfer disc for a centrifugal analyser.

Cross-section shown in figure 6.2

The analysis takes place in a circular disc called the transfer disc or rotor (Figure 6.1) made of an inert material such as Teflon or acrylic whose surface properties allow minimal adhesion of liquid. It contains radially arranged sets of cavities, each set composed of two or more interconnected cavities, one for the sample and the others for the reagents (Figure 6.2). Small measured volumes of sample and reagent, e.g. $5-200\,\mu l$, are automatically pipetted into the appropriate wells so that each set contains a different sample but the same

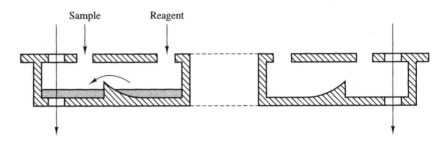

Figure 6.2
A cross-section of the transfer disc shown in Figure 6.1.

reagents. One set is usually prepared as a reagent blank. The wells are designed so that the sample and reagents remain in their separate compartments until centrifugation of the disc commences. During acceleration, mixing occurs and the chemical reaction is initiated. After a suitable time readings are taken while the disc is still spinning. These measurements may be absorbance, nephelometric or turbidimetric depending upon the design of the instrument. In some instruments the optical cuvettes, one for each set of wells, are incorporated into the perimeter of the transfer disc, while in other systems the fluid is expelled into individual cuvettes inside the centrifuge. Analysis is complete within a few minutes and, if the transfer disc is not disposable, it is automatically washed and dried ready for reloading.

A computer is incorporated in the instrument allowing the assay temperature and spinning programme, i.e. the speeds and duration of the mixing, and reaction stages, to be preset. It is also used to calculate the concentrations of the test samples by comparison of their photometric readings with those of standards treated identically. Centrifugal analysers are applicable not only to end-point analysis but, because of their continuous monitoring capability, are particularly useful for kinetic assays. Measurements of enzyme activity and enzyme immunoassays are particularly amenable to these types of analyser.

➤ Immuno methods –
see Section 7.4.

6.1.2   Dry chemistry analysers

➤ Immobilized enzymes
– see Section 8.7.

Advances in the technology associated with the production of immobilized reagents in thin layers has resulted in the development of a range of analytical systems in which the assays are carried out by the addition of a liquid sample to the reagents in a dry form. This method of analysis requires no reagent preparation and can be conveniently performed without the usual laboratory facilities. A major application is in the analysis of a range of blood serum components for clinical diagnostic purposes. Assays have been designed to be sensitive and rapid and operation of the instruments is simple. All the analytical

▶ Reflectance photometry – see Section 2.3.5.

parameters are stored within the instruments and these are activated by insertion of the bar-coded reagent strip or slide. There is a wide range of assay systems based on colour reactions with detection by reflectance photometry and involving such substances as enzymes and antibodies in addition to the more conventional reagents. The Boehringer Reflotron (Figure 6.3) is an example of a dry reagent strip system whereas the Vitros Chemistry System uses small

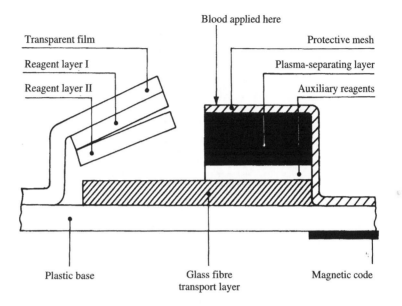

Figure 6.3 **Diagrammatic representation of the dry reagent strips known as reagent carriers,** which are used in the Boehringer Reflotron® System for the chemical analysis of blood samples. Red blood cells are removed in the separating layer and the plasma passing into the glass fibre transport layer is analysed. The magnetic code carries information about the analytical procedure.

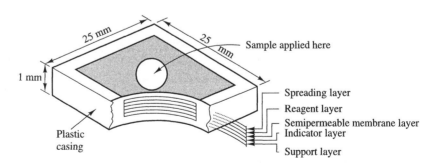

Figure 6.4 **Cross-section through a dry reagent slide for use in the Vitros Chemistry System,** previously known as the Kodak Ektachem analyser. A range of slides, which vary in the nature, number and composition of the layers, is available for a variety of analytes in blood serum. The sample (approximately 10 μl) is applied to the spreading layer and reactions take place as it permeates through the various layers. Detection is by reflectance photometry.

➤ Ion-selective electrodes – see Section 4.1.3.

multilayer slides (Figure 6.4), some of which incorporate miniature ion-selective electrodes for measuring ions such as sodium and potassium. In the Du Pont aca analyser the chemicals are held in pop-out pockets in a small transparent bag where the test reaction takes place (Figure 6.5).

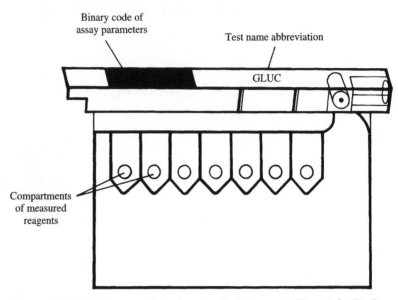

Binary code of assay parameters

Test name abbreviation

GLUC

Compartments of measured reagents

Figure 6.5 **Transparent analytical test pack** (100 mm × 80 mm) for Du Pont aca® analyser containing all the reagents for the assay of glucose. A wide range of test packs is available for use in the automated discrete analyser. Once the pack is loaded into the instrument, sample and diluents are automatically injected. The sealed reagent compartments are then broken open at various stages in the assay and the contents of the pouch are mixed and incubated. The transparent pouch then serves as a cuvette for absorbance measurements.

Self test questions

Section 6.1

1. Which of the following statements are true of discrete analysers?
    (a) They can measure only one analyte.
    (b) They can perform only one analysis at a time.
    (c) They may use dry reagent test strips.
    (d) They perform the analysis as the samples and reagents flow through tubes.
2. Which of the following statements are true of centrifugal analysers?
    (a) They use centrifugation to remove interfering substances.
    (b) They use centrifugal force to mix sample and reagents.
    (c) They cannot perform kinetic assays.
    (d) They incorporate a preparative ultracentrifuge.
3. Multitest analysers measure several analytes in a sample
    BECAUSE
    random access analysers have multitest capability.
4. Only solid samples are tested on dry chemistry analysers
    BECAUSE
    reagents in dry form are used with dry chemistry analysers.

6.2   Flow analysis

In automated flow analysis, samples are analysed as they are pumped in sequence through a series of tubing containing a continuously flowing reagent stream.

6.2.1   Continuous flow analysis

In 1957 Professor Skeggs described a method of continuous flow analysis which was developed and marketed by the Technicon Instruments Co. Ltd. under the name of 'AutoAnalyser'. The single- and dual-channel instruments had the flexibility of being able to perform a variety of different tests in laboratories where samples requiring a particular analysis could be batched and analysed together. A limited range of instruments is now available from several manufacturers and although they vary in design, showing several modifications from the original model, the concept of continuous flow analysis is common to them all.

The sequential multiple analysers (SMAs) were later developments and they were much more complex in design and included a computer. They could analyse each sample for several constituents simultaneously, the number of channels determining this capacity, i.e. 6, 12 or 20. These instruments were developed primarily for hospital clinical chemistry laboratories to allow an overall assessment of the chemical composition of blood samples. They have now been superseded by other types of analysers.

Continuous flow analysers can perform all the basic steps involved in a simple quantitative photometric procedure. These include pipetting of the sample and reagents, mixing, heating at specific temperatures, detection and presentation of results. In addition, sample digestion, phase separation, and the removal of interfering substances, particularly protein by dialysis, may be included when required. Although colorimetric detection is most frequently employed, fluorimetric, nephelometric, flame photometric, refractive index, radiochemical and electrochemical techniques have also been used, widening the range of analytical applications considerably.

Analysis takes place as the samples and reagents are pumped through a system of tubing instead of being retained in separate containers. The stream passes through the tubing to the different units of the analyser, each of which performs a specific analytical function. As distinct from manual methods and discrete analysers, absolute volumes are not pipetted but the liquids involved are mixed in carefully controlled proportions. These ratios are determined by using delivery tubes of varying internal diameter with a constant-speed pump and it is the different rates of flow (ml min^{-1}) within these tubes that is the important factor.

The samples are aspirated in sequence into the reagent stream with a small volume of water separating each one. Air bubbles are introduced into the flowing stream and in order to maintain a regular liquid–air pattern in the stream, a wetting agent, e.g. Brij 35, is usually added to the liquids.

The air bubbles have several important functions and are fundamental to

▶ Segmentation of the liquid streams by air bubbles is a feature of continuous flow analysis.

this type of analyser. They completely fill the lumen of the tubing and help to preserve the integrity of each individual aliquot of liquid by providing a barrier between them. They reduce contamination between succeeding samples because the pressure of the bubble against the inner wall of the tubing wipes away any droplets from the preceding sample which might contaminate the next. These features permit the use of a shorter water wash between samples and therefore save time. The bubbles also assist in mixing sample and reagents within each liquid segment. Whenever two air-segmented streams are brought together, mixing within each liquid segment is achieved by passing the stream through a horizontal glass mixing coil, whose circumference should be at least three times greater than the length of the liquid segments. During passage through the coil, each fluid segment is repeatedly inverted allowing the heavier layers to fall through the lighter ones, ensuring complete mixing.

Quantitation in continuous flow analysis is based on the direct comparison of the peak heights on a recorder trace of the samples and standards. It is not necessary for the reactions to go to completion, because all the measurements in a particular method are made after the same fixed reaction time, which is determined by the length of analyser tubing and relies on a constant pump speed.

Single- and dual-channel instruments
The design of all single-channel instruments stems from the original Technicon AAI system, which comprises several interconnected units, called modules. The liquid streams flow through the modules, each of which performs a different analytical function (Figure 6.6). The subsequently developed AAII analysers did not consist of separate modules but included an analytical cartridge which combined several functions in one unit.

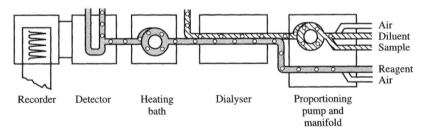

Figure 6.6
Diagrammatic representation of the function of the continuous flow analyser modules.

Recorder   Detector   Heating bath   Dialyser   Proportioning pump and manifold

Air
Diluent
Sample

Reagent
Air

Dual-channel instruments are very similar to their single-channel counterparts and differ only in the fact that they analyse each sample for two constituents simultaneously. The sample stream is divided into two fractions, each one flowing independently through its own system of tubing with the appropriate reagents and detectors.

Proportioning pump and manifold
The mixing of the sample and reagents in predetermined proportions rather than specific volumes is a fundamental principle of continuous flow analysis and is achieved by using a constant-speed roller bar peristaltic pump and flexible plastic delivery tubes of different internal diameters. These pump tubes are stretched between two plastic end-blocks, which hold them taut between a

spring-loaded platen and the pump rollers. The amount of air required to produce the bubbles is similarly introduced using a pump tube with a particular internal diameter. There is a choice of pump (manifold) tubes which deliver between 0.015 and 3.9 ml min^{-1} and are colour coded for easy identification.

Sampler
The sampler unit holds a tray (plate) which can be loaded with up to 40 cups containing the individual samples. The samples are aspirated in turn with water between each one. The length of time for which the sample is aspirated determines the volume of sample that is analysed.

> ➤ Sample and wash times are the dwell times of the sampling probe in the sample and water receptacle respectively.

Separation modules
A separation step is sometimes an essential part of an analytical method and may be as diverse as distillation, filtration, digestion, extraction, phase-separation or dialysis. These can all be performed by continuous flow analysers either by adding a specially designed glass fitting to the manifold or analytical cartridge or by the addition of a separate module to the analyser. Many biological samples contain protein and dialysis is often used to remove this protein, which would otherwise affect the analysis.

> ➤ Dialysis – see Section 3.4.1.

The dialyser unit consists of a thermostatically controlled water bath (37 $\pm$ 0.1 °C) which houses a matched pair of spirally grooved plastic (Lucite) plates, which are clamped together with a semipermeable cellophane membrane stretched between them. This in effect forms a long narrow tube which is divided lengthways by the membrane. As the air-segmented sample (donor) and recipient (usually saline or a reagent) streams flow side by side along the semicircular channels, small analyte molecules diffuse across the membrane into the recipient stream (Figure 6.7). The rate of dialysis is dependent on the concentration of the dialysable molecules if all the other factors which affect dialysis are kept constant. This makes it possible to compare standards and samples directly and although transfer is never complete, it is constant, usually between 5 and 25%.

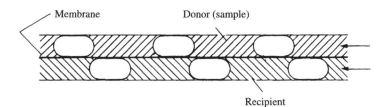

Figure 6.7 **Schematic representation of dialysis.** Small analyte molecules from the donor (sample) stream dialyse across the semipermeable membrane into the recipient stream. After emerging from the dialyser, the donor stream is usually pumped to waste while the analysis continues with the recipient stream.

Flow diagram
A diagrammatic representation of the components and layout of the system which is required to perform a particular method is known as a flow diagram (Figure 6.8). This is a convenient and well-recognized way to record information

about a particular method, e.g. the sampling rate, the temperature of the heating bath, the type of detection system and the size of pump tubes required for each reagent.

GLUCOSE

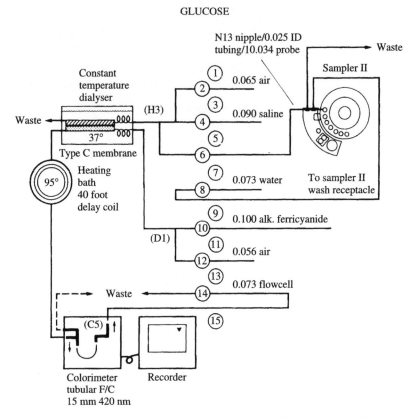

Figure 6.8 **Flow diagram for the measurement of glucose using alkaline ferricyanide.** The pump tubes are specified with respect to their internal diameters and position on the end-blocks. The colour coding of the tubes is:

Position 2	0.065	Blue/Blue
Position 4	0.090	Purple/Black
Position 6	0.020	Orange/Yellow
Position 8	0.073	Green/Green
Position 10	0.100	Purple/Orange
Position 12	0.056	Yellow/Yellow
Position 14	0.073	Green/Green

The odd-numbered positions on the end-blocks are those on the lower level and are unfilled. H3 and D1 refer to the glass connectors.

Theoretical aspects of continuous flow analysis

In continuous flow analysis, samples follow one another through the tubing and interaction of adjacent samples, which is called 'carry-over', does occur. Carry-over manifests itself on the recorder trace as peaks that are not completely differentiated. This is most noticeable when a sample of low concentration follows one of high concentration and is either seen as a shoulder on

> ➤ Carry-over occurs when there is some mixing of adjacent samples in the flowing stream.

the high peak or obliterated (Figure 6.9). However, carry-over may also result in the reduction of peak height when a high sample follows a low one, but this is not apparent from the trace and is therefore undetected. The introduction of water segments between the samples and of air bubbles into the liquid streams significantly reduces carry-over but does not eliminate it.

Appropriate sample and water volumes, expressed as aspiration times (seconds), must be determined for each method. This involves aspirating a standard solution of the analyte for varying times and observing the peak heights. A wash time may then be selected by keeping the sample time constant and varying the wash time. The highest sampling rate that results in the least amount of carry-over is usually sought.

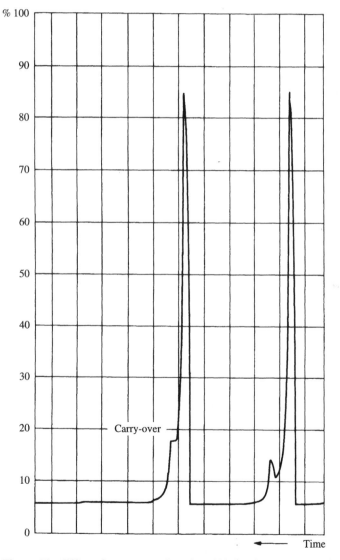

Figure 6.9 **Effect of carry-over.** Samples of high and low concentration were analysed with different wash times. Carry-over was more apparent with the shorter wash time.

Assessment of carry-over

An assessment of carry-over can be made by analysing a sample with a high concentration of the analyte (A) in triplicate, giving results a_1, a_2 and a_3, followed by a sample of low concentration (B) in triplicate giving results b_1, b_2 and b_3 (Fig 6.10). The carry-over between a_3 and b_1 is given by the equation:

$$\text{Percentage interaction } (k) = \frac{b_1 - b_3}{a_3 - b_3} \times 100$$

where b_3 is assumed to be the 'true' value for sample B since it is preceded by two samples of equal concentration; the effect of carry-over should be negligible.

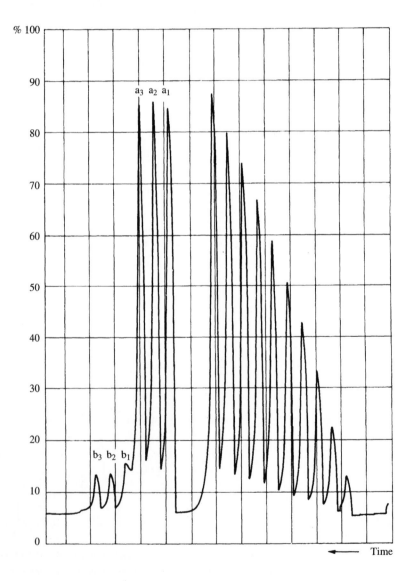

Figure 6.10
Assessment of carry-over.

6.2.2 Flow injection analysis

In this technique, which was developed in the 1970s, microlitre volumes of liquid sample are injected, at intervals, into a continuously flowing carrier stream which is not air-segmented. Various reagent streams are introduced as required and controlled mixing of reagents and sample occurs. The fact that flow injection analysis does not involve air-segmented streams makes it possible to include such separation steps as solvent extraction and gas diffusion.

When the merged stream flows through the detector, transient peaks are recorded in proportion to analyte concentration. Excessive dispersion of the sample zone into the carrier stream is reduced by the use of small sample volumes, narrow bore tubing and short residence times in the system. Rapid analysis and easy starting up procedures make this technique particularly useful for single samples which arrive infrequently and require immediate analysis (Table 6.1).

Table 6.1 Typical parameters for flow injection analysis

Sample volume	40–200 μl
Tube diameter	0.35–0.9 mm
Flow rate	0.5–2.5 ml min^{-1}
Coil lengths	10–200 cm
Flow cell volume	8–40 μl
Residence time	15–60 s
Sample through-put	60–200 h^{-1}

Instrumentation

➤ HPLC – see Section 3.2.2.

A precision injection device is required to minimize sample dispersion and keep the sample volume and length of sample zone reproducible. This is normally a rotary valve similar to that used for injection in HPLC. Exact timing from sample injection to detection is critical because of rapidly occurring reactions which are monitored before they reach completion. This demands a constant flow rate with low amplitude pulsing, normally achieved by a peristaltic

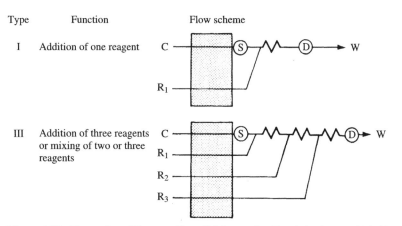

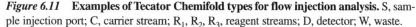

Figure 6.11 **Examples of Tecator Chemifold types for flow injection analysis.** S, sample injection port; C, carrier stream; R$_1$, R$_2$, R$_3$, reagent streams; D, detector; W, waste.

pump. Alternatively, a syringe or gas pressure may be used to propel the fluids. Correct proportions of sample and reagents are achieved by the use of tubes of varying diameters. Tecator produces a series of complete flow injection analysers. Their manifolds, which provide a means of bringing together the fluid lines and allowing rinsing and chemical reactions to take place in a controlled way, are marketed under the name Chemifolds (Figure 6.11) and a range is available for a variety of applications.

6.3   Robotics

A major shortcoming of the automated analysers developed in the 1960s and 1970s was that, to a large extent, sample preparation remained unautomated. Also their use was generally limited to quantitative photometric procedures on liquid samples. Robotic systems were developed to extend the range of tasks which could be performed. The Zymark Corporation based its robots on the fact that all laboratory procedures, regardless of complexity, can be broken down into a series of simple steps called Laboratory Unit Operations, LUOs (Table 6.2). The various modules or workstations required to perform these tasks can be linked by a robotic arm, which is programmed to transfer samples between them in any sequence. This provides a flexible, custom-built system by which a method can be fully automated irrespective of the number of steps and the sequence required. It combines robotics, computer control and instrumentation and produces a final report (Figure 6.12).

Less complex workstations are also available which perform more limited functions, either individually or in sequences, e.g. evaporation of organic solvents, small volume pipetting, reactions of hazardous reagents and auto-injection in column chromatographic techniques.

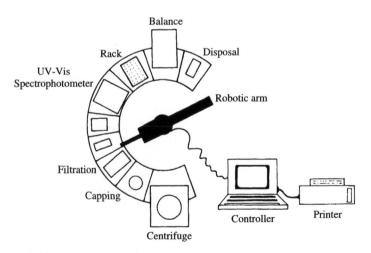

Figure 6.12 **Schematic representation of Zymate robotic system (Zymark Corporation).** The robotic arm is programmed to move between the various modules required for a particular analysis.

Table 6.2 Laboratory Unit Operations (LUOs)

Class	Definition	Examples
Weighing	Quantitative sample mass measurement	Using a balance
Homogenization	Reducing sample particle size and creating a uniform sample	Sonication, grinding
Manipulation	Physical handling of laboratory materials	Moving test-tubes, capping, uncapping
Liquid handling	Physical handling of liquid reagents	Pipetting, transferring, dispensing
Conditioning	Modifying and controlling the sample environment	Incubating, timing, shaking
Measurement	Direct measurement of physical properties	Absorbance, fluorescence, pH
Separation	Coarse mechanical or precision separation	Filtration, extraction, centrifugation
Control	Use of calculation and logical decisions in laboratory procedures	—
Data reduction	Conversion of raw analytical data to useable information	Peak integration, spectrum analysis
Documentation	Creating records and files for retrieval	Notebooks, listings, computers

Self test questions

Sections 6.2/3

1. Which of the following statements are true of flow injection analysis?
   (a) Liquid streams are segmented with air bubbles.
   (b) Centrifugation is used to mix reagent and sample.
   (c) Microlitre volumes of sample are typically used.
   (d) Reactions take place within lengths of tubing.
2. Which of the following statements are true of automated flow analysis?
   (a) Air bubbles are always introduced into the system.
   (b) The reagent volumes required are specified for each particular method.
   (c) Samples follow one another through a system of tubing.
   (d) Flexible tubes of varying diameters are used to regulate reagent delivery.

3. In automated flow analysis it is not necessary for the reactions to go to completion
   BECAUSE
   in automated flow analysis, quantitation is based on the comparison of recorder peaks of samples and standards analysed under the same conditions.
4. In air segmented continuous flow analysis, air bubbles help to reduce the interaction between adjacent samples (carry-over)
   BECAUSE
   it is not possible to assess the extent of carry-over for a particular continuous flow analysis method.

6.4   Further reading

Valcarcel, M. and Luque de Castro M.D. (1988) *Automatic methods of analysis*, Elsevier, Netherlands.

Stockwell, P.B. (1996) *Automatic chemical analysis*, 2nd edition, Taylor and Francis Ltd, UK.

Ruzicka, J. and Hansen, E.H. (1988), *Flow injection analysis*, 2nd edition, Wiley-Interscience, USA.

Sonntag, O. (1993) *Dry chemistry*, Elsevier, Netherlands.

7 Immunological methods

Key topics

- General processes of the immune response
- Antigen–antibody reactions
- Analytical techniques – precipitation reactions
- Analytical techniques – immunoassay

➤ An antigen is a molecule that can react with a preformed antibody.

➤ A hapten is too small to initiate an immune response but may do so when attached to a larger molecule.

➤ An epitope is that part of an antigen that interacts with an antibody.

➤ A paratope is the complementary portion of the antibody that interacts with an epitope.

Antibodies are proteins produced by an animal, in a process known as the immune response, as a result of the introduction of a foreign substance into its tissues. Such a substance is known as an antigen or immunogen and the antibodies that are produced are capable of binding with that antigen when allowed to react in an appropriate manner. Some substances, known as haptens, are not capable of initiating an immune response by themselves but when conjugated with a protein may act as an epitope, resulting in the formation of antibodies.

Antigen–antibody reactions have provided the basis of very useful methods of qualitative and semi-quantitative analysis for many years, particularly in microbiology. Antibodies have been used in the identification of bacteria and often antibody preparations were given names that described a demonstrable feature of their reaction with the antigen. Thus lysins were antibodies that caused the disruption of cell membranes, opsonins rendered the antigen susceptible to phagocytosis, while agglutinins and precipitins caused the flocculation of cellular and soluble antigens respectively. It was originally thought that the antibodies that caused these different effects were quite different from each other and it was not appreciated at the time that the same antibody might have different effects depending on the environmental conditions and the presence of other substances.

Major progress in the analytical use of antibodies occurred with the development of several fundamental techniques: the ability to detect cell-bound antibody (Coombs Test, 1945); immunoprecipitation in gels (Ouchterlony, 1953); radioimmunoassay (Yalow and Berson, 1960) and monoclonal antibodies (Kohler and Milstein, 1975). These, together with a considerably improved understanding of the immune response and antibodies, paved the way for development of many very sensitive quantitative methods of analysis.

The formation of antibodies is only one mechanism by which an animal may protect itself from substances or microorganisms that are potentially harmful. A **mechanical protection** against infection is provided by the presence of an intact skin surface and membranes together with the secretion of mucus from many internal membrane surfaces. The acids secreted by the stomach and skin have a bactericidal effect as does the presence in many body fluids of certain enzymes, particularly lysozyme.

➤ An inflammatory response is a non-specific reaction of the tissues against toxic substances.

Invasion of the tissues by an infective agent initiates an **inflammatory response** in the animal. This is non-specific and is mediated primarily by substances released from tissues that are damaged as a result of either trauma or the toxic effects of the infective agent. The major mediator is the vasoactive amine histamine, which causes an increased local blood flow and capillary permeability, resulting in local oedema. A major aspect of the inflammatory response is the involvement of large numbers of phagocytic cells, particularly the polymorphonuclear leucocytes. These are chemotactically attracted to the inflamed tissues and are mainly responsible for the elimination of particulate material. This often results in the destruction of many of these cells and the formation of pus.

One very important group of cells known as the lymphocytes, which are widely distributed throughout the tissues, appear in increased numbers during an inflammatory response and are primarily responsible for the **immune response**. This is a specific response to the invading substance or agent by the animal and involves the production of cells and antibodies with the ability to recognize and bind the invading substance.

7.1.1   Cells involved in the immune response

➤ Lymphocytes are small, mononuclear cells primarily associated with the immune processes.

The **lymphocytes** are a very heterogeneous group of cells, almost identical when studied using light microscopy methods and only showing slight differences by electron microscopy techniques and yet the group contains cells with many different roles. Although large numbers of lymphocytes can be detected in the circulating blood and body fluids, the majority of lymphocytes are to be found in the group of tissues known collectively as the reticulo-endothelial system. This includes such tissues as the liver, spleen, bone marrow, thymus and lymph nodes, all of which are important in the immune response. Experiments involving the removal of various tissues from experimental animals have indicated that there are two different features to the immune response. The removal of the thymus, a small gland located behind the sternum, impairs the ability of a young animal to reject skin transplants but does not affect to the same extent its ability to produce antibodies. This aspect of

➤ Cell-mediated immunity is that aspect in which cells are directly involved in the protective processes.
➤ T lymphocytes are cells produced in the thymus and responsible for cell-mediated immunity.
➤ Humoral immunity is that aspect that is mediated by antibodies.

➤ B lymphocytes are cells produced in the bone marrow and responsible for antibody production.

the immune response is known as **cell-mediated immunity** and is due to a sub-population of lymphocytes called T lymphocytes (thymus derived). Experimental work with fowls has demonstrated that the removal of a lymphoid tissue nodule in the gut known as the bursa of Fabricius results in a reduced ability to produce antibodies but does not significantly alter the response to skin grafts. It was subsequently demonstrated that the production of antibodies, a feature known as **humoral immunity**, is associated with another sub-population of lymphocytes known as B lymphocytes (bursa derived). The bone marrow acts as the source of B lymphocytes in man.

The introduction of an antigen into the tissues of a susceptible animal results initially in increased proliferation of lymphocytes in the tissues of the reticulo-endothelial system particularly the lymph nodes and the spleen (Figure 7.1). An animal will show different responses if the antigen is completely new to it (a primary response) or if the antigen has been encountered

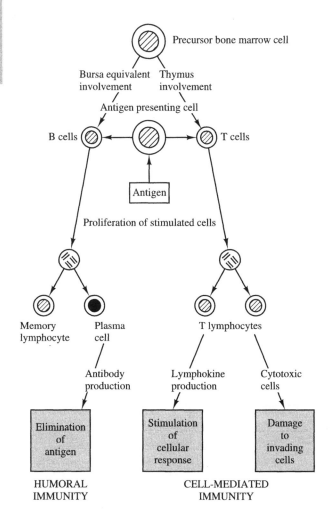

Figure 7.1 **Cellular processes of the immune response.** A simplified summary of the sequence of events in both the humoral and cell-mediated response to an antigen.

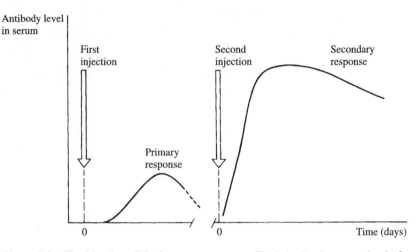

Figure 7.2   **The kinetics of the immune response.** The injection into an animal of a second dose of an antigen several weeks after the first injection will result in a response that is more rapid and more intense than the first.

➤ A primary response describes the type of reaction when an animal encounters an antigen for the first time.

➤ A secondary response describes the type of reaction when an animal encounters an antigen more than once.

➤ Plasma cells are B lymphocytes that are in the process of synthesizing antibody protein.

➤ Complement are non-immune plasma proteins that act in conjunction with antibodies.

➤ Lymphokines are substances produced by leucocytes which are involved in signalling between cells of the immune system.

➤ Clonal selection is the process by which only a limited number of lymphocytes are stimulated to replicate by an antigen.

on a previous occasion (a secondary response) (Figure 7.2). The secondary response shows a reduced lag period and a considerably increased rate of antibody synthesis compared with the primary response and the antibody persists for a longer period. The kinetics of the response vary depending upon the antigen and the animal but the relationship between the primary and secondary responses is characteristic.

There are three major types of lymphocyte which, when stimulated, are directly responsible for the effects of an immune reaction. **B lymphocytes** during proliferation develop into **plasma cells,** which are the antibody-producing cells, the process referred to earlier as humoral immunity. An antibody protects in a variety of ways. It can render the free antigen more susceptible to elimination by normal cellular processes such as phagocytosis or it may block the effects of a toxic substance. It can also function in conjunction with a series of plasma proteins known collectively as **complement** to cause lytic damage to the membrane of a target antigenic cell. There are two types of **T lymphocyte** involved: the cytotoxic cells (T_c cells) when stimulated are capable of binding to the antigen cell and causing irreversible lytic damage to the cell membrane; other T lymphocytes (T_d cells) release soluble substances known as lymphokines, which damage the invading antigen cells and stimulate other aspects of the host immune system. The effects of T_c and T_d lymphocyte activity are features of cell-mediated immunity.

Immunologically competent cells, whether they are T or B lymphocytes, have membrane receptors that are specific for an antigen. It is basically the binding of the antigen to the specific receptor on the appropriate lymphocyte which initiates the whole process, stimulating the cell to proliferate and producing a clone of identical cells, a process known as 'clonal selection'. The nature of the secondary response is due in the main to this large number of cells now available.

The processes of antigen recognition and immune stimulation are extremely complex and involve a sequence of steps. Upon entering the tissues, the antigen is taken up by macrophages or other 'antigen-presenting cells' (APCs) and, after modification, is exhibited on the outer membrane of the cells. This membrane-bound antigen can be recognised by lymphocytes via the antigen-specific receptors carried on their membranes. A key group of cells in the process are the T helper cells (T_h cells). These cells recognize the antigen presented by the macrophage and when stimulated by this recognition process they are capable of stimulating the specific B, T_c or T_d cells (Figure 7.3).

> ➤ Antigen-presenting cells are cells that take up antigen and present it, in recognizable form, to the lymphocytes.

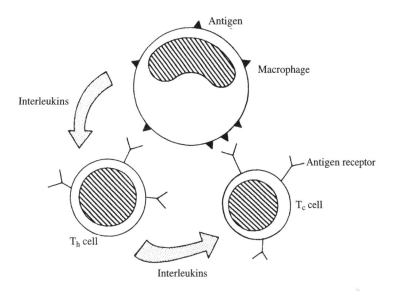

Figure 7.3 **Process of antigen recognition and lymphocyte stimulation.**

In all cellular interactions in the immune process, as well as antigen recognition being a major feature, there are complex stimuli resulting from the secretion of a range of cytokines by the different cells involved. In the immune process, because the interactions are between leucocytes, these cytokines are known as interleukins. They are specific for the type of cell on which they can act and are necessary, often in conjunction with antigen recognition, to stimulate the target lymphocyte to act in the required manner. This action may involve protein synthesis, replication or differentiation and is an essential step in mounting and controlling a full immune response. An individual lymphocyte is capable of being stimulated by only a limited number of antigens (probably only one but at the most only two or three) and in order to provide a large number of different antibodies it is essential that there are an equivalent number of different lymphocytes in the tissues.

> ➤ Interleukins are a group of lymphokines that act on other leucocytes to control the immune process.

7.1.2 Antibody structure

> ➤ Human serum proteins – see Figure 11.13.

Antibodies are members of a group of proteins known collectively as immunoglobulins. The name is derived from the observation that during electrophoresis of blood plasma the proteins associated with antibody activity migrate with the gamma globulin fraction.

Immunoglobulins are glycoprotein molecules composed of four polypeptide chains linked covalently by disulphide bonds (Figure 7.4). The four polypeptides consist of two identical chains of relative molecular mass 23 000 and 50 000–75 000 and are designated light and heavy chains respectively. In addition to the interchain disulphide bonds, each polypeptide contains a number of intrachain disulphide bonds which divide up the polypeptide chain into a series of domains. In addition, immunoglobulins possess an area known as the hinge region which allows for flexibility of the chains in relation to one another.

➤ Protein structure – see Section 11.1.

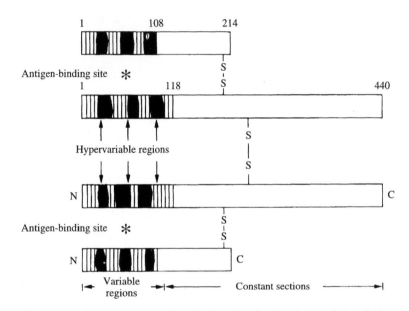

Figure 7.4 **Basic structure of an IgG molecule.** Two heavy chains (440 residues) and two light chains (214 residues) are joined by disulphide bonds and each shows a relatively constant amino acid sequence in one section (C-terminal end) and a variable sequence section (N-terminal end). The variable regions of both heavy and light chains are involved in the formation of the antigen-binding site.

Comparison of the amino acid sequence of the heavy chains of a particular class of immunoglobulin reveals that approximately three-quarters of each chain from the C-terminal end show very similar sequences (the constant section). The remaining quarter of the peptide chain (the variable section) shows considerable variation in the amino acid sequence and corresponds to that part of the chain associated with the antigen-binding site. Similar constant and variable regions are also demonstrable in the light chains although in this case each involves approximately half of the peptide. The variation is particularly noticeable at three distinct sections (hypervariable sections) in the heavy chains and it is suggested that these sections when associated with three similar sections in the light chains are responsible for the antibody activity and specificity of an immunoglobulin.

➤ Isotypes are variants that are present in all members of a species.

Five major sub-classes of immunoglobulin (isotypes) (Table 7.1) have been recognized and the differences between the classes lie in the heavy chains, which, although of approximately the same size for all classes, vary considerably in the amino acid sequence. The heavy chains of immunoglobulins are designated as γ, μ, α, δ or ϵ and when two identical heavy chains are combined in an immunoglobulin, the molecule is designated as being either IgG, IgM, IgA, IgD or IgE respectively. The heavy chain isotypes can be further subdivided into several subtypes. There are four sub-types of IgG (γ_1, γ_2, γ_3, γ_4), two of IgA (α_1, α_2) and two of IgM (μ_1, μ_2). The light chains do not show such variation and only two main types are demonstrable, known as the kappa (κ) and the lambda (λ) chains.

Table 7.1   Classes of immunoglobulins

Immunoglobulin	RMM	Number of basic four-chain units	Heavy chain	Heavy chain sub-classes	Antigenic valency	Percentage in normal serum
IgA	1.5×10^5	1 or 2	α	$\alpha_1\ \alpha_2$	2–4	13
IgD	1.8×10^5	1	δ		2	1
IgE	2.0×10^5	1	ϵ		2	0.002
IgG	1.6×10^5	1	γ	$\gamma_1\ \gamma_2\ \gamma_3\ \gamma_4$	2	80
IgM	1.0×10^6	5	μ	$\mu_1\ \mu_2$	10	6

7.1.3   The role of antibodies

IgG comprises some 80% of the total immunoglobulin in plasma and because it is relatively small it is capable of crossing membranes and diffusing into the extravascular body spaces. It can cross the placental membrane and provides the major immune defence during the first few weeks of life until the infant's own immune mechanism becomes effective.

IgM is a large molecule composed of five units, each one similar in structure to an IgG molecule. The tetramer contains an additional polypeptide, the J chain (relative molecular mass 15 000), which appears to be important in the secretion of the molecule from the cell. It is an effective agglutinating and precipitating agent and, although potentially capable of binding ten antigen molecules, it is usually only pentavalent. It does not cross membranes easily and is largely restricted to the bloodstream.

IgA is associated mainly with seromucous secretions such as saliva, tears, nasal fluids, etc., and is secreted as a dimer with both a J chain and a secretor piece (relative molecular mass 70 000), the latter apparently to prevent damage to the molecule by proteolytic enzymes. Its major role appears to be the protection of mucous membranes and its presence in blood, mainly as the monomer, may be as a result of absorption of the degraded dimer.

IgE is known as a cytophilic immunoglobulin because of its ability to bind to cells, which may account for its low concentration in body fluids. When IgE reacts with an antigen it causes degranulation of the mast cell to which it is bound, with the release of vasoactive amines such as histamine. This process may well be helpful in initiating the inflammatory response but in allergic individuals the reaction is excessive and leads to a hypersensitive or over-reactive state.

Antigen–antibody reactions

An antibody combines specifically with the corresponding antigen or hapten in a manner that is very similar to the binding of an enzyme to its substrate and involves hydrophobic and electrostatic interactions. The bonding between an antibody and antigen, however, involves no subsequent chemical reaction and its stability depends upon the complementary shape of the antigen and the binding site of the antibody. Because of the relative weakness of the forces that hold antibody and antigen together, these combinations are reversible and the complex will dissociate, dependent upon the strength of binding:

$$Ag + Ab \Leftrightarrow AgAb$$

Where the binding is strong the equilibrium will lie to the right; where weak, to the left. The strength of the binding of an antibody to an antigen is referred to as its affinity and is defined by the equilibrium constant K, where

$$K = \frac{[AgAb]}{[Ag][Ab]}$$

Where more than one antibody in an antiserum can combine simultaneously with an antigen, the sum of the binding strength is defined as the avidity of the antiserum.

> **Affinity is the strength of binding between a single antibody and its antigenic epitope.**

Most antigens are large and may have many antigenic characteristics or epitopes (determinants) and as a result serum taken from an animal which has been immunized against that antigen will contain several different antibodies against the different antigenic determinants. It is possible that another antigen may share some similar antigenic determinants with the original antigen with the result that some of the antibodies in an antiserum will bind to both antigens. Such an antiserum will show cross-reactivity between the two antigens and is said to lack specificity. Antiserums used for analytical purposes should be specific and it is essential that every antiserum should be thoroughly tested prior to its use.

> **Cross-reactivity occurs when antigens share epitopes such that an antibody generated against one antigen will also recognize another.**

7.2.1 Production of antibodies

The raising of a specific antiserum by the immunization of an animal usually involves intramuscular injection of the pure antigen, although intravenous injection may be appropriate for particular antigens. Because of the primary and secondary responses to antigens, a series of injections involving small amounts of antigen is likely to be more effective than a single large injection. The precise sequence and timing can significantly affect the quality of the antiserum produced. An initial injection is normally followed by several booster doses given at two- to four-week intervals. Too frequent injections, although possibly giving a quicker response, may result in an antiserum of reduced avidity.

> **Avidity is the total binding strength of an antiserum.**

The species of animal used should be as different as possible from the animal source of the antigen. It should be relatively easy to handle and yet provide enough serum to make the process worthwhile. The animals most frequently used for the production of antibodies are the guinea pig and rabbit but for larger supplies of serum, goats, sheep and horses are used.

> An adjuvant is a material designed to stimulate the immune system non-specifically.

Antigens vary considerably in their ability to initiate an immune response and it is usual to incorporate an adjuvant into the sample before injection. An adjuvant is a mixture of substances that stimulates an inflammatory response and prevents the rapid removal of the antigen from the tissues by the normal drainage mechanism. Freund's adjuvant consists of an emulsion of dead mycobacteria in mineral oil but simpler alternatives of aluminium phosphate or hydroxide have a similar effect. Although proteins are generally immunogenic, they do need to have a relative molecular mass of at least 4000 and some structural rigidity to be effective. In order to raise antibodies against a non-immunogenic molecule (a hapten) it is necessary to link it to a carrier protein which is capable of initiating a response. Bovine or human serum albumin is frequently used for this purpose as well as synthetic polypeptides such as poly-L-lysine. The hapten should be linked covalently with the carrier protein, a process readily achieved using a variety of chemistries and often involving the incorporation of a spacer molecule between the carrier protein and hapten.

> A polyclonal antiserum contains a heterogeneous mix of antibodies directed against an antigen.
> A monoclonal antibody is a homogeneous serum containing only one antibody directed against an antigen.

Antiserums produced in this manner are known as **polyclonal** because they contain many antibodies produced by different lymphocytes, each one responding either to different antigenic determinants of the original antigen or to other immunogenic substances in the injected material. Such a range of antibodies reduces the specificity of the method.

The development of techniques for the production of **monoclonal** antibodies by Kohler and Milstein has enormously expanded the potential of antibodies as analytical and therapeutic agents. A monoclonal antibody is one that is produced by a clone of cells all derived from a single lymphocyte. Any lymphocyte can probably produce only a single immunoglobulin and hence the antibody produced by a clone of identical cells is very restricted in the antigens to which it will bind, making it a very specific reagent.

The production of monoclonal antibodies starts with the immunization of an animal (usually a mouse) in the traditional manner. However, instead of allowing the immune system of the mouse to generate antibodies, lymphocytes are separated from the spleen of the mouse and fused *in vitro* with myeloma cancer cells growing in cell culture. A fusogen, usually polyethylene glycol, is used and the resulting fused cell is known as a hybridoma. The cell suspension is diluted and distributed among a large number of sub-cultures in order to achieve single-cell distribution. The original cancer cell has the ability to synthesize immunoglobulin non-specifically but when fused with a lymphocyte, which was stimulated by the injected antigen, produces the immunoglobulin for which the lymphocyte has the genetic information.

> A myeloma is a cancer of the antibody-producing plasma cell and as such is immortal.

The hybridoma cells are initially grown in a medium that will not maintain the growth of the cancer cells; these therefore die, as do non-fused lymphocytes, leaving only the fused cells. As the hybridoma cells grow, the supernatant fluid is tested for the presence of antibodies. Those cultures producing the desired antibody are further cloned and either grown in bulk or as tumours in animals and the monoclonal antibodies harvested.

> A hybridoma is a cell line created by the fusion of two different lymphocyte cell lines, one of which is derived from a tumour.

In addition to their improved specificity, monoclonal antibodies offer other significant advantages over polyclonal antiserums: there is an indefinite supply of antibodies with constant characteristics together with relative ease in purification.

7.2.2   The effects of antigen–antibody complex formation

The combination of an antigen and an antibody usually results in the formation of a lattice-type structure (Figure 7.5). If this structure is large enough it will sediment and be measurable in some way. If the antigen was originally soluble, this process is known as precipitation but if the antigen was cellular or particular, the process is known as agglutination.

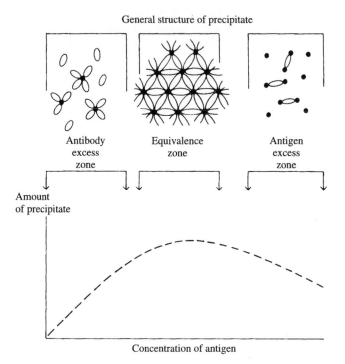

Figure 7.5 **Antigen–antibody reaction.** Maximum precipitation occurs when the antigen and antibody are present in equivalent amounts and is due to the formation of large lattice structures. On either side of the equivalence zone the amount of precipitation is reduced because the aggregates are smaller and more soluble.

It is extremely difficult to measure the amount of antibody present in absolute terms but its activity or concentration can usually be related to a demonstrable aspect of the antigen–antibody reaction. A study of the effect of the concentration of the two components indicates that the proportion of antibody to antigen is often critical in order to produce the detectable effect. Figure 7.5 shows the effect of varying amounts of antigen in the presence of a fixed amount of antiserum on the extent of precipitation. As the amount of antigen increases from zero, so the amount of precipitate increases in an approximately proportional manner up to a maximum value. However, if the amount of antigen is further increased, the amount of precipitate often diminishes. This means that instead of working with excess reagents, as is often the case in chemical methods of analysis, it is often essential to work with optimal or equivalent proportions of antigen and antibody in order to produce the maximum effect.

In some cases, particularly if the antigen is cellular, the combination of antigen and antibody will result in a biological effect rather than a physical effect. Cells may be damaged, resulting in either lysis or the inhibition of their natural processes, such as motility. Viruses may be neutralized making them non-infective. Many tests involving such mechanisms are qualitative and do not readily lend themselves to quantitative analysis. Table 7.2 summarizes the major types of analytical methods involving antibody reactions.

Table 7.2   Immunological methods of analysis. Demonstrable reactions between antibodies and antigens

Nature of reaction	Nature of antigen	Analytical value	Example
Agglutination	Particulate or cellular	Mainly qualitative	Blood grouping Detection of antibodies
Cytotoxicity	Cellular	Qualitative or semi-quantitative	Cell typing Complement fixation tests
Precipitation	Soluble	Qualitative or quantitative	Double diffusion Single radial diffusion
Complex formation	Labelled molecules	Quantitative or qualitative	Radioimmunoassay Enzyme immunoassay Immunohistochemistry Immunoblotting

Self test questions

Sections 7.1/2

1 What is antibody production a feature of?
   (a) Cell-mediated immunity.
   (b) Humoral immunity.
   (c) T lymphocytes.
   (d) B lymphocytes.

2 Which of the following describes an IgG immunoglobulin?
   (a) It is a pentamer.
   (b) It has two antigen-binding sites.
   (c) It has an RMM of 1.0×10^6 dalton.
   (d) It is the major immunoglobulin of the blood.

3 The production of an antibody response requires an antigen to be taken up by antigen-presenting cells (APCs) in order to be recognized by lymphocytes
   BECAUSE
   T lymphocytes are the major group of cells involved in cell-mediated immunity.

4 The nature of an antigen–antibody complex depends upon the ratio of the antigen to antibody rather than their concentrations
   BECAUSE
   IgM has ten antigen binding sites but IgG has two.

7.3 Analytical techniques – precipitation reactions

Many qualitative and quantitative methods for soluble antigens are based on their precipitation by antibodies. It is important that the conditions for each assay are carefully optimized in order to achieve the desired end.

7.3.1 Immunoprecipitation in solution

The number and size of the immune complexes formed by the reaction of antigen with antibody is dependent on the relative concentrations of the reactants (Figure 7.5). The formation of antigen–antibody complexes may be observed by measuring the apparent absorption of light when incident light is scattered by the complexes or by direct measurement of the scattered light.

> ➤ Turbidimetry is the measurement of the apparent absorbance when light is scattered by a suspension of particles.
> ➤ Nephelometry is the direct measurement of light scattered by a suspension of particles.

The reaction is carried out in an excess of antibody and a calibration curve is prepared by measuring the turbidity of a series of standard solutions of the antigen. The measurement of the turbidity of a solution containing antigen–antibody complex, in which the amount of incident light lost by scattering is measured, has the advantage of simplicity but generally shows poor levels of sensitivity and precision. The method may be considerably improved by the use of a nephelometer rather than a simple photometer. A nephelometer is similar in design to a fluorimeter in that the detector is placed at right angles to the incident light beam and thus measures the light scattered by the complex. The extent of light scattering by a turbid solution increases as the wavelength of incident radiation decreases and this, together with a requirement to minimize the internal reflection of incident light, has led to the increasing use of lasers as a light source.

Immunoprecipitation is an extremely valuable technique because it permits the quantification of a specific protein in the presence of many other similar proteins. It is capable of a good level of sensitivity, the lower limit being mainly determined by the clarity of the sample blanks. In order to achieve maximum sensitivity the antiserum must show a high avidity for the antigen and be capable of being used at low concentration (high titre), the latter being important not only from a cost perspective but also to minimize turbidity due to the reagents. In addition, the development of particle-enhanced light-scattering assays in which the antibodies are immobilized onto latex particles has overcome many of the limitations of basic immunoprecipitation in terms of improved sensitivity and assay range. A major advantage of immunoprecipitation is that it may be readily automated, a feature that has been particularly exploited using centrifugal analysis instrumentation.

7.3.2 Immunoprecipitation in gels

Gels are used in immunoprecipitation techniques to stabilize the precipitate, enabling both the position and the area of the precipitate to be measured. The point has already been made that maximum precipitation occurs when the equivalent proportions of both antigen and antibody are available. Hence, if a high concentration of antigen is permitted to diffuse into a gel that contains a uniform concentration of antibody, at some point in the concentration gradient of antigen that is formed there will be optimum concentrations of both reactants and a precipitate will form. The dimensions of the gradient will depend on the original concentration of the antigen and hence the distance between the precipitate and the original starting point of the antigen will be proportional to its initial concentration.

This principle forms the basis of single radial immunodiffusion (SRID), a technique first developed in 1965 by Mancini. SRID involves pipetting a measured volume of antigen into holes cut into a buffered agar gel containing the antibody. The loaded gel is placed in a moist chamber at room temperature for at least 18 hours to permit diffusion. Rings of precipitate form around each well (Procedure 7.1), the precipitate being maximal at the periphery and less

> ➤ Single radial immunodiffusion (SRID) is a quantitative immunoprecipitation technique.

Procedure 7.1: Quantitation of albumin by single radial immunodiffusion

Reagents
Agar gel containing 10% (v/v) of a polyclonal antibody to human albumin.
Standard solutions of human albumin with concentrations ranging from 10 to 100 mg l^{-1}.

Method
Small wells were cut into the gel.
10 μl volumes of albumin solutions were pipetted into the wells.
Allowed to stand at 4°C for 24 h.
The zones of precipitation were measured in two directions at right angles.

Results

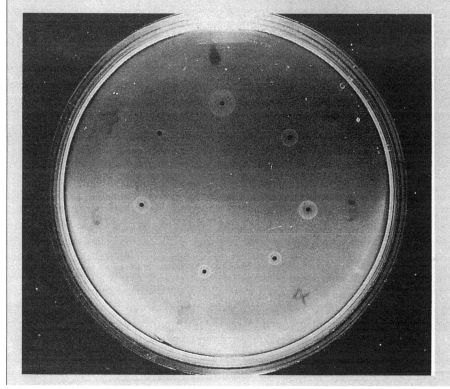

Well	Albumin concentration (mg l^{-1})	Ring diameter (m)
1	100	9.0
2	Sample A	6.5
3	60	7.0
4	30	5.0
5	20	4.0
6	Sample A	6.5
7	10	3.0

Calculation

A calibration graph was prepared from the results of the standard solutions by plotting the square of the diameter against the albumin concentration. The concentration of the test sample can then be determined from the graph and gives a value of 54 mg l^{-1}.

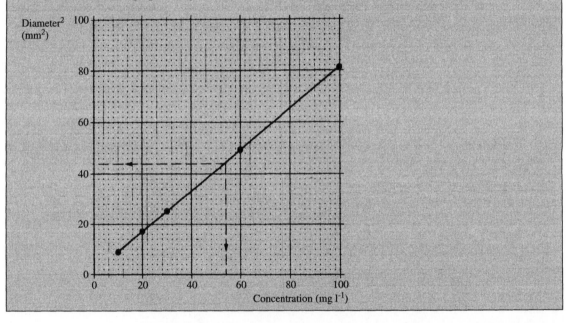

intense towards the well. The time taken for samples containing a high antigen concentration to arrive within the zone of equivalence will be longer than for lower concentrations of antigen and so the diameter of the formed rings will be larger. The radial nature of the diffusion is such that the diameter of the precipitation ring will be related to the concentration of the antigen. In practice, the diameter of the ring is measured and a plot of the square of the diameter against concentration generally gives a straight line relationship (Procedure 7.1). Two measurements of the diameter are taken at right angles in order to allow for any slight irregularities in the shape of the rings. SRID techniques provide useful and specific methods for the quantitation of individual proteins. The limit of sensitivity is about 5 mg l^{-1} and it is dependent on the ability to detect and measure very small precipitation rings. The concentration of antibody in the gel has to be carefully selected; reducing the amount of antibody in the gel will increase the average ring size but will result in less precipitate and a compromise has to be made between ring diameter and precipitate intensity. The visualization of the precipitate may be improved by staining the protein with a suitable dye.

One of the drawbacks to SRID is the time for the reaction to take place. Laurell, in 1966, sought to overcome this difficulty by using electrophoresis rather than diffusion to produce the concentration gradient. In electro-immunoassay the antibody is incorporated into the gel in a similar way to SRID. The electrophoresis is usually performed at pH 8.6, conditions under which the antibodies do not migrate significantly but the test proteins do. Because the samples are driven towards the equivalence zone by electrophoresis rather than by diffusion the precipitation lines characteristically appear as peaks or 'rockets' in the gel (Figure 7.6). A calibration curve for the quantitation of unknown samples is constructed by plotting peak height against the concentration of antigen.

➤ Rocket immunoelectrophoresis is a quantitative immunoprecipitation technique.

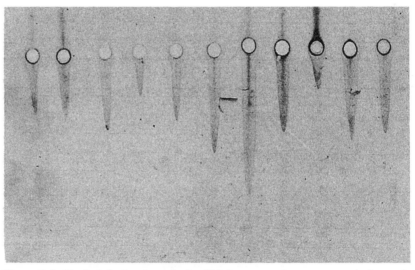

Figure 7.6   **Rocket electrophoresis.** At pH 8.6 most proteins move towards the anode and precipitation occurs where the antigen and the antibody (in the gel) are in equivalent proportions. The size of the resulting 'rocket'-shaped pattern of precipitate is proportional to the original concentration of antigen. The immunoglobulins show very little electrophoretic mobility at pH 8.6 and so remain in the gel during the process.

Although the buffer systems are designed to restrict the migration of antibodies during electrophoresis, immunoglobulins may still be measured by this technique if they are first carbamylated or formylated. These procedures are designed to increase the ionic character of the immunoglobulins without significantly altering their immunological properties. Carbamylation involves a reaction with potassium cyanate, and formylation a reaction with formaldehyde.

These techniques are notable for the fact that they rely on the diffusion of only one component of the antigen–antibody reaction; in the technique of

➤ Double diffusion immunoprecipitation is a qualitative technique.

double diffusion both antigen and antibody are placed in wells and allowed to diffuse. Precipitation also occurs at the equivalence point but in this case the precipitation is in the form of a line or arc in the gel. The position of the line is characteristic of the antigen and allows its identification in a complex mixture of similar proteins (Figure 7.7). A variety of similar qualitative techniques (immunoelectrophoresis, crossed immunoelectrophoresis) have been developed.

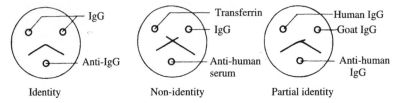

Figure 7.7 **Double diffusion in gels** showing reactions of identity, non-identity and partial identity.

7.3.3 Immunohistochemistry

The specificity of antibodies can be exploited in order to probe the *in situ* organization of cells and tissues. Cellular antigens can be identified both in viable cells and in frozen or fixed tissue sections. Antibodies are used to identify the appropriate antigen in the section and then the position of this primary antibody may itself be detected either directly if it was initially labelled or indirectly using another secondary antibody or molecule to attach to the antibody (Figure 7.8). Samples need to be carefully washed after addition of the primary or labelled antibody in order to prevent any non-specific reactions. Labels that have been successfully linked to antibodies include the following:

1. *Fluorescent dyes*. Immunofluorescence using fluorescein or rhodamine has been very successfully employed for immunohistochemistry. Fluorescein, when illuminated with UV light, emits a characteristic green fluorescence while rhodamine gives an orange colour.
2. *Enzymes*. Enzymes too are useful labels and several have been employed, with peroxidase and alkaline phosphatase being the most popular to date. One important feature of this technique is for the enzyme to be able to convert a soluble substrate into an insoluble product in order to localize the antigen properly. Several suitable substrates are available with 3′3′-diaminobenzidine (DAB) and 3-amino-9-ethylcarbazole (AEC) being used with peroxidase and 5-bromo-4-chloro-3-indoylphosphate/nitroblue tetrazolium (BCIP/NBT) with alkaline phosphatase.
3. *Colloidal gold*. This is a useful label for both direct and indirect staining methods since it requires no further reagent additions. Gold probes may be readily seen by light microscopy when coupled with silver enhancement and in addition, being electron dense, offer excellent sensitivity for the electron microscope.

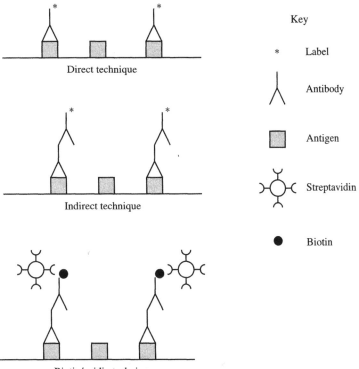

Key

*   Label

    Antibody

    Antigen

    Streptavidin

●   Biotin

Figure 7.8 **Some examples of antibody detection techniques.** (a) Direct labelling of the primary antibody, (b) indirect using labelled secondary antibody, (c) indirect using biotinylated secondary antibody and labelled streptavidin.

7.3.4 Immunoblotting

Antibodies may also be used to determine the presence or identity of soluble antigens by a process known generally as immunoblotting. In a technique known as 'dot blotting', soluble antigens are applied to a nitrocellulose membrane in the form of 'dots', antibody is then applied to the membrane, the membrane is washed to remove unbound antibody, and the presence or absence of the antigen is determined by similar means to that employed during immunohistochemistry.

Alternatively, as in Western blotting (Figure 7.9), soluble antigens may be separated by electrophoresis on polyacrylamide gel either with or without prior treatment with SDS. After electrophoresis the antigens are transferred from the gel to a nitrocellulose membrane. Once on the membrane antibodies may be used to probe for the presence of particular antigens either directly or indirectly as in immunohistochemistry. Non-specific binding sites may be 'blocked' using other non-specific proteins such as bovine serum albumin or casein before washing to remove unbound antibody. The labelled molecule may be visualized by the techniques used for immunohistochemistry or by autoradiography.

➤ Blotting techniques – see Section 11.3.3.

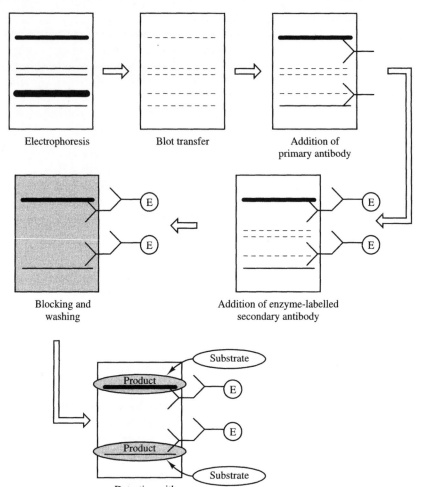

Figure 7.9   Immunoblotting. A representation of the stages in the technique.

Section 7.3

1 Which of the following techniques are quantitative?
   (a) Single radial immunodiffusion (SRID).
   (b) Laurell rocket immunoelectrophoresis.
   (c) Immunohistochemistry.
   (d) Immunoblotting.

2 Which of the following techniques require a labelled antibody?
   (a) Single radial immunodiffusion (SRID).
   (b) Laurell rocket immunoelectrophoresis.
   (c) Immunohistochemistry.
   (d) Immunoblotting.

3 Double immunodiffusion techniques permit the quantitation of one
   protein in the presence of other proteins

BECAUSE

the position of the precipitation line in double immunodiffusion is characteristic of the antigen.

4 In Laurell rocket immunoelectrophoresis, the height of the rocket is proportional to the concentration of the antigen

BECAUSE

in Laurell rocket immunoelectrophoresis, the movement of the larger molecules is retarded by the structure of the gel.

7.4   Analytical techniques – immunoassay

Immunoassay as an analytical technique was introduced by Rosalind Yalow and Solomon Berson in 1960 with their use of anti-insulin antibodies to measure the concentration of the hormone in plasma. This advance, for which Rosalind Yalow was awarded the Nobel prize, was probably the most important single advance in biological measurement of the following two decades. Examples of the use of immunoassay may now be found in almost all areas of analytical biochemistry.

Despite many novel developments in immunoassay design the principles are confined to two broad approaches: those that rely on the competition between antigens labelled with a molecule which may be readily observed (for example, a radioisotope) and unlabelled antigens for a limited number of antibody binding sites; and those in which the antibody is available in excess and for which there is no competition for binding sites.

➤ Competitive immunoassay relies on the competition between labelled and unlabelled antigens for a fixed and limited number of antibody-combining sites.

➤ Non-competitive immunoassay employs antibodies in excess.

➤ Kits – see Section 1.1.3.

These principles have been exploited by commercial manufacturers to provide 'kits', which are packages of reagents, designed to analyse samples for a wide variety of analytes. The commercial exploitation of immunoassay technology and the concurrent development of suitable automated instrumentation has had a profound effect on the analysis of biological samples. Kits can provide a wide range of tests offering complex technology and consistent reagents allowing even small laboratories to perform a full range of analyses.

7.4.1 Competitive binding immunoassay

The basis for this technique lies in the competition between the test antigen and a labelled antigen for the available binding sites on a fixed amount of antibody. While the binding sites are traditionally associated with an antibody, any source of specific reversible binding sites may be used to create an assay in this format. Examples of such are specific transport proteins such as thyroxine-binding globulin and certain cellular receptors such as opiate or benzodiazepine receptors. Under these circumstances the equilibrium mixture may be represented thus:

$$
\begin{array}{ccc}
\text{Ag} & & \text{Ag [Ab]} \\[1em]
& + \quad \text{[Ab]} \quad \underset{K_2}{\overset{K_1}{\rightleftharpoons}} & \\[1em]
\text{[Ag*]} & & \text{Ag* [Ab]}
\end{array}
$$

Given that the quantities of labelled antigen [Ag*] and antibody [Ab] are fixed when no unlabelled test antigen (Ag) is present then the labelled antigen has free access to the binding sites and the subsequent equilibrium state will represent the maximum amount of labelled antigen that may be bound. When unlabelled test antigens are present there will be competition between the labelled and the unlabelled antigens for the available binding sites with the result that at equilibrium the amount of labelled antigens bound will be reduced by an amount proportional to the amount of unlabelled antigens in the system. The concentration of antigens in unknown samples may be determined by comparison of the proportion of bound labelled antigens with the proportion bound when a series of standards of known antigen concentration is used.

7.4.2   Non-competitive or immunometric immunoassay

This approach to immunoassay is characterized by the fact that the antibody is present in excess and is generally also labelled. Because the labelled antibody is in excess there is no requirement for the setting up of an equilibrium since all of the test antigen may be sequestered by the excess of antibody.

$$Ag \quad + \quad Ab* \blacktriangleleft \text{-----------} \blacktriangleright AgAb* \quad + \quad Ab*$$

In immunometric assays, unlike competitive systems, the amount of labelled antibody bound is directly proportional to the amount of unlabelled antigen present rather than inversely proportional.

For each of these types of immunoassay, in order to observe the ratio of bound to free in the final reaction mixture, a method has to be employed which will separate these fractions and various ingenious ways have been devised to do this. Techniques which require the separation of free from bound fractions are referred to as heterogeneous assay systems.

➤ Heterogeneous assays are those that require separation of the bound fraction from the free.

7.4.3   Components of immunoassay systems

The antibody

The properties of the antibody used in an immunoassay will in large measure define its usefulness as an analytical technique. Both polyclonal and monoclonal antibodies have been used in immunoassays.

It is essential to determine the required optimum antibody concentration when using a competitive assay. By convention this is the dilution of antiserum which will bind 50% of the labelled antigen in the absence of unlabelled antigen (Figure 7.10) although this convention has no sound theoretical basis and many workers question its validity. This dilution of antiserum is known as its titre but this is a proxy unit because in a polyclonal antiserum it is not generally possible to determine the mass of immunoglobulin involved in antigen. Figure 7.10 also illustrates the effect that decreasing the mass of labelled antigen has on the assessment of the titre.

Once the titre of an antiserum is known, its specificity may be assessed.

➤ The titre of an antibody in the context of immunoassay is the dilution of serum used in the assay.

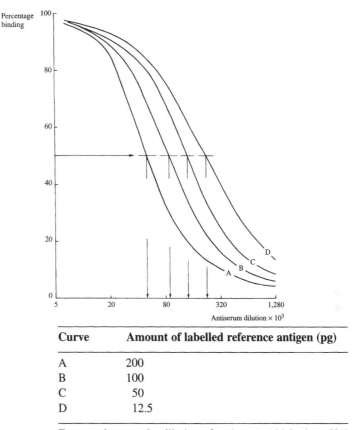

Figure 7.10
Radioimmunoassay of insulin – titration of antiserum.

Curve	Amount of labelled reference antigen (pg)
A	200
B	100
C	50
D	12.5

From each curve the dilution of antiserum which gives 50% binding of that amount of labelled antigen can be determined.

> ➤ Specificity is the ability of an antiserum to recognize only the antigen for which it was generated.

Where related molecules are recognized by an antibody, they are said to cross-react and such antibodies are likely to be unsuitable for some assays. For example, where measurements of a particular drug are made, metabolites of the drug having similar structures may also react with the antibody. The extent of cross-reactivity may be assessed by comparing the concentration at which a 50% displacement of the related compound is obtained with that required for the antigen (Figure 7.11).

Table 7.3 Detector molecules commonly used in immunoassay

Detector system	Example
Radioisotopes	Iodine-125, carbon-14, tritium
Enzymes	Horseradish peroxidase, alkaline phosphatase, β-galactosidase
Chemiluminescence	Luminol, acridinium esters, adamantyl dioxetane
Bioluminescence	Luciferase/luciferin
Fluorescence	Fluorescein, rhodamine, europium chelates

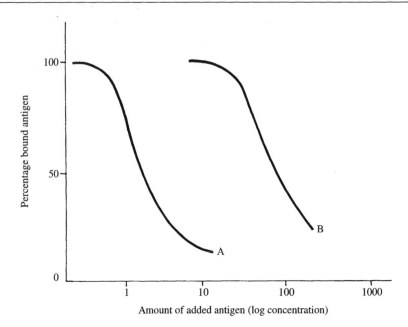

Figure 7.11 **Cross-reactivity.** Compound B displaces 50% of the label at 100-fold greater concentration than compound A and is said to have a 1% cross-reactivity.

The label

Immunoassays require a pure sample of either antibody or antigen for labelling with an appropriate molecule. Such a molecule should retain a high signal efficiency (i.e. be readily detected at low concentration) while its incorporation into the antigen or antibody should have no effect on their subsequent immunoreactivity.

The choice of label is important and the most common are listed in Table 7.3.

Radioisotopes

➤ Radioisotopes – see Section 5.3.1.

The most commonly used isotopes are carbon-14, tritium (hydrogen-3) and iodine-125. Carbon-14 and tritium are both beta-emitting isotopes while iodine-125 emits both beta particles and gamma radiation. Carbon-14 and tritium are examples of internal labels since the radioactive atom replaces an existing atom within the antigen, while iodine-125 is described as an external label because it is usually necessary to attach the iodine covalently to the antigen. There are advantages and disadvantages to each of these labels.

The beta-emitting carbon-14 and tritium have the advantage that the labelled form of the molecules is identical to the unlabelled antigen, but suffer from the disadvantage that the efficiency of the measurement of the beta emission, which uses the technique of scintillation counting, is less than that of the gamma emissions associated with iodine-125. Equally, iodine-125 suffers from the problem that the covalent attachment of the isotope to the antigen often means that there is a significant structural difference between the unlabelled and the labelled antigen. The gain in signal measurement over the potential loss of immunoreactivity, however, is sufficient to make iodine-125 an isotope of choice for immunoassay systems. Iodine may be readily substituted onto the aromatic side-chain of the amino acid tyrosine by mild oxidation using a variety of agents; chloramine T, the enzyme lactoperoxidase (EC 1.11.1.7) and the sparingly soluble agent 'iodogen' (1,3,4,6-tetrachloro-3,6-diphenyl-glycouril) have all been used successfully to yield a stable and efficient label. In applications where there is no suitable tyrosine residue available, a carrier molecule containing both a phenol or imidazole group suitable for iodination and an amine group suitable for coupling to a carboxylate group on the antigen may be used.

After iodination the label is usually purified to remove damaged antigen and unreacted iodine and this may be conveniently accomplished using either gel permeation chromatography or HPLC.

➤ Chromatography –
see Section 3.2.

Enzymes

Enzyme labels are usually associated with solid-phase antibodies in the technique known as enzyme-linked immunosorbent assay (ELISA). There are several variants of this technique employing both competitive and non-competitive systems. However it is best used in combination with two monoclonal antibodies in the 'two-site' format in which an excess of antibody is bound to a solid phase such as a test-tube or microtitre plate; the test antigen is then added and is largely sequestered by the antibody (Figure 7.12). After washing

1. Attachment of antibody to solid phase

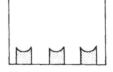

2. Wash

3. Incubate with sample containing antigen

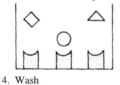

4. Wash

5. Incubate with antibody–enzyme conjugate

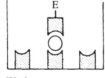

6. Wash

7. Incubate with enzyme substrate and measure product

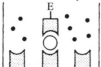

Figure 7.12
The two-site assay employing two monoclonal antibodies directed against two distinct epitopes.

to remove the remaining biological material a second, enzyme-labelled, monoclonal antibody is added which recognizes a different epitope on the antigen (the second site), thus sandwiching the antigen between the enzyme-labelled and the solid-phase antibodies. The excess enzyme–antibody is washed away before addition of a suitable substrate and development of a colour which is measured either kinetically or as an end-point reaction.

The enzymes commonly used as labels include alkaline phosphatase (EC 3.1.3.1), horseradish peroxidase (EC 1.11.1.7) and beta-galactosidase (EC 3.2.1.23). The enzymes used should be capable of being covalently linked to the antigen or antibody without loss of either catalytic activity or immunoreactivity. Glutaraldehyde may be used as a linking agent, while glycoprotein enzymes such as peroxidase may be linked via the carbohydrate group using periodic acid to form a reactive aldehyde group.

The choice of enzyme is governed by the availability of substrate and the type of detector. Several substrates have been developed for use with horseradish peroxidase: o-phenylene diamine (OPD), 2,2-azino-di(3-ethylbenzothiazoline-6-sulphonate) (ABTS) and 5,5′-tetramethylbenzidine hydrochloride (TMB).

> Enzymatic cycling – see Section 8.5.1.

Enzymes may also be used to amplify the signal from the label, using cycling systems. Amplification of the signal from the label has the potential to increase the sensitivity of the immunoassay system.

> Luminescence – see Sections 2.1.2 and 8.3.4.

Luminescent labels
There are several different types of luminescence differing only in the source of energy used to excite the molecules to a higher energy state; radioluminescence occurs when energy is supplied from high energy particles; in chemiluminescence, energy is derived from a chemical reaction; in bioluminescence the excitation is performed by a biological molecule such as an enzyme; and in fluorescence or photoluminescence the excitation is derived from light energy. Each has been successfully used as a label system for immunoassay.

In chemiluminescence immunoassay the antigen is tagged with a molecule such as luminol or an acridinium ester which emits light with a high quantum yield on oxidation. Alternatively, the antigen may be labelled with a bioluminescent molecule such as luciferin, which emits light when oxidized by the enzyme luciferase.

> Fluorescence – see Section 2.4.2.

Successful immunoassays have been developed using photoluminescent labels such as fluorescein and rhodamine but there are significant drawbacks to the use of these compounds. A background signal can be generated by compounds present in the sample which themselves fluoresce when excited by light of the same wavelength as the fluorophore employed as the label. In addition, many biological molecules will act to decrease the emitted light by absorbing the energy, a process known as quenching. Biological molecules can also scatter the exciting light energy in such a way as to reduce the efficiency of excitation. Each of these effects has contributed to the limited usefulness of fluorescent labels in immunoassay.

The problem of background fluorescence has been largely overcome with the introduction of time-resolved fluorimetry. This technique relies on the use of fluorophores with a long-lived fluorescence which can be measured after the background fluorescence has decayed. The usual decay time for background fluorescence is about 10 ns while that of long-lived fluorophores such as chelates of the lanthanide metal europium is of the order of 103–106 ns (Figure 7.13). Instead of continuous excitation of the fluorophore, the exciting light is pulsed and readings made after the background emissions have decayed. Once a measurement has been taken, the fluorophore is pulsed again allowing the accumulation of signal from the label. Assays based on this principle have been shown to be both as precise and as sensitive as radioisotopic assays.

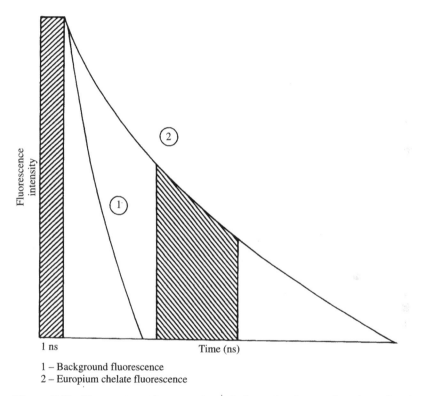

1 – Background fluorescence
2 – Europium chelate fluorescence

Figure 7.13 **Decay curve for europium chelates** showing reading time after the decay of background fluorescence.

Standard materials

The quantitation of antigen will ultimately rely on comparison of test antigen responses with those of a series of standard solutions, and the type of standard solution will depend upon the nature of the assay. The matrix of the standard solution should resemble the sample matrix (blood, urine, saliva, etc.) as closely as possible and should of course be antigen free. This may be difficult to achieve in practice and only reputable suppliers should be used.

Once the assay reagents have been optimized, a standard curve can be prepared. It is customary to express the binding of labelled antigen as a percentage of the binding of the zero standard (B/B_0), although this is by no means the only way of expressing the final result. The responses of the unknown samples may now be noted on the curve and their concentrations conveniently read off (Figure 7.14).

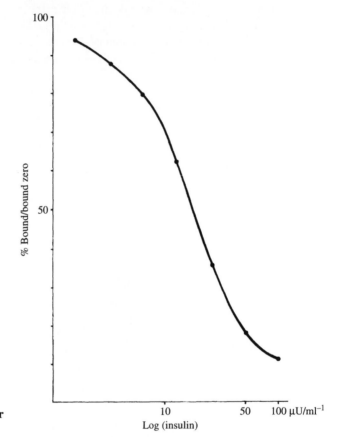

Figure 7.14
Typical standard curve for competitive insulin assay.

Separation of bound from free antigen

The whole quantitative basis of heterogeneous competitive immunoassay relies upon the physical separation of the bound and the free fractions in order that the relative proportions of each fraction may be assessed. Since the components are in solution, a variety of techniques have been developed in order to separate them (Table 7.4).

Some separation techniques rely on the physical removal of one of the fractions: charcoal will strongly adsorb the free fraction allowing its ready removal by centrifugation; the addition of dextran reduces the tendency of charcoal to 'strip' bound antigen from the complex; alternatively, the bound fraction may be precipitated by the addition of suitable concentrations of various protein precipitants such as alcohol, ammonium sulphate and polyethylene glycol (PEG).

Amongst other techniques are those involving the immunoprecipitation of the bound fraction using a second antibody, which reacts with the proteins of the first antibody. This second antibody may be produced by immunization:

Table 7.4 Techniques used to separate the bound and free fractions in immunoassay

Principle	Example
Adsorption of free fraction	Dextran-coated charcoal
Precipitation of bound fraction	Ethanol, ammonium sulphate, PEG
Immunological	Use of second antibody directed against the primary antibody species
Solid phase	Using antibodies coupled to plastic; tubes, beads, micro-plates
	Using antibodies coupled to magnetic particles
Specific binding	Use of other specific binding properties; staphylococcal protein A, avidin–biotin

> ➤ Double antibody techniques use a second antibody to bind a complex between the antigen and the primary antibody.

> ➤ Immobilized proteins – see Section 8.7.

immunoglobulins from the species in which the primary antibodies were raised are injected into a different species. Such a separation procedure is often called a double antibody technique. The concentration of the second antibody is adjusted such that the soluble bound fraction is converted into an insoluble matrix which can be easily separated from the free antigen by centrifugation. It is possible to accelerate the formation of the insoluble matrix by adding the second antibody in combination with a protein precipitant such as polyethylene glycol or ammonium sulphate (assisted second antibody technique).

Many solid-phase systems have been developed in which either the primary or secondary antibody is immobilized onto surfaces such as plastics. Antibodies may be attached to plastic surfaces both by simple adsorption at high pH and by covalent linkage. Various formats have been employed such as tubes, beads or microtitre plates. Once equilibrium has been reached the bound fraction is removed from the free by simple decantation of the tube contents, leaving the bound antigen attached to the solid phase. A novel use of this idea is the attachment of antibodies to magnetizable particles which, once the reaction is complete, allows the simple separation of bound from free fractions by placing the tubes above a strong magnetic field. When the particles have migrated to the bottom of the tube the free fraction may be decanted or aspirated away from the particles.

Other substances that exhibit specific binding may be used to separate the free and the bound fractions: when attached to a solid phase the ability of staphylococcal protein A to bind to the FC fragment of certain isotypes of IgG can be utilized; the strong binding of the vitamin biotin to tetravalent avidin may also be employed. Biotin may be readily incorporated into antibody molecules and these molecules may be subsequently captured by an avidin solid phase. Alternatively avidin may be used to provide a link between a biotinylated antibody and a biotinylated solid phase.

7.4.4 Homogeneous assays

Homogeneous assays require no separation step in order to observe the ratio between the labelled and unlabelled antigen and the antibody.

The most common of these systems is the enzyme-multiplied immuno-assay technique or EMIT, which is particularly suited to the measurement of small molecules (haptens) such as drugs. EMIT is a trade mark of the Syva Corporation of Palo Alto, California. Although it does not involve the separation of bound fraction from free it is nevertheless a competitive assay system. The antigen is labelled with an enzyme in such a way that the enzyme retains its catalytic activity. When the antigen binds to the antibody the enzyme becomes inhibited, probably by an induced conformational change or by steric hindrance of the enzyme active site (Figure 7.15).

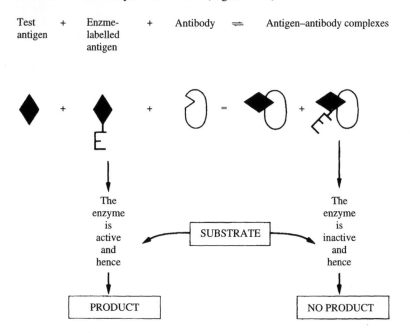

| Test antigen | + | Enzme-labelled antigen | + | Antibody | ⇌ | Antigen–antibody complexes |

Figure 7.15 **Enzyme-multiplied immunoassay (EMIT).** The three reactants, test (or standard) antigen, enzyme-labelled antigen and a limited amount of antibody are allowed to react and reach an equilibrium position. The unbound labelled antigen which remains is the only source of enzyme activity, the bound enzyme being inactivated. This free enzyme can be quantitated using a direct kinetic assay method and is proportional to the amount of unlabelled antigen originally present.

This inhibition will be reduced by the presence of free antigen in the test sample competing for the antibody-binding sites. The more antigen that is present, the more activity is retained and the more coloured product can be formed. The responses of the unknown are compared with the responses of standard doses of the antigen. The enzymes lysozyme (EC 3.2.1.17), malate dehydrogenase (EC 1.1.1.37) and glucose-6-phosphate dehydrogenase (EC 1.1.1.49) have all been used for this purpose. The dehydrogenase enzymes are particularly useful, owing to the additional ease of measuring their activity by monitoring the absorption of NADH at 340 nm.

Other successful homogeneous assays have been developed such as the fluorescence polarization technique made available by Abbott Laboratories of North Chicago, USA. This competitive immunoassay technique relies upon the principle that, in solution, small molecules generally rotate more quickly than larger molecules. The hapten label is fluorescent and excited by plane polarized light. If the hapten label is unbound then during the time between excitation and emission the molecule has rotated sufficiently to alter the plane of polarized light. Where the hapten label is bound, the larger complex does not rotate sufficiently during excitation and emission to alter the plane of polarized light and the emission may be measured. Thus the degree of binding of labelled hapten may be determined without separation of bound and free fractions. This technique is particularly well suited to the measurement of small molecules such as therapeutically administered drugs.

7.4.5 Membrane-based immunoassay devices

➤ Point-of-use devices are designed to be used in situations other than a laboratory.

➤ Dry chemistry analyser – see Section 6.1.2.

A particularly interesting advance in immunoassay technology has been the development of novel devices which contain all the necessary reagents, usually in a dry form. Many of these devices are designed for 'point-of-use' testing and are qualitative in nature, usually being assessed visually. Areas covered include pregnancy testing, drugs, microbiological antigens and environmental molecules such as pesticides and antibiotics. However a significant number are quantitative in design and involve dedicated instruments, some designed to be portable. These immunoassay devices are directly comparable to the dry chemistry systems described earlier.

The devices usually contain antibody immobilized on a membrane surface above a filter and absorbent pad (Figure 7.16). The sample is placed in the device and passes through the membrane into the absorbent pad. Where analyte is present in the sample it is sequestered by the antibody on the membrane and is then visualized by addition of a labelled second antibody. Labelled second antibody not bound to analyte also passes into the absorbent pad, sometimes with the aid of a wash solution. The labelled molecule may be an enzyme, which would then require addition of a suitable substrate, or a label such as colloidal gold, which has the virtue of being visible without the aid of a second reagent.

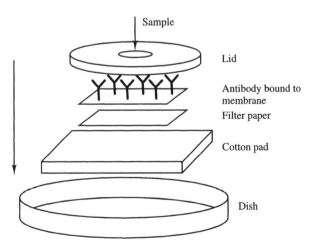

Figure 7.16 **Membrane-based device.** Antibody is bound to the membrane and removes any test antigen as the sample is drawn through into the absorbent pad. A second labelled antibody is then applied and, if antigen is trapped in the membrane, this second antigen will also be held in the membrane and can be demonstrated by means of the label.

This type of approach is essentially non-competitive and usually requires the use of two monoclonal antibodies directed against two distinct epitopes on the analyte. Other devices have employed a two-stage competitive system in which analyte and labelled analyte compete for antibody in one part of the device. This is followed by transfer of the equilibrium mixture to a separate part of the device where membrane-immobilized antibody removes the unbound labelled material and allows the bound to go through the membrane into the absorbent pad.

Self test questions

Section 7.4

1 Which of the following immunoassay techniques are competitive in nature?
   (a) Radioimmunoassay (RIA).
   (b) Enzyme-linked immunosorbent assay (ELISA).
   (c) Enzyme-multiplied immunoassay (EMIT).
   (d) Fluorescence polarization immunoassay (FPIA).
2 Separation of the free antigen from the bound antigen is achieved in different immunoassays by which of the following techniques?
   (a) Adsorption of the free antigen.
   (b) Immunoprecipitation of the free antigen.
   (c) Immunoaffinity chromatography.
   (d) Immobilized second antibodies.
3 Competitive binding immunoassays require a pure, labelled sample of the antigen
   BECAUSE
   the basis of competitive binding immunoassays lies in the competition between the antibody and the free label for unlabelled antigen.
4 The presence of a cross-reacting antigen in a sample will result in falsely increased test values when using competitive binding immunoassays
   BECAUSE
   a cross-reacting antigen will compete with test antigen for the available label.

7.5 Further reading

Kirkwood, E. and Lewis, C.J. (1991) *Understanding medical immunology*, 2nd edition, Wiley-Liss, USA.

Coleman, R.M. (1992) *Fundamental immunology*, 2nd edition, William C. Brown Co., USA.

Clausen, J. (1989) *Immunological techniques for the identification and estimation of macromolecules*, 3rd edition, Elsevier Science, Netherlands.

Goding, J.W. (1996) *Monoclonal antibodies: principles and practice*, 3rd edition, Academic Press, UK.

Price, C.P. and Newman, D.J. (eds) (1996) *Principles and practice of immunoassay*, 2nd edition, Grove Dictionaries Inc., USA.

Johnstone, A.P. and Thorpe, R. (1987) *Immunochemistry in practice*, 3rd edition, Blackwell Science, UK.

8 Enzymes

Enzymes occupy an important place in analytical biochemistry and many investigations require their detection and quantitation. Studies of the enzyme content of blood plasma are particularly useful in clinical biochemistry both in the monitoring of normal metabolic processes and in the detection of abnormal levels of enzyme production or release. Enzyme assays also provide convenient methods for assessing the quality of foodstuffs and checking the efficiency of sterilization and pasteurization processes.

Enzymes are valuable analytical tools and offer sensitive and specific methods of quantitation for many substances. The increasing availability of highly purified enzyme preparations, both in solution and in immobilized forms, permits the development of a wide range of methods.

Because of the difficulties in measuring the amount of enzyme in the conventional units of mass or molar concentration, the accepted unit of enzyme activity is defined in terms of reaction rate. The **International Unit (IU)** is defined as that amount of enzyme which will result in the conversion of 1 μmol of substrate to product in 1 minute under specified conditions. The SI unit of activity, which is becoming more acceptable, is the **katal** and is defined as that amount of enzyme which will result in the conversion of 1 mol of substrate to product in 1 second. A convenient sub-unit is the nanokatal, which is equal to 0.06 International Units.

The **specific activity** of an enzyme preparation is expressed as the catalytic activity per milligram of protein (units mg^{-1} protein) and is a convenient

➤ An International Unit (IU) is the amount of enzyme that will convert 1 μmol of substrate to product in 1 minute.
➤ A katal is the amount of enzyme that will convert 1 mole of substrate to product in 1 second.
➤ Specific activity is the enzyme activity per milligram of protein.

way of comparing the purity of enzyme preparations. The turnover number expresses the catalytic activity in terms of units per mole of pure enzyme rather than milligrams of protein. This allows direct comparison of the catalytic activity of different enzymes, i.e. to say that one enzyme is X times more active than another. It can only be used if the relative molecular mass of the enzyme and protein content of the sample are known and the sample is known to be a pure preparation.

➤ Turnover number is the enzyme activity per mole of pure enzyme.

Enzymes are classified under six main headings, which relate to the chemical reactions involved. Each class is then subdivided in order to classify the other features of the reaction. The main headings are listed below but further details may be found in the Appendix.

1. **Oxidoreductases**. Enzymes that catalyse the transfer of hydrogen or oxygen atoms or electrons.
2. **Transferases**. Enzymes that catalyse the transfer of specific groupings.
3. **Hydrolases**. Enzymes that catalyse hydrolytic reactions.
4. **Lyases**. Enzymes that catalyse the cleaving of bonds by reactions other than hydrolysis.
5. **Isomerases**. Enzymes that catalyse intramolecular rearrangements.
6. **Ligases**. Enzymes that catalyse the formation of bonds and require ATP.

Each enzyme has been given a four-digit number by the Enzyme Commission of the International Union of Biochemistry. The first three digits relate to the reaction catalysed by the enzyme and the final one is required if several enzymes with different protein structures catalyse the same reaction.

Name of enzyme (EC W.X.Y.Z)

EC – Enzyme Commission number system
W – indicates the reaction catalysed (1−6)
X – indicates the general substrate or group involved
Y – indicates the specific substrate or coenzyme
Z – the serial number of the enzyme.

| 8.1 | The nature of enzymes |

All enzymes are proteins although many are conjugated proteins and are associated with non-protein groups. Their catalytic activity depends on the maintenance of their native structure and slight variations may result in significant changes in this activity.

A common feature of enzymes is the presence of a cleft or depression in the structure which is lined with mainly hydrophobic amino acid residues and into which the substrate fits. Certain amino acid residues which are concerned with either the orientation of the substrate, and hence the specificity of the enzyme, or are involved in the catalysis of the reaction, are located in this cleft. Those amino acid residues that are associated with the latter role form the active site of the enzyme and are often located towards the base of this cleft. In most cases they are ionic or reactive and include histidine, lysine, cysteine and serine as well as glutamic and aspartic acids. In addition, the binding of ions from the solution, particularly cations, may also aid either the location of the substrate or the catalysis of the reaction.

8.1.1 Factors affecting enzyme activity

Temperature

An increase in temperature increases the rate of all chemical reactions, including those catalysed by enzymes, but it also increases the rate of denaturation of enzyme protein (Figure 8.1). Denaturation also occurs more readily with pure solutions of enzymes than with impure.

It is sometimes suggested that enzymes show an optimum operating temperature but the most suitable temperature for a particular reaction is a compromise between maximal activity for a short period of time and a falling activity due to denaturation for a longer period of time. Many assays are performed at 37°C either because it is mistakenly assumed that the human body temperature is the optimum temperature for an enzyme or, more realistically, because above this temperature the rate of inactivation of the enzyme becomes far more significant. The International Union of Biochemistry originally recommended that 25°C should be regarded as a standard temperature but subsequently raised it to 30°C because of difficulties in keeping to the lower temperature in hot climates. It is possible to convert enzyme activities quoted for one temperature to equivalent activities at another temperature by using predetermined conversion factors, but the validity of such methods is questionable. Currently no standard temperature is specified but it is recommended that the assay temperature should be quoted in all references to enzyme activities.

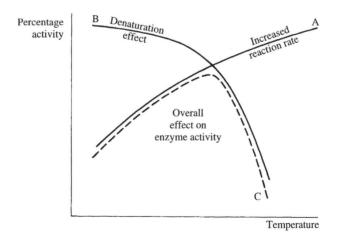

Figure 8.1 **The effect of temperature on enzyme-catalysed reactions.** The velocity of a chemical reaction increases with increasing temperature (A) but because of the increasing denaturation of the protein, the proportion of active enzyme falls (B). These two processes result in the characteristic temperature profile of an enzyme (C).

pH

Enzymes are very sensitive to changes in pH and function best over a very limited range with a definite pH optimum. The effects of pH are due to changes in the ionic state both of the amino acid residues of the enzyme and of the substrate molecules. These alterations in charge will affect substrate binding and the resulting rate of reaction. Over a narrow pH range, these effects will be reversible but extremes of acidity or alkalinity often cause serious distortion of protein structure and result in permanent denaturation (Figure 8.2). It is important to appreciate that the pH optimum of the enzyme reaction may differ for different substrates. Thus a quoted value may not necessarily hold true for every assay method and this should always be determined when designing enzyme assays.

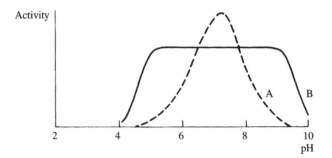

Figure 8.2   **The effect of pH on the enzyme lactate dehydrogenase** (EC 1.1.1.27). The enzyme shows maximum activity at pH 7.4 (A). When stored in buffer solutions with differing pH values for 1 h before re-assaying at pH 7.4, it shows complete recovery of activity from pH values between 5 and 9 but permanent inactivation outside these limits (B).

Substrate concentration

Experimental studies on the effect of substrate concentration on the activity of an enzyme show consistent results. At low concentrations of substrate the rate of reaction increases as the concentration increases. At higher concentrations the rate begins to level out and eventually becomes almost constant, regardless of any further increase in substrate concentration. The choice of substrate concentration is an important consideration in the design of enzyme assays and an understanding of the kinetics of enzyme-catalysed reactions is needed in order to develop valid methods.

8.1.2   Kinetics of enzyme-catalysed reactions

The law of mass action states that the rate of a chemical reaction is proportional to the product of the concentrations of the reactants. This means that the rate of a reaction which has a single component will increase in direct relation to the increasing concentration but for a two-component reaction it will increase in proportion to the square of the concentration. These relationships may be expressed by the following equations:

$$\text{Rate} = k_1 \text{(concentration)} \qquad \text{(single reactant)}$$
$$\text{Rate} = k_2 \text{(concentration)} \times \text{(concentration)} \quad \text{(two reactants)}$$

where k_1 and k_2 are the reaction velocity constants or rate constants for the reactions. The reactions are said to show first- and second-order kinetics respectively. Occasionally situations arise where increases in the concentration of a reactant do not result in an increase in the reaction rate. Such reactions are said to show zero-order kinetics.

The effect of increasing the concentration of the substrate (Figure 8.3) can be explained most satisfactorily by the formation of an enzyme–substrate complex as a key stage in the reaction. It is the breakdown of this single component, the ES complex, which results in the formation of the products as illustrated by equation [1] and hence first-order kinetics apply.

$$\text{E} + \text{S} \rightleftharpoons \text{ES} \rightleftharpoons \text{E} + \text{P} \qquad\qquad [1]$$

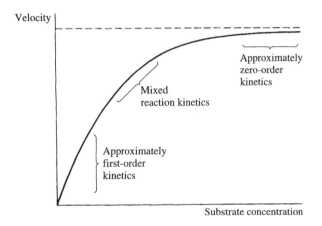

Figure 8.3 **The effect of substrate concentration on the activity of an enzyme.** At low substrate concentrations, the rate of reaction resulting from a fixed amount of enzyme is proportional to the concentration of the substrate. However, at high concentrations, the reaction is almost constant and independent of the substrate concentration. The range of substrate concentrations between these two extremes results in mixed reaction kinetics.

Michaelis–Menten equation

The concept of an enzyme–substrate complex is fundamental to the appreciation of enzyme reactions and was initially developed in 1913 by Michaelis and Menten, who derived an equation that is crucial to enzyme studies. Subsequent to Michaelis and Menten several other workers approached the problem from different viewpoints and although their work is particularly useful in advanced kinetic and mechanistic studies, they confirmed the basic concepts of Michaelis and Menten.

The ES complex is formed when the enzyme and substrate combine but it may also be formed by the combination of the enzyme and product. The complex normally breaks down to form free enzyme and product but it may also revert to the free enzyme and substrate. These four reactions are summarized by equation [2]:

$$E + S \underset{k_2}{\overset{k_1}{\rightleftharpoons}} ES \underset{k_4}{\overset{k_3}{\rightleftharpoons}} E + P \qquad [2]$$

The dissociation constant of the complicated equilibrium involving the enzyme–substrate complex is known as the Michaelis constant, K_m, and involves the rate constants for each of the four reactions involving the ES complex:

$$K_m = \frac{k_2 + k_3}{k_1 + k_4} \qquad [3]$$

When concentration of the complex is constant, the rate of formation of the complex is balanced by the rate of disappearance of the complex. If measurements of the reaction rate for any substrate concentration are made before the product is present in appreciable amounts the rate of the reverse reaction will be negligible and the rate constant k_4 can be ignored. Under these conditions:

$$K_m = \frac{k_2 + k_3}{k_1} \qquad [4]$$

It is extremely difficult to determine the relative amounts of free and complexed enzyme but it is possible to measure the total activity of the enzyme. The proportion of free enzyme can be represented as the difference between the total enzyme (E) and that complexed with substrate (ES) Hence:

Rate of ES formation $= k_1$ [E − ES] [S]
Rate of ES removal $= k_2$ [ES] $+ k_3$ [ES]

where the square brackets indicate the concentration of the reactant. Therefore:

$$\frac{k_2 + k_3}{k_1} = \frac{[E - ES][S]}{[ES]} = K_m$$

Rearrangement of the equation results in:

$$K_m = \frac{[E][S]}{[ES]} - \frac{[ES][S]}{[ES]}$$

$$K_m = \frac{[E][S]}{[ES]} - [S]$$

$$[ES] = \frac{[E][S]}{K_m + [S]} \qquad [5]$$

By definition, the observable rate of product formation (v) is proportional to the concentration of the enzyme substrate complex, ES:

$$v = k_3 [ES]$$

Therefore substituting in equation [5]:

$$v = \frac{k_3 [E][S]}{K_m + [S]} \qquad [6]$$

The true concentration of enzyme is difficult to measure especially in terms of molar concentration but if the substrate concentration is large compared with that of the enzyme, all of the enzyme will be present as the ES complex and the reaction will proceed at maximum velocity. Under these conditions of excess substrate and maximum velocity (V_{max}):

$$V_{max} = k_3 [E]$$

➤ The Michaelis equation describes the relationship between substrate concentration and the rate of an enzyme-catalysed reaction.
➤ Michaelis constant (K_m) is a measure of the catalytic effectiveness of an enzyme.

Substituting in equation [6] results in the common form of the Michaelis equation:

$$v = \frac{V_{max} \times [S]}{K_m + [S]} \qquad [7]$$

The equation gives a measure of the Michaelis constant (K_m) in terms of the measured velocity of the reaction (v) which results from a substrate concentration ([S]) and the maximum velocity (V_{max}) which can be achieved using very high concentrations of substrate.

The value for the maximum velocity is related to the amount of enzyme used but the Michaelis constant is peculiar to the enzyme and is a measure of the activity of the enzyme. Enzymes with large values for K_m show a reluctance to dissociate from the substrate and hence are often less active than enzymes with low K_m values. The substrate concentration required for a particular enzyme assay is related to K_m and when developing an assay, the value for K_m should be determined.

Determination of the Michaelis constant

The reaction velocity produced by a fixed amount of enzyme with varying concentrations of substrate is determined and a plot of the two variables shows a characteristic shape (Figure 8.4). The substrate concentration that results in

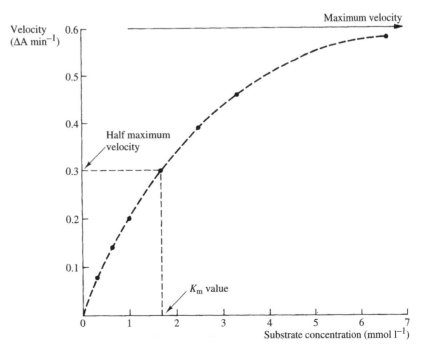

Figure 8.4
Determination of the Michaelis constant (K_m).

a velocity that is half the maximum velocity is numerically equal to the Michaelis constant:

$$\frac{V_{max}}{2} = \frac{V_{max} \times [S]}{K_m + [S]}$$

$$[S] = K_m$$

While this method is extremely simple it is also experimentally inaccurate. Because of the hyperbolic nature of the relationship, the curve approaches maximum velocity asymptotically making the deduction of a value for V_{max} difficult. Any error in assessing this value will be reflected in the value ascribed to K_m.

Lineweaver and **Burk** (1934) described a method for the determination of K_m which uses the reciprocal form of the Michaelis equation converting it to a linear relationship (Procedure 8.1):

> ➤ The Lineweaver–Burk equation uses the reciprocal form of the Michaelis equation to give a linear relationship with intercepts of $1/K_m$ and $1/V_{max}$.

$$\frac{1}{v} = \frac{K_m}{V_{max}} \times \frac{1}{[S]} + \frac{1}{V_{max}} \qquad [8]$$

A plot of the reciprocal of velocity against reciprocal of the substrate concentration gives a straight line graph with intercepts of

$$\frac{1}{[S]} = -\frac{1}{K_m}$$

and

$$\frac{1}{v} = \frac{1}{V_{max}}$$

An alternative method known as the **Hofstee** plot uses the Michaelis equation in the form

> ➤ The Hofstee equation uses another reciprocal form of the Michaelis equation to give a linear relationship with intercepts at V_{max} / K_m and V_{max}.

$$v = V_{max} - K_m \frac{v}{[S]}$$

in which v is plotted against $v/[S]$ and gives intercepts at V_{max} and V_{max}/K_m.

Both of these methods are extremely useful because the linear relationships make graphical treatment and also the statistical and subsequent computer handling of the data simple. The Lineweaver–Burk method is more frequently used, although it does suffer from the disadvantage that the experimental values which are the least precise (i.e. those involving very low substrate concentrations and resulting in very low velocities) are the furthest from the origin and hence tend to exert the most influence on the graph unless a weighted linear regression analysis is used. It is for this reason that the Hofstee plot is felt by some to be a more reliable method.

8.1.3   Enzyme mechanisms

The initial steps in enzyme-catalysed reactions involve the binding of the reactants to the enzyme surface and one of the functions of the enzyme is to orientate these reactants relative to each other. This idea was suggested by Fischer as a 'lock-and-key' hypothesis, where the enzyme is the lock and the

Procedure 8.1: Determination of the Michaelis constant (K_m) for the enzyme D-amino acid oxidase (EC 1.4.3.3) using the Lineweaver–Burk method

Data

Final substrate concentration (mmol l^{-1})	Rate of reaction (ΔA_{340} min^{-1})
0.33	0.08
0.66	0.14
1.00	0.20
1.66	0.30
2.50	0.39
3.33	0.46
6.66	0.58

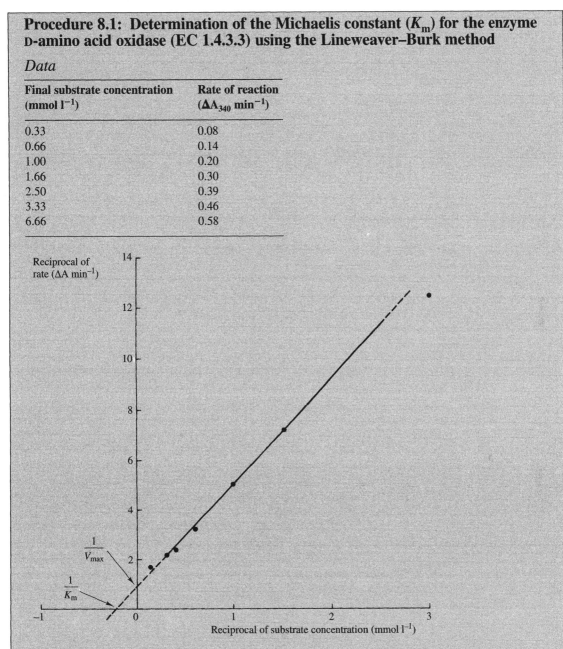

Method
Plot the reciprocal of the reaction velocity against the reciprocal of the substrate concentration.

Calculation
The values for the Michaelis constant (K_m) and the maximum velocity are calculated from the intercepts on each of the axis which give the reciprocals of the K_m and the V_{max}.

Results
Michaelis constant (K_m) is 4.4 mmol l^{-1}
Maximum velocity is 1.1 absorbance units per minute

substrate is the key. Although this explained the idea of the specificity of an enzyme, the idea of a rigid protein structure was difficult to accept. Koshland proposed an alternative 'induced-fit' hypothesis which says that the binding of the substrate causes alteration in the geometry of the enzyme which results in the correct orientation of the appropriate groups in both the substrate and the enzyme.

Procedure 8.2: Determination of the Michaelis constant for the enzyme D-amino acid oxidase (EC 1.4.3.3) using the Hofstee method

Data
The same as for Procedure 8.1.

Method
Plot the velocity at the specified substrate concentration against the value, $V/[S]$.

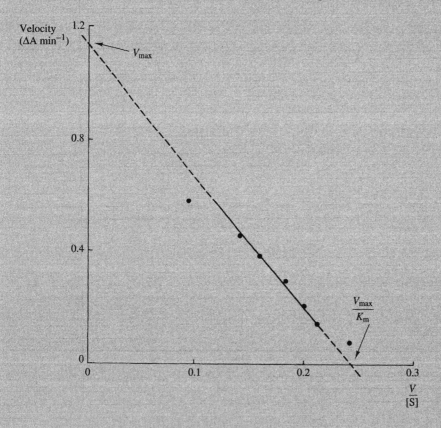

Calculation
The values for the Michaelis constant (K_m) and the maximum velocity are calculated from the intercepts on each axis, which give the value for V_{max} and the fraction V_{max}/K_m.

Results
   Michaelis constant (K_m) is 4.7 mmol l^{-1}
   Maximum velocity is 1.14 absorbance units per minute

Another hypothesis suggests that the binding of a substrate to an enzyme causes a strain or deformation of some of the bonds in the substrate molecule, which are subsequently broken. The effectiveness of this mechanism depends upon the strength of the binding force and does not necessarily involve any movement of the protein but suggests the idea of a flexible enzyme.

Many enzymic reactions are a consequence of either nucleophilic or electrophilic attack on the substrate, the former being the more common. If such a species approaches another group it will tend to push electrons away from the 'positive centre'. If the reaction to be catalysed is

> ➤ A nucleophile is a 'positive-centre seeking' species capable of donating electrons.

$$A—X \rightarrow A^+ + X^-$$

then an approach by a nucleophile (C^-, the catalyst) will make the reaction more feasible:

$$C^- \frown A—X \rightarrow C—A + X^-$$

> ➤ An electrophile is a 'negative-centre seeking' species capable of accepting electrons.

When a bond is broken the nature of the leaving group is important in determining the energy of activation. The basicity of the group is related to its nucleophilicity and if the basic nature of the leaving group is diminished so also will be the tendency to re-form the broken bond:

$$H^+ + A—X \rightarrow A—X—H^+$$

$$C^- \frown A—X—H^+ \rightarrow C—A + H^+X$$

This is an example of acid catalysis and the effect is to pull electrons away from the leaving group. Often both acid catalysis and nucleophilic attack are involved in enzyme-catalysed reactions in what are known as 'push–pull' mechanisms.

8.1.4 Activation of enzymes

> ➤ A prosthetic group is a non-protein moiety which is either covalently or ionically linked to a protein molecule.
> ➤ An apoenzyme is the part of a conjugated enzyme without the prosthetic group.

Many enzymes require additional substances in order to function effectively. Conjugated enzymes require a prosthetic group before they are catalytically active, such groups being covalently or ionically linked to the protein molecule and remaining unaltered at the end of the reaction. Catalase (EC 1.11.1.6), for instance, contains a haem group while ascorbate oxidase (EC 1.10.3.3) contains a copper atom.

The purification of some enzymes inactivates them because substances essential for their activity but not classed as a prosthetic group are removed. These are frequently inorganic ions which are not explicit participants in the reaction. Anionic activation seems to be non-specific and different anions are often effective. Amylase (EC 3.2.1.1), for example, is activated by a variety of anions, notably chloride. Cationic activation is more specific, e.g. magnesium is particularly important in reactions involving ATP and ADP as substrates. In cationic activation it seems very likely that the cation binds initially to the substrate rather than to the enzyme.

In addition to these activators there is a group of substances which are not simple activators but are often essential co-substrates for many different enzymes; the most common examples are given in Table 8.1.

Table 8.1   Cofactors as transporting systems

Cofactor	Group transported	Type of reaction
NAD$^+$, NADP$^+$	Redox carrier	Oxidation
Flavins	Redox carrier	Oxidation
Cytochromes	Redox carrier	Oxidation
AMP, ADP	Phosphate	Energy transfer
Pyridoxal phosphate	Amino	Transamination
UDP	Glycosyl	Carbohydrate metabolism
Coenzyme A	Acyl	Lipid metabolism

8.1.5   Inhibition of enzymes

A substance that decreases the rate of an enzyme-catalysed reaction is known as an inhibitor and its effects may be permanent or transient. The inhibition of some reactions by substances which may be products of either that reaction or a subsequent reaction provides a control mechanism for cellular metabolism, while the selective inhibition of enzymes is the basis of many aspects of pharmacology and chemotherapy.

Some inhibitors structurally resemble the substrate and are bound by the enzymes but cannot be converted to the products. Because the formation of the complex between the enzyme and the inhibitor is a reversible reaction, the inhibitor can be displaced by high concentrations of the normal substrate. The

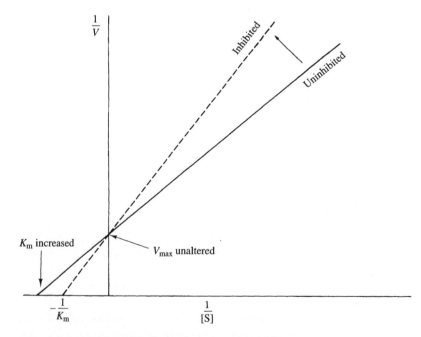

Figure 8.5   **The kinetic effects of a competitive inhibitor.** The effects of a competitive inhibitor can be reversed by high concentrations of substrate and the increased amount required to achieve maximum velocity results in an increased value for the Michaelis constant.

inhibition causes an increase in the K_m value for the enzyme but the maximum velocity for the reaction remains unaltered (Figure 8.5). Such substances are known as **competitive inhibitors** and their effects can be reduced and even eliminated by high concentrations of substrate. The classical example of competitive inhibition is the action of the sulphonamide antibiotic drugs due to their structural similarity to the natural substrate *p*-aminobenzoic acid (Figure 8.6).

> ➤ A competitive inhibitor competes with the substrate for the enzyme's active site.

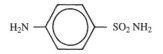

p-Aminobenzoic acid

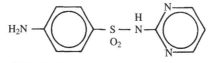

Sulphanilamide

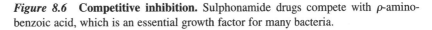

Sulphadiazine

Figure 8.6 **Competitive inhibition.** Sulphonamide drugs compete with *p*-amino-benzoic acid, which is an essential growth factor for many bacteria.

Another type of inhibitor combines with the enzyme at a site which is often different from the substrate-binding site and as a result will inhibit the formation of the product by the breakdown of the normal enzyme–substrate complex. Such **non-competitive inhibition** is not reversed by the addition of excess substrate and generally the inhibitor shows no structural similarity to the substrate. Kinetic studies reveal a reduced value for the maximum activity of the enzyme but an unaltered value for the Michaelis constant (Figure 8.7). There are many examples of non-competitive inhibitors, many of which are regarded as poisons because of the crucial role of the inhibited enzyme. Cyanide ions, for instance, inhibit any enzyme in which either an iron or copper ion is part of the active site or prosthetic group, e.g. cytochrome *c* oxidase (EC 1.9.3.1).

> ➤ A non-competitive inhibitor binds to the enzyme even in the presence of the substrate.

Not all inhibitors fall into either of these two classes but some show much more complex effects. An **uncompetitive inhibitor** is defined as one that results in a parallel decrease in the maximum velocity and the K_m value (Figure 8.8). The basic mode of action of such an inhibitor is to bind only to the enzyme–substrate complex and not to the free enzyme and so it reduces the rate of formation of products. Alkaline phosphatase (EC 3.1.3.1) extracted from rat intestine is inhibited by L-phenylalanine in such a manner.

> ➤ An uncompetitive inhibitor binds to the ES complex rather than the free enzyme.

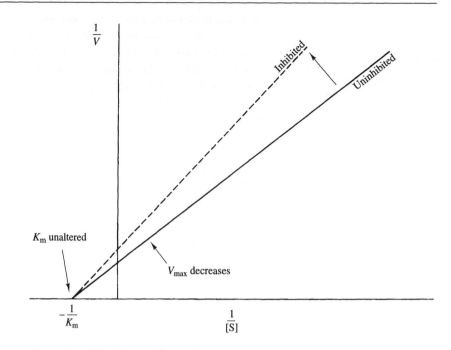

Figure 8.7   The kinetic effects of a non-competitive inhibitor. The effect of a non-competitive inhibitor is not reversed by high concentrations of substrate and the enzyme reaction shows a reduced value for the maximum velocity. The enzyme remaining is unaltered and gives the same value for the Michaelis constant as originally shown by the uninhibited enzyme.

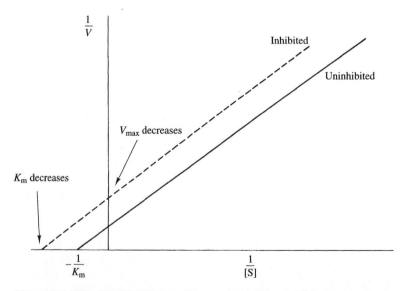

Figure 8.8   The kinetic effects of an uncompetitive inhibitor. The effects of an uncompetitive inhibitor are very complex and the reaction usually shows a parallel decrease in both the maximum velocity and the Michaelis constant.

Another different class of inhibitors binds covalently to specific amino acids in the enzyme and these are referred to as **irreversible inhibitors**. The organophosphorus compounds, of which nerve gases are examples, inactivate enzymes which rely on the hydroxyl group of serine residues for their activity, e.g. cholinesterase (EC 3.1.1.8).

8.1.6 Allostery

➤ Allosteric enzymes have sites other than the catalytic or active site which are associated with the activation and inhibition of the enzyme.

Allosteric enzymes show various activation and inhibition effects which are competitive in nature and related to conformational changes in the structure of the enzyme. Such allosteric enzymes are often crucial enzymes in metabolic pathways and exert control over the whole sequence of reactions. The name allostery refers to the fact that inhibition of the enzyme is by substances that are not similar in shape to the substrate.

The basic concept of allosteric enzymes initially proposed by Monod is an extension of Koshland's induced-fit theory. An allosteric enzyme in order to be catalytically active must be in a conformation that allows the binding of the substrate. Its structure is normally flexible and may only be stabilized by the binding of other molecules. Hence the binding of an activator at the activator site locks the enzyme in a form in which the substrate-binding site is available. The binding of an inhibitor at a separate inhibitor site causes distortion of both the substrate binding site and the activator site (Figure 8.9). 6-Phosphofructokinase (EC 2.7.1.11), a key enzyme in the glycolytic pathway, is inhibited by ATP and citrate but is activated by AMP, ADP, inorganic phosphate and the normal substrate, fructose-6-phosphate. These effects mean that its activity is reduced when the concentration of high

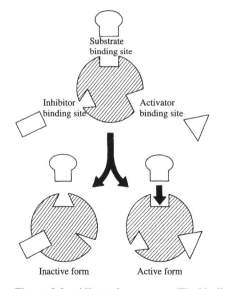

Figure 8.9   Allosteric enzymes. The binding of an activator stabilizes the enzyme in an active form while the binding of an inhibitor distorts the active site, causing a loss of activity.

energy compounds is high or there is an adequate source of energy from the citric acid cycle. Conversely, the enzyme is activated by compounds whose presence in increased amounts suggests a lack of high energy compounds in the cell.

8.1.7   Iso-enzymes

> ➤ Heteroenzymes catalyse the same reaction as each other but they are derived from different sources.
> ➤ Iso-enzymes are different molecular forms of an enzyme and are from the same organism.

Enzymes that perform the same catalytic function are known as homologous enzymes and fall into two classes. Heteroenzymes are derived from different sources and although they catalyse the same reaction they show different physical and kinetic characteristics. The hydrolytic enzyme α-amylase (EC 3.2.1.1) is found in the pancreatic secretion in man and is different from the enzymes of the same name which are derived from bacteria or malt. Iso-enzymes, sometimes referred to as isozymes, are different molecular forms of the same enzyme and are found in the same animal or organism although they often show a pattern of distribution between tissues.

Many iso-enzymes are hybrids of a limited number of sub-units. Some enzymes show multiple forms owing to increasing levels of polymerization of a single sub-unit; these should not really be called iso-enzymes, because they do not have any genetic difference between the different forms. Cholinesterase, for instance, shows five forms consisting of a single sub-unit existing as monomers, dimers, trimers, tetramers and pentamers.

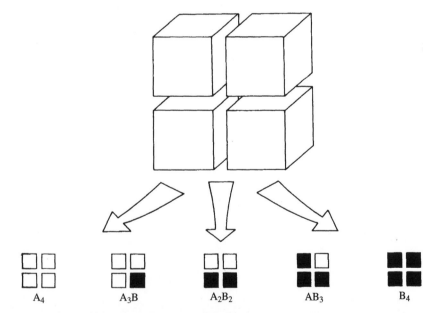

Figure 8.10   **The quaternary structure of proteins.** The enzyme lactate dehydrogenase (EC 1.1.1.27) has a relative molecular mass of approximately 140 000 and occurs as a tetramer produced by the association of two different globular proteins (A and B), a characteristic that results in five different hybrid forms of the active enzyme. The A and B peptides are enzymically inactive and are often indicated by M (muscle) and H (heart). The A_4 tetramer predominates in skeletal muscle while the B_4 form predominates in heart muscle but all tissues show most types in varying amounts.

Lactate dehydrogenase (EC 1.1.1.27) shows a classical iso-enzyme pattern. In man it is composed of two different protein sub-units known as A and B arranged into tetramers (Figure 8.10) giving five different hybrid forms of the two basic units. The A and B sub-units are enzymically inactive but all the tetramers are active and because of their different composition can be separated from each other by electrophoresis or ion-exchange chromatography. They also show different catalytic properties, e.g. substrate specificity and inhibition kinetics, and these properties often provide the basis for relatively specific iso-enzyme assays. Enzymes with iso-enzyme variants often show a characteristic distribution of the iso-enzymes in different tissues and the B_4 form of lactate dehydrogenase is found mainly in heart muscle and the A_4 mainly in skeletal muscle.

While iso-enzymes show regular variations in their molecular structure, it is the difference in their catalytic activity that is most significant. They often show different inhibition and activation effects, permitting one iso-enzyme to function under conditions that would reduce the activity of another and it is likely that such variations aid the control of the same reaction under different cellular or tissue conditions.

Self test questions

Section 8.1

1. The following will always cause a reduction in the activity of an enzyme. Select true or false for each of the statements:
   (a) A reduction in temperature.
   (b) A reduction in substrate concentration.
   (c) An increase in product concentration.
   (d) An increase in pH.
2. What is high catalytic activity of an enzyme suggested by?
   (a) A low specific activity.
   (b) A low Michaelis constant (K_m).
   (c) A low turnover number.
   (d) A low maximum velocity (V_{max}).
3. At low concentrations of substrate, any slight increase in concentration will result in an increase in enzyme activity
   BECAUSE
   at low substrate concentrations not all the available enzyme is incorporated into the enzyme–substrate complex.
4. The effect of a competitive inhibitor can be reversed by increasing the substrate concentration
   BECAUSE
   a competitive inhibitor normally shows no structural similarity to the substrate.

8.2   Enzyme assay methods

Enzyme assays in which either the substrate or the product of the test enzymes is measured are known as direct assays. For many enzymes direct assays are either not convenient or not feasible, and coupled assays are used.

The reaction mixture for a coupled assay includes the substrates for the initial or test enzyme and also the additional enzymes and reagents necessary to convert the product of the first reaction into a detectable product of the final reaction. The enzyme aspartate aminotransferase (EC 2.6.1.1), for instance, results in the formation of oxaloacetate, which can be converted to malic acid by the enzyme malate dehydrogenase (EC 1.1.1.37) with the simultaneous conversion of NADH to NAD$^+$, a reaction which can be followed spectrophotometrically at 340 nm:

$$\text{aspartate } + \text{ oxoglutarate } \xrightarrow{\substack{\text{aspartate} \\ \text{aminotransferase}}} \text{glutamate } + \text{ oxaloacetate}$$

$$\text{oxaloacetate } + \text{ NADH } \xrightarrow{\text{malate dehydrogenase}} \text{malate } + \text{ NAD}^+$$

In this example, malate dehydrogenase is known as the indicator enzyme.

Many assays have been described in which the initial product forms the substrate of an intermediary reaction involving auxiliary enzymes. The assay of creatine kinase (EC 2.7.3.2), for example, involves hexokinase (EC 2.7.1.1) as the auxiliary enzyme and glucose-6-phosphate dehydrogenase (EC 1.1.1.49) as the indicator enzyme:

> ➤ An indicator enzyme is the enzyme that catalyses the final measurable reaction in a coupled enzyme assay.
> ➤ An auxiliary enzyme is the enzyme or enzymes that catalyse the intermediary reactions in a coupled enzyme assay.

$$\text{creatine phosphate } + \text{ ADP } \xrightarrow{\text{creatine kinase}} \text{creatine } + \text{ ATP}$$

$$\text{ATP } + \text{ glucose } \xrightarrow{\text{hexokinase}} \text{ADP } + \text{ glucose-6-phosphate}$$

$$\text{glucose-6-phosphate } + \text{ NADP}^+ \xrightarrow{\substack{\text{glucose-6-phosphate} \\ \text{dehydrogenase}}}$$
$$\text{6-phosphogluconate } + \text{ NADPH}$$

8.2.1 Optimized assays

Prior to any experimental work in developing an assay, information about the enzymes should be obtained from the literature, such as pH optima, activators, inhibitors and K_m values. It is important that all such information is confirmed experimentally using the selected method.

The pH optimum of an enzyme will often vary from one substrate to another and must be determined for each substrate. The buffer system used will often affect the overall activity of an enzyme and may alter its pH optimum. In general, the amino buffers such as glycylglycine and tricine, etc. (Table 8.2) result in a greater enzyme activity than do the simple inorganic buffers such as phosphate and carbonate. Buffers are most effective over a narrow pH range (approximately two units) which centres on their pK_a value. Those buffers with pK_a values similar to the known optimum pH of the enzyme should be tested for their effect on the activity of the enzyme over a limited pH range.

> ➤ A buffer system is a mixture of a weak acid and its conjugate base and resists pH change.

> ➤ pK_a – see Section 10.1.3.

It is difficult to determine the optimum pH of each enzyme when designing a coupled assay because of the presence of other enzymes and it is necessary to use a direct assay technique for each one. The effect of buffers, activators and inhibitors must also be checked using the same direct method.

The selection of the overall assay pH may not be easy because of the differences in optima between the various enzymes. In general, it is advisable to select the pH optimum of the test enzyme in order to achieve maximum sensitivity and the proportionate loss of activity for each of the additional enzymes can be determined from their pH profiles (Figure 8.2) and compensated for by adding more of the enzyme to the assay mixture.

Table 8.2   Buffers

Common name	Base compound	pK_a
ACES	Aminosulphonic acid	6.8
ADA	Iminoacetic acid	6.6
BES	Aminosulphonic acid	7.1
BICINE	Hydroxyglycine	8.3
CAPS	Aminosulphonic acid	10.4
EPPS	Piperazinesulphonic acid	8.0
GG	Glycylglycine	8.2
HEPES	Piperazinesulphonic acid	7.5
MES	Morpholinosulphonic acid	6.1
MOPS	Morpholinosulphonic acid	7.2
TAPS	Aminosulphonic acid	8.4
TES	Aminosulphonic acid	7.5
TRICINE	Hydroxyglycine	8.1
Tris	Hydroxyaminomethane	8.1

It is essential to identify the need for any activators and to determine their optimal concentrations. In many coupled assays it is not uncommon to find that an activator for one enzyme can inhibit another. The selected optimum concentration of all activators must be checked against all the other enzymes for any possible inhibition effects. Problems can often be solved by careful control of concentrations and by increasing the amounts of any inhibited supplementary enzymes.

To achieve maximum velocity, a substrate concentration which is at least ten times greater than the K_m value for the enzyme should be used. Although maximum velocity is only theoretically achieved at an infinite substrate concentration, it is possible using the Michaelis–Menten equation to calculate the percentage of maximum velocity given by any concentration of substrate. For a substrate concentration of ten times greater than the K_m value the velocity (v) achieved will be:

$$v = \frac{V_{max} \times 10}{1 + 10}$$

$$v = \frac{10}{11} = 91\% \text{ maximum velocity}$$

While it is desirable to use as high a concentration as possible, some enzymes are subject to substrate inhibition and give a characteristic

Lineweaver–Burk plot (Figure 8.11). In such cases the chosen substrate concentration must give the highest reaction velocity possible. It is important, when describing any enzyme assay, to report the percentage maximum velocity which the method will give.

The only rate-limiting factor in a coupled assay should be the concentration of the initial and linking products and all other reagents should be in excess. The role of the auxiliary and indicator enzymes is essentially that of a substrate assay system and under optimum assay conditions the rate of the indicator reaction should be equal to the rate of formation of the initial product. The indicator reaction must be capable of matching the different test reaction rates and its velocity can be defined by the Michaelis–Menten equation in the usual way:

$$\text{test velocity} = \text{indicator velocity} = \frac{V_{max} \times [P]}{K_m + [P]}$$

where [P] is the concentration of the test product.

The product concentration is normally very low and almost constant due to a steady state being established between its formation and use in the indicator reaction. Hence:

$$\text{velocity} = \frac{V_{max} [P]}{K_m}$$

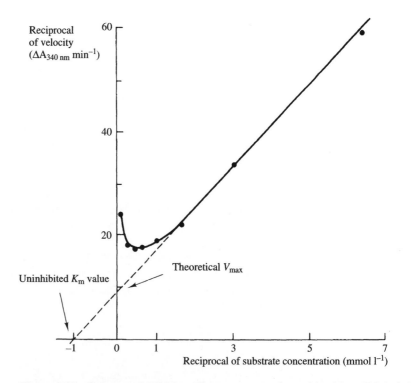

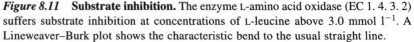

Figure 8.11 **Substrate inhibition.** The enzyme L-amino acid oxidase (EC 1. 4. 3. 2) suffers substrate inhibition at concentrations of L-leucine above 3.0 mmol l^{-1}. A Lineweaver–Burk plot shows the characteristic bend to the usual straight line.

Table 8.3 Examples of kinetic spectrophotometric methods of enzyme assay

Enzyme	EC number	Direct or coupled	Substrate used	Product detected	Linking substrate	Linking enzymes
Lactate dehydrogenase	1.1.1.27	D	Pyruvate	NADH		
D-Amino acid oxidase	1.4.3.3	C	D-Alanine	NADH	Ammonia	Glutamate dehydrogenase
γ-Glutamyl transpeptidase	2.3.2.2	D	γ-Glutamyl-p-nitroanilide	p-Nitroaniline		
Aspartate aminotransferase	2.6.1.1	C	L-Aspartate	NADH	Oxaloacetate	Malate dehydrogenase
Alkaline phosphatase	3.1.3.1	D	p-Nitrophenyl phosphate	p-Nitrophenol		
5'-Nucleotidase	3.1.3.5	C	5'-AMP	NADH	Adenosine Ammonia	Adenosine deaminase Glutamate dehydrogenase
Argininosuccinate lyase	4.3.2.1	D	Argininosuccinate	Fumarate		
Aldolase	4.1.2.3	C	Fructose-1,6-diphosphate	NADH	Glyceraldehyde-3-phosphate	Glycerophosphate dehydrogenase
Glucose phosphate isomerase	5.3.1.9	C	Fructose-6-phosphate	NADP$^+$	Glucose-6-phosphate	Glucose-6-phosphate dehydrogenase
Glutamine synthetase	6.3.1.2	C	L-Glutamate	NADH	ADP Pyruvate	Pyruvate kinase Lactate dehydrogenase

In order for the indicator reaction to parallel the test reaction the ratio, V_{max}/K_m, for the indicator enzyme should be significantly greater than the same ratio for the test enzyme. The value for V_{max} is the only variable in the equation and is determined by the amount of enzyme present, usually quoted in units per assay volume. An increase by a factor of 100 is usually recommended for each stage and will permit concentrations of approximately 1/100 of the K_m value to give a rate comparable to the test reaction rate. It may be necessary, if pH or inhibition effects reduce the activity of the enzyme by a known proportion, to increase further the amount of enzyme added to the assay. Table 8.3 lists some examples of coupled and direct assays for a range of enzymes.

8.2.2 Measurement of enzyme activity

In order to follow the progress of an enzyme-catalysed reaction it is necessary to measure either the depletion of the substrate or the accumulation of the product. This demands that either the substrate or the product show some measurable characteristic which is proportional to its concentration. This is not always the case and a variety of techniques have been developed in order to monitor enzyme reactions. In order to illustrate some of the methods and also to give an appreciation of the technical details and the calculations, three examples are given (Procedures 8.4 to 8.6) that use the enzyme D-amino acid oxidase (Table 8.4).

Table 8.4 D-Amino acid oxidase (EC 1.4.3.3)

Source	Mammalian kidney
Relative molecular mass	125 000
Reaction catalysed	Amino acid + O_2 + H_2O = H_2O_2 + oxoacid + NH_3
Michaelis constant (D-alanine)	4.4×10^{-3} mol l^{-1}
Optimum pH (D-alanine)	8.3
Cofactor	FAD

The rate at which the substrate is converted to the product by the action of an enzyme is dependent upon the concentrations of both the enzyme and the substrate. The initial rate of such a reaction is maximal and it is this initial rate of reaction that reflects the enzyme activity (Figure 8.12). The major factor in this decline is the depletion of the substrate but the increasing amount of product competing for the enzyme also reduces the rate of the forward reaction. Additionally, inactivation of the enzyme may occur, particularly if an appreciable reaction time is involved.

A typical reaction curve of a direct assay with a maximal initial velocity is often modified in a coupled reaction where there may be an initial lag period (Figure 8.13) during which the linking products build up to a steady-state concentration and the maximum velocity is determined from the slope of the steepest section of the curve.

The measurement of reaction velocity is most frequently done in one of three ways. The velocity of the reaction in terms of substrate depletion or

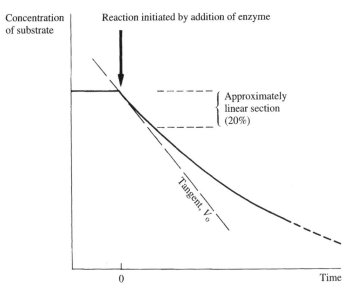

Figure 8.12   Characteristic progress curve of an enzyme-catalysed reaction. The rate of reaction declines from a maximum initial velocity, V_0, which can be represented by the tangent to the curve at zero time. The rate falls least during the first 15–20% of the total reaction change.

▶ Kinetic enzyme assays are designed to measure the velocity of the reaction.

product accumulation may be monitored by recording the concentration change and plotting the values as in Figure 8.12. Such a technique is known as a **kinetic** or **continuous assay**. The most valid measure of the rate of the reaction is the initial velocity, V_0, and this may be determined either by drawing a tangent to the curve at zero time or with the use of electronic equipment to measure the change in the first few seconds. Hence:

$$\text{enzyme concentration } \alpha \text{ velocity } (V_0)$$

It is not always possible to measure initial velocities in this way owing to difficulties in detecting either substrate or product, in which case an alternative method can be used in which the amount of substrate used over a relatively long but specified period is measured. While this may be prone to error owing to the declining reaction rate, it can be a valid analytical method because the first section of a reaction progress curve is relatively linear. This linear section (Figure 8.12) usually comprises the first 10−20% of the total change possible and provided that the assay time has been demonstrated to be within these limits the enzyme activity can be related to the amount of substrate or product converted. Such methods are known as **fixed time assays**. Hence:

▶ Fixed time enzyme assays measure the amount of substrate used or product produced in a fixed time.

$$\text{enzyme concentration } \alpha \frac{1}{\text{amount of substrate used}}$$

$$\alpha \text{ amount of product formed}$$

A technique that depends upon the same basic assumption as fixed time

> ➤ Fixed change enzyme assays measure the time taken for a fixed amount of substrate to be used or product to be formed.

assays but relates the enzyme activity to the time taken for a fixed amount of product to be formed is called a **fixed change assay**. It is particularly useful for reactions which result in a pH change and can be monitored potentiometrically. Hence:

$$\text{enzyme concentration } \alpha \ \frac{1}{\text{time}}$$

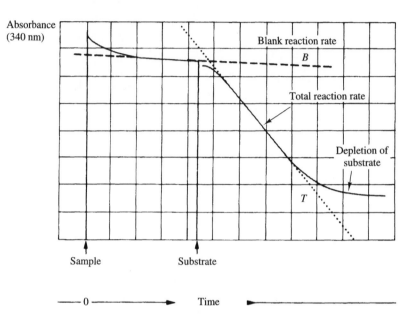

Figure 8.13 **Typical reaction trace of a coupled enzyme assay.** The indicator reaction in many coupled assays will often show a demonstrable change after the addition of the sample but before the addition of the substrate for the test enzyme. This blank reaction may be due to the presence of endogenous substrates in the sample and its rate (B) must be measured in order to be able to calculate the activity of the test enzyme ($T - B$) from the total rate of reaction (T) which results from adding the substrate.

8.2.3   Quantitation of enzyme activities

It is normally not possible to use standard preparations of enzymes and although pre-assayed samples of enzymes are commercially available, they are usually used in quality assurance programmes and not as standards and, as a consequence, all test values have to be calculated from measured reaction rates or product concentrations. Because such calculations assume that there is a direct relationship between the rate of reaction and the concentration of the enzyme it is essential, when setting up an assay, to determine the effective range of activity for which this relationship is linear. This is most conveniently done by assaying carefully prepared dilutions of a sample containing a high enzyme activity and plotting the resulting reaction rates against the percentage dilution of the sample (Figure 8.14).

The enzyme activity in katals is the number of moles of substrate converted by the enzyme in 1 second. Considerable difficulty is often experienced in calculating the activity from first principles and it is important to appreciate the necessary stages in the calculation. Almost invariably all of these steps can be condensed into a single conversion factor which may be applied to the measured reaction rate or instrument reading in order to calculate the enzyme activity in a sample. Procedures 8.4 to 8.6 give details of the calculations.

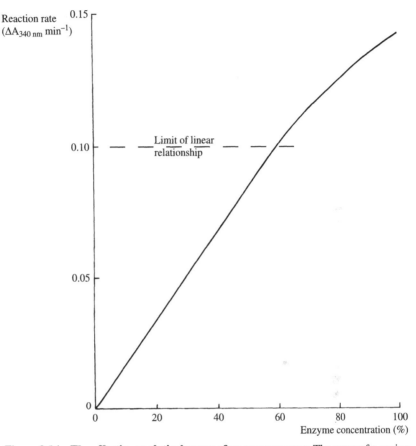

Figure 8.14 **The effective analytical range of an enzyme assay.** The assay of D-amino acid oxidase (EC 1.4.3.3), using the method detailed in Procedure 8.5, shows a valid analytical range up to a maximum reaction rate of 0.10 absorbance change per minute.

Procedure 8.3: Sequence of steps for calculating enzyme activity

1   *Calculate the concentration of product formed in 1 second, expressed in moles per litre*

For fixed time assays this most frequently involves the use of standards and a calibration graph. Some methods, e.g. the use of the molar absorbance coefficient in spectrophotometry, do not require standards and gasometric methods permit the calculation of molar concentration from the volume of gas (1 gram mole of gas occupies 22.4 litres at standard temperature and pressure, STP).

2   *Calculate the actual amount of product formed in 1 second*
Knowing the total assay volume the actual amount of product can be calculated from the concentration value, giving the value for the number of units of enzyme activity in the assay.

3   *Calculate the enzyme activity in a standard volume of the sample*
Knowing the volume of sample used in the assay, the number of enzyme units per litre or millilitre can be calculated.

Consult the calculations for Procedures 8.4 to 8.6 to see these steps illustrated.

Self test questions

Section 8.2

1. How much product will be formed in two minutes by 10 mg of an enzyme preparation with a specific activity of 2.0 millikatal mg^{-1}?
   (a) 0.04 mol.
   (b) 0.48 mol.
   (c) 2.4 mol.
   (d) 40 mol.
2. If 1.0 ml of enzyme preparation in a total assay volume of 5.0 ml results in the formation of a concentration of 0.3 mmol l^{-1} of product in 10 minutes, what is the activity of the enzyme preparation in katal ml^{-1}?
   (a) 2.5×10^{-7} katal ml^{-1}.
   (b) 1.5×10^{-5} katal ml^{-1}.
   (c) 1.2×10^{-3} katal ml^{-1}.
   (d) 3.0×10^{-3} katal ml^{-1}.
3. It is the initial rate of a reaction that most validly reflects the activity of an enzyme
   BECAUSE
   the initial rate of an enzyme-catalysed reaction is affected by the ambient temperature.
4. In coupled assays, it is essential that the concentration of the indicator enzyme be kept low
   BECAUSE
   the amount of product formed by the test reaction in a coupled assay is initially very low.

8.3   **Monitoring techniques**

8.3.1 Gasometric methods

Some enzyme-catalysed reactions result in the production or uptake of a gas and some of the earliest assay methods used this phenomenon as a basis for monitoring the reaction. Classically, Krebs, in elucidating the metabolism of glucose, used such methods. They measure either the pressure changes that result in a fixed volume (Warburg manometry) or the volume changes required to keep a constant pressure (Gilson respirometry). The latter is potentially more useful because the molar concentration of the gas can be more easily calculated from volume than from pressure changes but both types of method are extremely tedious and difficult to perform and, as a result, are rarely used nowadays. They have been largely superseded by electrochemical gas measuring devices.

The uptake of oxygen by oxidative enzymes and the generation of carbon dioxide by decarboxylating enzymes are the most frequent gaseous reactions used (Table 8.5). Some reactions involved both gases and it is possible to measure the uptake of oxygen in the presence of carbon dioxide by including a compartment in the reaction vessel containing a sodium hydroxide solution to absorb the carbon dioxide. If ammonia is generated it is normally retained in aqueous solution provided that the pH is not too high. It may be released subsequently by the addition of a concentrated solution of sodium hydroxide.

Table 8.5 Examples of gasometric methods of enzyme assay

Enzyme	EC number	Substrate	Product	Gas exchanged
D-Amino acid oxidase	1.4.3.3	D-Amino acids/oxygen	Ammonia	Oxygen uptake
Malate dehydrogenase	1.1.1.38	Pyruvate/carbon dioxide	Malate	Carbon dioxide uptake
Catalase	1.11.1.6	Hydrogen peroxide	Oxygen	Oxygen release
Cholinesterase	3.1.1.8	Acetylcholine	Acetic acid	Carbon dioxide release from bicarbonate buffer
Histidine decarboxylase	4.1.1.22	L-Histidine	Carbon dioxide	Carbon dioxide release
Pyruvate carboxylase	6.4.1.1	Oxaloacetate/ADP	Carbon dioxide	Carbon dioxide release

The actual volume of gases exchanged during such reactions is very small and therefore the measuring devices must be very sensitive. Because slight variations in temperature significantly affect either the volume or pressure of a gas, strict temperature control is necessary. Reaction vessels are usually the characteristically shaped Warburg flasks with varying numbers of side arms. The majority of the reactants are contained in the main vessel while those that will initiate the reaction are held in the side arms. The centre well is used to hold a small filter paper wick soaked in a solution of sodium hydroxide to absorb any carbon dioxide generated during the reaction. The reaction is started by tilting the flask to mix the contents of the side arms and the main vessel.

With all gasometric methods it is vital that all the connections and stoppers are gas-tight and usually these glass-to-glass connections are lubricated with lanolin grease. However, it requires very careful technique to ensure that the system is leak-proof and it is desirable to test for leaks once a system has been set up and prior to initiating the reaction.

Warburg manometry
The basic design of a Warburg manometer is illustrated in Figure 8.15 and consists of a U-tube manometer connected by a side arm to the reaction

flask, which is immersed in a constant temperature bath. The manometer limb is calibrated from a mid-point of zero over a 300 mm range and the composition of the fluid is such that the pressure exerted by 10 000 mm is equal to 1 atm.

The readings obtained reflect the rate of the reaction and in many cases are sufficient although they may not give the quantitative information regarding the amount of gas evolved. For quantitative work, it is necessary to determine the flask constant (k), which takes into account the total gas volume of the system, the solubility of the gas in water and the absolute temperature of the reaction. This factor can then be used to convert the pressure readings into gas volume.

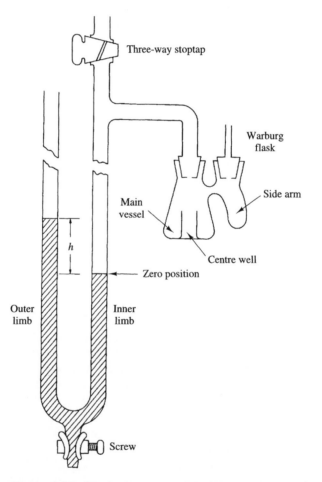

Figure 8.15 **Warburg manometer.** Prior to any readings being taken, the manometer fluid in the inner limb is adjusted to the zero position by means of the pressure screw at the base of the manometer. Changes in the height (h) will then reflect the pressure changes in the flask which result from any gaseous exchange during the reaction.

Gilson respirometry

The Gilson respirometer (Figure 8.16) uses a constant pressure system, and variations in the gas volume are compensated for by altering the volume of the flask by means of a calibrated syringe. The pressure generated by the test flask and its contents is balanced against a reference volume of gas using a small indicating manometer. As gaseous exchange occurs, the manometer fluid is kept at the same level by altering the syringe piston (volumeter), which is calibrated in microlitres. The reading of the volumeter gives a direct measure of the volume of gas exchanged but it is still necessary to correct the value for standard temperature and pressure if critical results are required:

$$\text{Gas volume} \times \frac{T_0}{T} \times \frac{P}{P_0}$$

where T is the absolute temperature of the reaction,

T_0 is absolute zero temperature ($-273\,^\circ$C),

P is the test atmospheric pressure,

P_0 is the standard atmospheric pressure (760 mm mercury).

The volume of gas can be converted to moles of gas using the fact that 1 mole of gas at standard temperature and pressure occupies 22.4 litres.

The method described for the assay of D-amino acid oxidase in Procedure 8.4 illustrates the use of the Gilson respirometer in a gasometric assay.

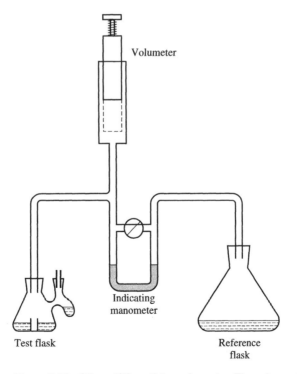

Figure 8.16 **Gilson differential respirometer.** The volume of any gas exchanged can be measured by altering the volume of the test flask using a calibrated syringe system (volumeter) until the original balance in pressure between the test and reference flasks is restored.

Procedure 8.4: Gasometric assay of D-amino acid oxidase using a Gilson differential respirometer

Reaction – catalysed by D-amino acid oxidase

$$CH_3CHNH_2COOH + O_2 \rightarrow CH_3COCOOH + H_2O_2 + NH_3$$

Reagents

Glycylglycine buffer	0.1 mol l^{-1} adjusted to pH 8.3
FAD	0.3 mmol l^{-1}
Sodium hydroxide	100 g l^{-1}
D-Alanine	0.5 mol l^{-1}

Method

1. Into a Warburg flask pipette:

Main compartment	1.2 ml buffer solution
	0.1 ml D-alanine
	0.1 ml FAD
Side arm	0.1 ml sample
Well	0.1 ml sodium hydroxide with a filter paper wick.

2. Seal the joints of the flasks with lanolin and equilibrate at 37 °C for 15 min with the system open to the air.
3. Close the system and monitor the reading for a further 15 min.
4. Initiate the reaction by mixing the contents of the main vessel and the side arm. Monitor the change in volume due to the uptake of oxygen for 30 min and record the results as microlitres of gas per second.

Calculation

The enzyme activity in katals per litre can be calculated from the volume change after correcting for temperature and pressure:

$$\text{Recorded volume} \times \underbrace{\frac{273}{310} \times \frac{P}{760}}_{A} \times \underbrace{\frac{1}{22.4 \times 10^6}}_{B} \times \underbrace{\frac{1000}{0.1}}_{C}$$

The various steps in the calculation are:

A. Correction for temperature and pressure (*P* mm Hg).
B. Converting microlitres of gas to moles.
C. Correction for volume of sample used.

8.3.2 Spectrophotometric methods

The use of spectrophotometry to monitor enzyme-catalysed reactions (Table 8.6) is a very convenient and popular method owing to the simplicity of the technique and the precision that is possible. The technique lends itself readily not only to temperature control using water-jacketed or electrically heated cell holders but also to the measurement of initial velocities by continuous monitoring and recording techniques or by automated analysis systems.

Although only a small number of substrates or products show absorption maxima in the visible region of the spectrum a considerable number show

➤ Absorption maximum – see Section 2.1.

maxima in the ultraviolet. However, in order to be able to measure the rate of substrate depletion or product formation, it is necessary that these two components of the reaction show different absorption characteristics to permit the measurement of one without interference from the other.

The natural substrates which are most frequently used are the nucleotide coenzymes NAD^+ and $NADP^+$, which are reversibly reduced by many enzymes:

➤ Nucleotides – see Section 13.1.

$$NAD(P)^+ + H_2 \rightleftharpoons NAD(P)H + H^+$$
(oxidized)   (reduced)

The two forms of these coenzymes (oxidized and reduced) have different spectral characteristics (Figure 8.17), which permit the reduced form to be detected at 340 nm in the presence of the oxidized form.

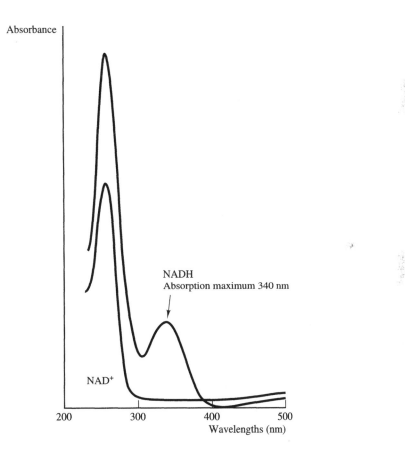

Absorbance

NADH
Absorption maximum 340 nm

NAD$^+$

200   300   400   500
Wavelengths (nm)

Figure 8.17
Absorption spectra of NAD$^+$ and NADH.

➤ A substrate analogue is a synthetic (rather than natural) compound that can act as a substrate for a particular enzyme.

The natural substrates for many enzymes do not show any significant spectral properties and alternative artificial substrates (analogues) may be used. Many hydrolytic enzymes can most easily be detected by using substrate analogues; the glycosidases, for instance, may be measured using p-nitrophenyl derivatives of the appropriate carbohydrates. Hydrolysis results in the production of p-nitrophenol, which can be measured at 404 nm.

Table 8.6 Examples of fixed time spectrophotometric methods of enzyme assay

Enzyme	EC number	Substrate	Product	Measurement
Glucose oxidase	1.1.3.4	Glucose	Hydrogen peroxide	Oxidation of a chromogen
Aspartate aminotransferase	2.6.1.1	L-Aspartate	Oxaloacetate	Formation of a hydrazone
Cholinesterase	3.1.1.8	Acetylcholine	Acetate	Conversion to hydroxamate
Aldolase	4.1.2.3	Fructose-1,6-diphosphate	Dihydroxyacetone phosphate	Formation of a hydrazone
Glucose phosphate isomerase	5.3.1.9	Glucose-6-phosphate	Fructose-6-phosphate	Reaction with resorcinol
L-Glutamate synthetase	6.3.1.2	L-Glutamate/ATP	Phosphate	Determination of phosphate

Table 8.7 Examples of fluorimetric methods of enzyme assay

Enzyme	EC number	Substrate	Product	Excitation wavelength (nm)	Emission wavelength (nm)
Monoamine oxidase	1.4.3.4	Kynurenine	**4-Hydroxyquinoline**	315	380
UDP-glucuronyltransferase	2.4.1.17	**4-Methylumbelliferone**	4-Methylglucuronide	340	440
Cholinesterase	3.1.1.8	Indoxylacetate	**Indoxyl**	395	470
Lipase	3.1.1.3	Dibutyrylfluorescein	**Fluorescein**	490	520

Note: All NAD$^+$- and NADP$^+$- linked enzyme assays are also capable of being monitored fluorimetrically. The fluorescent compounds in each assay are shown in bold type.

Table 8.8 Examples of electrochemical methods of enzyme assay

Enzyme	EC number	Substrates	Products	Device
D-Amino acid oxidase	1.4.3.3	D-Amino acids/oxygen	Ammonia	Ammonia-selective electrode
Urate oxidase	1.7.3.3	Uric acid/oxygen	Allantoin	Oxygen electrode
Cholinesterase	3.1.1.8	Acetylcholine	Acetic acid	pH electrode
Lipase	3.1.1.3	Triacylglycerols	Fatty acids	pH electrode
Carbonic anhydrase	4.2.1.1	Carbon dioxide	Bicarbonate	pH electrode
Glutamate decarboxylase	4.1.1.15	L-Glutamate	4-Aminobutyrate/CO_2	Carbon dioxide electrode
Arginase	3.5.3.1	L-Arginine	L-Ornithine/urea	Urea-selective electrode

Many methods have been developed in which a product of the reaction is chemically modified to produce a substance with a particular spectral property. The inorganic phosphate released by the hydrolysis of phosphate esters may be measured by simple chemical methods (Fiske and Subbarow) after the enzyme reaction has been stopped. Such techniques are often convenient but do not lend themselves to the measurement of initial velocity.

Coupled enzyme assays provide a good alternative to chemical modification and permit a kinetic technique to be employed. In a coupled assay, the rate at which the product is formed is measured by using the product of the reaction as a substrate for a second enzyme reaction which can be monitored more easily. Coupled assays offer great flexibility in enzyme methodology while still retaining all the advantages of continuous monitoring techniques and are illustrated by Procedure 8.5.

PROCEDURE 8.5: Spectrophotometric assay of D-amino acid oxidase (kinetic)

Test reaction – catalysed by D-amino acid oxidase

$$CH_3CHNH_2COOH + O_2 \rightarrow CH_3COCOOH + H_2O_2 + NH_3$$

Indicator reaction – catalysed by glutamate dehydrogenase

$$NH_3 + HOOCCH_2CH_2COCOOH + NADH \rightarrow HOOCCH_2CH_2CHNH_2COOH + NAD$$

Assay reagent

Glycylglycine buffer pH 8.3	$(0.1 \text{ mol } l^{-1})$	2.05 ml
2-Oxoglutarate	$(0.2 \text{ mol } l^{-1})$	0.20 ml
NADH	$(2.25 \text{ mmol } l^{-1})$	0.20 ml
FAD	$(0.3 \text{ mmol } l^{-1})$	0.10 ml
Glutamate dehydrogenase	(20 μkatal or 1200 units)	0.05 ml
ADP	$(15.0 \text{ mmol } l^{-1})$	0.10 ml

Method

1. Pipette the assay reagent (2.7 ml) into a clean dry cuvette and allow to attain a temperature of 37 °C in a thermostated cell holder.
2. Monitor the absorbance of the solution at 340 nm continuously.
3. Add 0.2 ml sample and mix carefully.
4. When the absorbance shows either no further fall or a slow but steady fall (the blank reaction) initiate the reaction by adding 0.1 ml 0.5 mol l^{-1} D-alanine.
5. Monitor the reaction for at least 5 min or until a linear section of the reaction trace can be clearly defined.
6. The rate of the blank reaction and the total reaction are determined graphically and reported as absorbance change per second. Under conditions described the reaction gives 84% maximum velocity for any enzyme preparation.

Calculation

The enzyme activity in katals per litre may be calculated from the molar absorption coefficient for NADH at 340 nm (6.22×10^3 l mol^{-1} cm^{-1}).

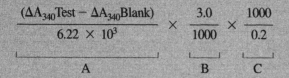

$$\underbrace{\frac{(\Delta A_{340}\text{Test} - \Delta A_{340}\text{Blank})}{6.22 \times 10^3}}_{A} \times \underbrace{\frac{3.0}{1000}}_{B} \times \underbrace{\frac{1000}{0.2}}_{C}$$

The various steps in the calculation are:

A. The use of the molar absorption coefficient for NADH to calculate the concentration change (mol l^{-1} s^{-1}).
B. Calculation of the amount of NADH oxidized in the 3.0 ml reaction volume in 1 second.
C. Correction for the sample volume used.

8.3.3   Fluorimetric methods

The use of fluorescent substrates or products permits sensitive kinetic measurement of enzyme reactions to be undertaken and although there are relatively few natural fluorescent substrates, analogues can sometimes be used (Table 8.7). Some products can be converted to fluorescent compounds and can be used in fixed time assays.

> ➤ Fluorimetry – see Section 2.4.

Fluorimetric methods are useful for monitoring reactions involving the nucleotide coenzymes. The natural fluorescence of the reduced forms in the region of 460 nm can be used in kinetic assays. However, this fluorescence is destroyed at pH values below 2.0, whereas any oxidized forms of the coenzymes present are stable. If the pH of the solution is then raised above 10.5 and heated, the oxidized forms are themselves converted to fluorescent derivatives. This latter procedure lends itself to fixed time assays such as is illustrated in Procedure 8.6.

Fluorimetric methods, while being inherently sensitive, have significant disadvantages which relate to the technical difficulties in fluorescence measurements. Many fluorescent compounds are unstable particularly in ultraviolet light, although it is sometimes possible to increase the stability by including additional reagents, e.g. imidazole. The presence of other fluorescent substances in either the sample or the reagents will give a background fluorescence which will reduce the sensitivity of the measurements. This is a problem particularly when analysing tissue extracts, etc., in which aromatic amino acids in the proteins will contribute to the fluorescence.

A major problem in fluorimetric measurements is caused by the fact that many substances, often present in small amounts as impurities in reagents and solvents, can quench the radiation emitted by the fluorescent substance. For this reason it is necessary to use only reagents of the highest purity and to ensure that all equipment is scrupulously clean.

PROCEDURE 8.6: Fluorimetric assay of D-amino acid oxidase (fixed time)

Test reaction – catalysed by D-amino acid oxidase

$$CH_3CHNH_2COOH + O_2 \rightarrow CH_3COCOOH + H_2O_2 + NH_3$$

Indicator reaction – catalysed by glutamate dehydrogenase

$$NH_3 + HOOCCH_2CH_2COCOOH + NADH \rightarrow HOOCCH_2CH_2CHNH_2COOH + NAD$$

Reagents

Assay reagent	as for Procedure 5
Sodium hydroxide	10 mol l^{-1} containing imidazole (10 mmol l^{-1})
Standard solution of NAD$^+$	1.5 mmol l^{-1}
D-Alanine	0.5 mol l^{-1}

Method

1. Pipette the assay reagent (2.7 ml) into each of the two test-tubes labelled TEST and BLANK. To each add 0.2 ml of sample.
2. After incubating at 37 °C for about 5 min, initiate the reaction in the TEST by adding 0.1 ml of D-alanine.
3. Incubate both tubes for a further 60 min. Add 0.1 ml D-alanine to the BLANK tube and immediately transfer 1.0 ml from each tube to appropriately labelled tubes containing 0.1 ml HCl (2.0 mol l^{-1}). Allow to stand at room temperature for 15 min to destroy the remaining NADH.
4. To both tubes, add 2.0 ml of the sodium hydroxide/imidazole reagent and heat at 56 °C for 15 min in the dark.
5. Cool and measure the intensity of fluorescence (excitation wavelength, 365 nm; emission wavelength, 455 nm).
6. The fluorescence produced by 1.0 ml dilutions of the standard NAD$^+$ are also measured in the same manner after heating with sodium hydroxide reagent.

Calculation

From the calibration curve, the concentration of NAD$^+$ in both the test and blank can be determined. From the difference in the results the amount of NAD$^+$ generated in 1 second can be calculated (mol l^{-1} s^{-1}) and from this the enzyme activity in katals per litre of sample:

$$\text{Concentration change} \times \underbrace{\frac{3}{1000}}_{A} \times \underbrace{\frac{1000}{0.2}}_{B}$$

The various steps in the calculation are:

A. Calculation of the amount of NAD$^+$ produced in the 3.0 ml reaction volume in 1 second.
B. Correction for the volume of sample used.

8.3.4 Luminescence methods

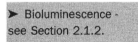

➤ Bioluminescence -
see Section 2.1.2.

Bioluminescence provides the basis for sensitive enzymic assay methods both for substrate assays and coupled enzyme assays. Firefly luciferase (EC 1.13.12.5) catalyses the production of light (540–600 nm) by the oxidation of luciferin (D-LH$_2$) (Figure 8.18).

$$D\text{-}LH_2 + ATP + O_2 \rightarrow AMP + PP + D\text{-}L + CO_2 + \text{light}$$

The enzyme is activated by magnesium ions, has an optimum pH of about 7.5 and offers a very sensitive assay method for ATP but has much wider applications when used in coupled assays with ATP-converting enzymes.

Luciferase from bacterial sources catalyses the oxidation of long chain aliphatic aldehydes and requires the coenzyme FMN. The wavelength of the emitted radiation in this reaction is approximately 490 nm:

$$FMNH_2 + RCHO + O_2 \rightarrow FMN + RCOOH + H_2O + light$$

The reaction can be coupled to the oxidation of either NADH or NADPH by the presence of a flavin dehydrogenase, which is usually present complexed with the luciferase:

$$NAD(P)H + FMN \rightarrow NAD(P)^+ + FMNH_2$$

A wide range of coupled assays involving ATP, NAD(P)H and FMN can be developed using these two enzymes and provide increased levels of sensitivity over other coupled assays (Procedure 8.7).

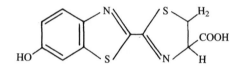

D-Luciferin

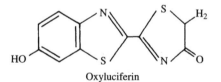

Oxyluciferin

Figure 8.18
Luciferin, a substrate for bioluminescence.

8.3.5 Electrochemical methods

The measurement of pH using the glass electrode provides a very convenient method for monitoring reactions in which either an acid or a base is produced, notably certain hydrolytic enzymes (Table 8.8). The change in pH resulting from the reaction has an effect on the activity of the enzyme and hence on the rate of the reaction, and to maintain the reaction at the optimum pH and yet measure the amount of acid or alkali formed a pH-stat system can be used. In such an instrument, slight changes from an initial pre-selected pH activate a solenoid valve to allow the addition of either an acid or an alkali to maintain the pH at the selected value. The reaction is monitored not as a change in pH but as the volume of titrant required to maintain the pH. Such a method also avoids the difficulties of making direct measurements of pH change in the presence of the buffers in the reaction mixture.

The use of ion-selective electrodes opens up the wider use of electrochemical methods for monitoring the formation of particular products, e.g. ammonia, while simple platinum electrodes can be used to monitor redox

► Oxygen electrode –
see Section 4.4.3.

reactions catalysed by the oxidoreductases or hydrolytic enzymes using substrate analogues which cause redox changes. Oxygen exchange may be monitored with an oxygen electrode as an alternative to manometry, and other electrodes, e.g. carbon dioxide, are also available.

Procedure 8.7: Bioluminescence assay of FMN using bacterial luciferase (EC 1.14.14.3)

Reaction – catalysed by luciferase

$$FMNH_2 + C_{10}H_{23}CHO + O_2 \rightleftharpoons FMN + C_{10}H_{23}COOH + H_2O + light$$

Assay reagent

Glycylglycine buffer pH 7.0	(0.1 mol l^{-1})	1.0 ml
Luciferase	(see below)	0.1 ml
Decylaldehyde	(0.1% in buffer)	0.1 ml
NADH solution	(10 mg ml^{-1} in buffer)	

Luciferase (*Vibrio fischeri*) 1×10^4 light units per ml in buffer containing 10 mg l^{-1} albumin and 6 mol l^{-1} dithiothreitol.

Method

1. Pipette the reagent into a tube appropriate for the model of luminometer being used.
2. Add 0.1 ml of the FMN standard or sample solution.
3. Mix and measure any background luminescence.
4. Add 0.1 ml NADH solution.
5. Quickly mix and monitor the luminescence for 5 min, recording the maximum level of emission.

Calculation

Using a series of standard solutions of FMN, a calibration graph can be drawn and used to determine the concentration of FMN in the test samples.

8.3.6 Microcalorimetric methods

Almost all enzyme-catalysed reactions are exothermic and the enthalpy change (ΔH) involves the release of energy as heat. The resulting increase in temperature, although very small, provides a non-specific method of monitoring the reaction. Most calorimeters used for this purpose measure the increase in temperature in an insulated container (adiabatic calorimeters) and although the temperature change may be as small as $1 \times 10^{-3}\,°C$, it can be detected and measured using thermistors.

The amount of heat released during a reaction is proportional to the amount of substance involved but the relationship is complicated in enzyme studies by secondary reactions. Although the use of entropy constants means that calorimetry theoretically does not require standardization, in many instances this will be necessary. The initial energy change can often be enhanced, giving an increase in the sensitivity of the method. Hydrogen ions released during a reaction, for instance, will protonate a buffer with an evolution of more heat.

$$\text{Glucose} + \text{ATP} = \text{glucose-6-phosphate} + \text{ADP} + \text{H}^+ \quad \Delta H = -27 \text{ kJ mol}^{-1}$$
$$\text{Tris} + \text{H}^+ \quad\quad = \text{Tris H}^+ \quad\quad\quad\quad\quad\quad \Delta H = -47 \text{ kJ mol}^{-1}$$

$$\text{Total } \Delta H = -74 \text{ kJ mol}^{-1}$$

Similarly, the use of coupled reactions can provide further energy release, e.g. the coupled determination of glucose using glucose oxidase (EC 1.1.3.4) and catalase (EC 1.11.1.6).

$$\text{Glucose} + O_2 = \text{gluconic acid} + H_2O_2 \quad\quad\quad \Delta H = -83 \text{ kJ mol}^{-1}$$
$$H_2O_2 \quad\quad = H_2O + \tfrac{1}{2}O_2 \quad\quad\quad\quad\quad \Delta H = -125 \text{ kJ mol}^{-1}$$

$$\text{Total } \Delta H = -208 \text{ kJ mol}^{-1}$$

Because the energy changes are so small, microcalorimetry lends itself to substrate assays in the presence of excess enzyme rather than to enzyme assays and is particularly useful when immobilized enzymes are used.

Self test questions

Section 8.3

1. What can enzyme-catalysed reactions that involve the nucleotide co-enzyme NAD be monitored by?
   (a) Fluorimetry.
   (b) Absorbance at 450 nm.
   (c) Luminometry.
   (d) Gaseous exchange.
2. What do bioluminescence reactions involve?
   (a) An oxidative process.
   (b) The enzyme, firefly amylase.
   (c) Emission of ultraviolet radiation.
   (d) Inhibition by proteins.
3. Spectrophotometry at 340 nm lends itself to kinetic enzyme assays
   BECAUSE
   NADH is a component of many oxidative enzymic reactions.
4. Enzyme reactions involving the nucleotide coenzymes can often be followed fluorimetrically
   BECAUSE
   almost all enzyme-catalysed reactions are exothermic.

8.4   Treatment of samples for enzyme assay

Studies of the enzyme content of cells frequently involve the use of coarse tissue samples of either animal or plant origin. In such cases some preliminary dissection of the tissue may be necessary to isolate the relevant tissue components and remove unwanted structural material such as collagen, cellulose, etc., before moving on to the more critical disruption of the cells. Sometimes it is possible to use the technique of tissue culture to provide pure cell preparations for subsequent studies.

8.4.1 Tissue culture

Tissue culture, more frequently used as cell culture, enables animal and plant cells to be cultured in large numbers by techniques comparable to those used in microbiology but, because of the fragile nature of the cells, does require special cultural conditions. The culture media used must supply all the essential factors for growth, such as a wide range of amino acids, nucleotides, enzyme co-factors as well as indeterminate factors that can only be supplied in special products, e.g. foetal bovine serum. The environmental conditions must be carefully controlled, particularly pH, and this is frequently maintained by culturing in a bicarbonate buffer system and a carbon dioxide saturated atmosphere.

The build-up of waste products inhibits growth and it is necessary to change the culture medium at regular intervals, usually every few days. As with all cell culture techniques, asepsis is essential but very difficult to maintain. Antibiotics may therefore be incorporated into the culture media to assist in maintaining sterility.

8.4.2 Storage of samples for enzyme assay

All samples must be stored appropriately to minimize the loss of activity due to protein denaturation, lack of stabilizers or presence of inhibitors. Optimal storage conditions will vary for different enzymes and the nature of the sample, blood, tissue, etc. Such information would be sought from specialist textbooks.

In general, storage at low temperature, with protecting molecules present, is often to be recommended, e.g. storage at $-20°C$ in bovine serum albumin or glycerol. A 50% (v/v) solution of glycerol is particularly useful because it is still fluid at $-20°C$ and, being non-protein, can subsequently be removed very easily. For enzymes that are prone to oxidative effects, the presence of mild reducing substances such as the amino acid cysteine or similar substances may be beneficial.

8.4.3 Treatment of tissue and cell samples

Very few enzymes are secreted by cells and it is often necessary to disrupt the cells in order to demonstrate enzyme activity. When the total enzyme content of cells is required, it is necessary to disrupt them completely and to remove all particulate material by centrifugation, leaving the dispersed enzyme protein in the supernatant fluid (Figure 8.19). Many enzymes are membrane bound and are only released from fragmented membranes if a suitable detergent is incorporated into the medium. The detergents used include Triton X100 and the bile salts, e.g. sodium deoxycholate. Enzymes that fail to exhibit maximum activity without the releasing effects of detergents are said to show 'latency'. Many cellular studies, however, require the careful disruption of the cell and the subsequent separation of the various cellular organelles such as mitochondria, nuclei, etc., by a process known as cell fractionation.

Gross homogenization of tissues may be achieved with one of the high

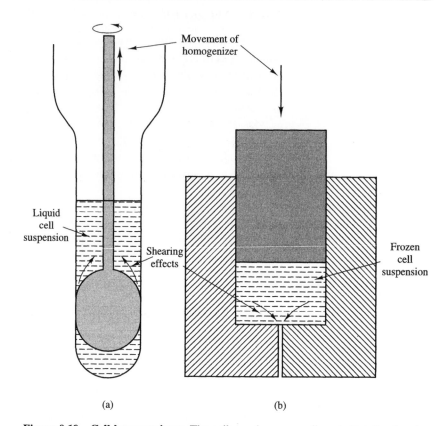

Figure 8.19 **Cell homogenizers.** The cell membranes are disrupted by the shearing effects of being forced through narrow apertures. In the Potter homogenizer (a) the gap between the plunger and the walls of the tubes is critical in determining the extent of the disruptive effect and in the cell press systems, (b) disruption is caused by rapidly forcing a frozen cell suspension through a small hole.

speed bladed instruments, usually known as blenders, and while they are generally used to produce cellular dispersions from chopped tissues, they also cause considerable damage to the cells when used at high speeds. They are therefore mainly used in preliminary homogenization of tissue samples prior to their complete cellular disruption using an alternative technique. For smaller samples of tissue, there are various cylindrical rotating blade instruments which, while causing the general disruption of tissues, can be controlled to produce more specific cellular damage. As well as showing the cutting effects of the rotating blade, such instruments often generate ultrasonic effects due to the castellated design of the blades.

Many cells, especially bacteria, yeast, etc., require very drastic measures to disrupt them and various cell disintegrators are available for this purpose. Ultrasonic vibrations cause the formation of minute bubbles, a phenomenon known as cavitation, which is caused by the extreme variations in pressure generated by the sound waves, although the generation of heat may cause problems unless the samples are cooled frequently during the treatment. An additional hazard lies in the potential damage to the hearing and it is necessary to use some sound-proofing around the instrument.

Shearing effects are often used to disrupt cells and probably the most frequently used is the Potter homogenizer, which consists of a glass tube with a close-fitting PTFE pestle which may be either used manually or driven by a variable speed motor. The clearance between the pestle and the walls of the tube is critical in order to achieve the degree of fragmentation required. A gap of approximately 0.25 mm is frequently used for the preparation of mitochondrial suspensions (Figure 8.19). Alternative disruptive shearing techniques include forcing a frozen sample through a very small orifice under high pressure or shaking with small glass beads.

➤ Cell fractionation – see Section 3.4.3.

Many methods for the fractionation of cellular organelles have been described and it is probably not possible to have a method that is equally effective at preserving each individual component. For each type of cell and organelles the most appropriate method must be selected and evaluated.

Sub-cellular enzymes

It is often necessary to assess the efficiency of cell fractionation procedures. Electron microscopy of the prepared fractions is very informative but gives no quantitative indication of the purity of the fraction. It is often easier to measure the relative concentrations of marker enzymes in each fraction (Table 8.9).

➤ Marker enzymes can be used to assess the presence of specific organelles in sub-cellular preparations.

These marker enzymes are assumed to be associated with only one of the cellular organelles and hence their relative concentrations in each fraction will give some indication of the degree of cross-contamination in the preparations. There is considerable debate about the validity of this assumption and more precise information may be gained from specialist texts.

Table 8.9 Organelle markers

Structure	Enzyme	EC number
Cytoplasm	Lactate dehydrogenase	1.1.1.27
Membrane	Adenylate cyclase	4.6.1.1
Endoplasmic reticulum	Glucose-6-phosphatase	3.1.3.9
Lysosomes	Acid phosphatase	3.1.3.2
Peroxisomes	Catalase	1.11.1.6
Mitochondria		
Inner membrane	Cytochrome *c* oxidase	1.9.3.1
Outer membrane	Amine oxidase	1.4.3.4
Matrix	Glutamate dehydrogenase	1.4.1.3
Nucleus	NMN adenyltransferase	2.7.7.1

8.5 **Substrate assay methods**

Enzymes are extremely useful analytical tools, primarily because of their specificity. Currently a very large number of enzymes are available commercially in various forms of purity and, although their cost is often high, only small amounts are usually required. Procedure 8.8 describes a fixed time method for the quantitation of D-alanine using D-amino acid oxidase.

Procedure 8.8: Spectrophotometric determination of D-alanine using D-amino acid oxidase

Test reaction – catalysed by D-amino acid oxidase

$$CH_3CHNH_2COOH + O_2 \rightarrow CH_3COCOOH + H_2O_2 + NH_3$$

Indicator reaction – catalysed by glutamate dehydrogenase

$$NH_3 + HOOCCH_2CH_2COCOOH + NADH \rightarrow HOOCCH_2CH_2CHNH_2COOH + NAD$$

Reagents

Glycylglycine buffer pH 8.3	0.1 mol l^{-1}
2-Oxoglutarate	0.2 mol l^{-1}
NADH	2.5 mmol l^{-1}
Glutamate dehydrogenase	$1.5 \text{ } \mu\text{katal or } 100 \text{ units ml}^{-1} \text{ in ADP solution } (0.5 \text{ mmol ml}^{-1})$
D-amino acid oxidase	$1.5 \text{ } \mu\text{katal or } 100 \text{ units ml}^{-1}$
Standard solution D-alanine	5.0 mmol l^{-1}

Method

1. Prepare a bulk assay mixture as follows:
   19.0 ml buffer solution
   1.0 ml NADH solution
   1.0 ml 2-oxoglutarate solution
   1.0 ml glutamate dehydrogenase preparation
   1.0 ml D-amino acid oxidase
2. Into the cuvettes pipette 2.9 ml of this mixture and monitor the absorbance for several minutes to ensure a steady reading or at least a very slow fall in absorbance.
3. To cuvette (BLANK) add 0.1 ml distilled water and note absorbance B_1. To the other cuvette (TEST) add 0.1 ml amino acid solution and again record absorbance T_1.
4. Incubate both cuvettes for 15 min at 37 °C and again record absorbance values B_2 and T_2.
5. The amount of NADH oxidized and hence the amount of amino acid is proportional to the absorbance change.

$$(T_1 - T_2) - (B_1 - B_2)$$

Calculation

The concentration of D-alanine in the sample (mol l^{-1}) can be calculated from the molar absorption coefficient for NADH:

$$\underbrace{\frac{(T_1 - T_2) - (B_1 - B_2)}{6.22 \times 10^3}}_{A} \times \underbrace{30}_{B}$$

The steps in the calculation are:

A. The use of the molar absorption coefficient for NADH to calculate the concentration change (mol l^{-1}).
B. Corrections for the dilution of the original sample.

8.5.1   End-point methods

In the most frequently used method of substrate quantitation, the total amount of change resulting from the presence of the substrate is measured and used to calculate the original amount of substrate present. There must be an adequate enzyme activity to ensure that the conversion of substrate to product takes place in a reasonable period of time, but towards the end of a reaction the concentration of the substrate will be very low and the time taken to reach the equilibrium position will be relatively long. The ratio V_{max}/K_m is useful in assessing the amount of enzyme that is required and numerical values of about 1.0 are recommended. Provided that the concentration of the substrate is less than the K_m value, such conditions will result in a reaction time of just over 3 min for approximately 99% conversion of the substrate.

Reactions do not necessarily go to completion and regardless of the amount of enzyme used, the equilibrium position of the reaction will not change. It is important for quantitative measurements that the reaction goes as near to completion as possible and this may be achieved by a variety of methods. The equilibrium position may be altered by changing the pH away from the optimum for the enzyme. For example, the equilibrium position for the reaction in which pyruvate is converted to lactate by lactate dehydrogenase (EC 1.1.1.27) lies very much towards pyruvate at the normal pH of 7.6 but at pH 9.0 the equilibrium is altered towards lactate.

In bi-substrate and tri-substrate reactions the use of high concentrations of the second and third substrates will displace the equilibrium of the test reaction towards the formation of the product and the use of a second reaction which results in the regeneration of one of the initial substrates from its product will similarly force the conversion of the test substrate. The assay of glutamate, for instance, can be achieved by the enzyme glutamate dehydrogenase (EC 1.4.1.3), in which oxoglutarate is the product, and the reaction can be forced almost to completion by incorporating pyruvate and lactate dehydrogenase (EC 1.1.1.27) to regenerate the NAD^+ used in the test reaction (Figure 8.20). The amount of oxoglutarate formed can subsequently be determined and the initial amount of glutamate calculated. The ADP/ATP systems are also very amenable to this approach using the enzyme pyruvate kinase (EC 2.7.1.40) and phosphoenolpyruvate to regenerate ATP.

End-point assays can also use coupled enzyme systems in which the product of the test reaction provides the substrate for the subsequent auxiliary and indicator reactions. The choice of pH is less critical than with the measurement of enzyme activity: frequently a compromise pH is used and

Figure 8.20
The recycling of NAD^+ in an assay of L-glutamate.

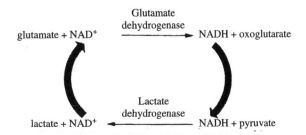

▶ Carbohydrate assay –
see Section 9.3.

increased amounts of enzyme used to compensate for any loss in activity. Alternatively, the reaction may be completed in two stages, changing the pH with the careful use of buffer systems for the second stage of the reaction. The quantitation of sucrose uses β-D-fructofuranosidase (invertase) (EC 3.2.1.26) at pH 4.6 and after a suitable period of time the pH is changed to 7.5 for the second stage of the reaction to be completed:

$$\text{sucrose} \xrightarrow{\beta\text{-D-fructofuranosidase}} \text{glucose} + \text{fructose} \quad \text{pH 4.6}$$

$$\text{glucose} \xrightarrow{\text{glucose oxidase}} \text{gluconic acid} + H_2O_2 \quad \text{pH 7.0}$$

$$H_2O_2 + \text{chromogen} \xrightarrow{\text{peroxidase}} \text{colour} \quad \text{pH 8.0}$$

Enzymic cycling

Very low concentrations of substrates may be assayed by recycling the test substrate for an appreciable but definite period of time and measuring the amount of product formed. The coenzyme NADPH, for instance, may be assayed using the two enzymes glutamate dehydrogenase (EC 1.4.1.3) and glucose-6-phosphate dehydrogenase (EC 1.1.1.49):

$$\text{NADPH} + \text{oxoglutarate} + \text{NH}_3 = \text{NADP}^+ + \text{L-glutamate}$$

$$\text{NADP}^+ + \text{glucose-6-phosphate} = \text{NADPH} + \text{6-phosphogluconate}$$

The reaction mixture contains oxoglutarate, ammonia, glucose-6-phosphate and the two enzymes. The reaction is started by adding the test substrate (NADPH) and allowed to continue for a defined time. The reaction is stopped and the enzymes are inactivated, usually by heat. The amount of 6-phosphogluconate which has accumulated is determined using an appropriate method and this relates to the amount of NADPH originally present. In this case the product may be determined using glucose-6-phosphate dehydrogenase and monitoring the oxidation of excess NADPH. Such methods do require calibration with known amounts of the test substrate.

8.5.2 Kinetic methods

▶ Automated analysis –
see Section 6.1.

Direct kinetic methods comparable to those used for enzyme assays are generally feasible only with automated instrumentation because of the difficulty in measuring the rapidly falling reaction rate as the low concentration of substrate is further depleted.

However, kinetic assays using coupled systems in which the concentration of the substrate is held constant by a recycling process may lend themselves to more general use. In such a situation the rate of the indicator reaction will also remain constant and can be related to the original substrate concentration. Small amounts of coenzyme NAD^+ can be measured by coupling its reduction to NADH to the simultaneous reduction of cytochrome c, a process that can be monitored at 550 nm (Figure 8.21). The assay reagent contains ethanol and cytochrome c, together with the enzymes alcohol dehydrogenase (EC 1.1.1.1) and cytochrome c reductase (EC 1.6.99.3). On initiating the reaction with the sample containing NAD$^+$, a steady state rapidly develops in which the rate of reduction of cytochrome c reflects the concentration of NAD$^+$.

Direct kinetic assays are the only valid methods for the measurement of activators and inhibitors and calibration plots of the percentage activation or inhibition by known amounts of the substance can be made. Examples of inhibition assays include the quantitation of organophosphorus pesticides using the inhibition of cholinesterase (EC 3.1.1.7) while manganese can be measured in amounts as low as 1×10^{-12} mol using its activating effect on isocitrate dehydrogenase (EC 1.1.1.41).

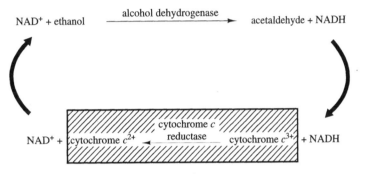

Figure 8.21   **A kinetic assay of NAD$^+$.** The rate of increase in absorbance at 550 nm as cytochrome c is reduced is a measure of the steady-state concentration of NAD$^+$.

8.6   Automated analysis

There are many instruments designed for either enzyme assays or substrate assays using enzymes. Information on the analytical capabilities of these instruments will be supplied by the manufacturers. This will often include protocols for specified assays using kits of commercially available pre-prepared reagents. These may be in liquid or dry form and may, for substrate assays, include immobilized enzymes. The facility to be able to develop additional automated methods on a particular instrument will depend upon its design and some instruments are dedicated solely to specified analyses.

➤ Centrifugal analyser – see Section 6.1.1.

Many automated instruments measure enzyme activity using fixed time colorimetric methods. Some, however, can be classed as reaction rate analysers, e.g. the centrifugal analysers, and these instruments determine the reaction rate from either the initial slope of the reaction curve or from repeat measurements at fixed intervals. In both methods the slope of the line is taken to represent the activity of the enzyme.

The main technical function of a reaction rate analyser is the automated addition and mixing of the reagents and sample under carefully controlled temperature conditions and the monitoring of the reaction immediately after initiation. The data generated are presented in various forms but all involve some degree of computerization. The most versatile types of instrument are those using spectrophotometric detection systems and using NADH-linked assays. Some instruments are dedicated to specific methodologies which often involve ion-selective electrodes or commercially prepared reagents and as a result may be restricted for general use.

The presence of a lag period in many coupled assays and difficulties in determining the linear portion of a curve present the main problems in the calculation of enzyme activity using reaction rate analysers. In the simplest instruments the slope of the curve in the first few seconds of the reaction is extrapolated into a straight line or, if the reaction is known to show a lag period, the rate of reaction after a defined period of time can be measured. The more sophisticated instruments use microcomputers to determine the linear portion of the curve and calculate the enzyme activity directly from the slope. The second derivative of the reaction progress curve (rate of change of the slope) can be monitored by the computer and when a value of zero is held for a period of time (10−15 seconds) this indicates a linear section of the graph. From the value for the slope, the enzyme activity can be calculated.

An alternative approach involves the measurement of the time taken for small fixed changes in absorbance and the calculation of the rate for each single measurement. The computer stores and compares all the values, finally selecting the slope of the linear section of the curve and again calculating the enzyme activity.

8.7   Immobilized enzymes

It is possible to bind enzymes to an insoluble matrix by a variety of methods and still retain their catalytic activity. The reusable nature of immobilized enzymes can significantly reduce costs and provides a convenient source of enzymes for performing substrate assays. Such preparations often show a greater stability and reduced inhibition effects than do soluble enzymes, although occasionally optimum pH values may be altered slightly.

➤ Immobilization of proteins – see Section 3.5.1.

Covalent linking to a variety of polymers using non-essential amino acid residues is a convenient method of immobilization. Enzyme inactivation during the process can be appreciably reduced by the presence of a competitive inhibitor of the enzyme. Physical immobilization techniques (Figure 8.22) are less drastic than covalent bonding but some changes in the properties of the enzyme may occur, e.g. pH optima and activation effects. Enzymes may be adsorbed on a variety of polar materials such as charcoal, silica gel, alumina and ion-exchange resins and although this is a simple technique, a major problem lies in the ease with which enzymes may be desorbed as a result of variations in pH and buffer concentration, etc. The formation of a cross-linked poly-

➤ Polyacrylamide – see Section 3.3.2.

mer, often polyacrylamide, in the presence of the enzyme results in trapping of the enzyme molecules within the matrix of the polymer but still permits the access of small substrate molecules to the enzymes. Alternatively, enzymes may be encapsulated by the production of an emulsion of an aqueous solution of the enzyme in an organic solvent. The droplets are stabilized by the addition of a suitable polymer which, after forming a membrane around the droplets, is allowed to harden and the capsules can then be washed free from the other reagents. Suitable semipermeable membranes can be produced with a variety of reagents including collodion and polystyrene.

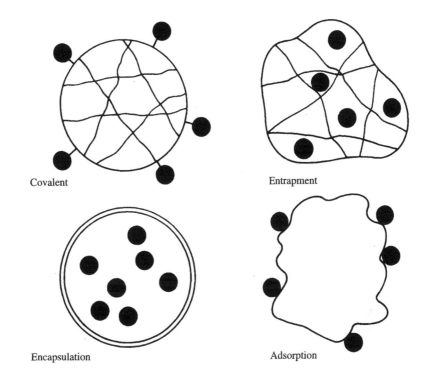

Figure 8.22
General methods of immobilizing enzymes.

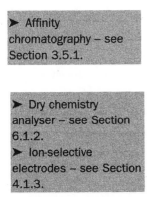

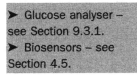

➤ Affinity chromatography – see Section 3.5.1.

➤ Dry chemistry analyser – see Section 6.1.2.
➤ Ion-selective electrodes – see Section 4.1.3.

➤ Glucose analyser – see Section 9.3.1.
➤ Biosensors – see Section 4.5.

Immobilized enzymes may be used in affinity chromatographic methods but their use as catalysts may be in either the production or removal of compounds in chemical processes or as analytical tools. Many substrate assays can be performed using enzymes immobilized on a variety of surfaces, e.g. glass beads, plastic or nylon tubing. Alternatively they may be incorporated into gel or microparticulate layers on dry strips or slides.

Immobilized enzymes used in conjunction with ion-selective electrodes provide very convenient methods of analysis. The immobilized enzyme may be held in a gel or membrane around the electrode and the substance to be measured diffuses into the enzyme gel. Its conversion to the product alters the ionic equilibrium across the ion-selective membrane (Figure 8.23). It is important that the enzyme layer is thin, to minimize any problems caused by slow diffusion rates through the layer.

The main advantage of enzyme electrodes lies in the simplicity of the method of analysis although the presence of interfering ions may present a problem, particularly when using cation-sensitive electrodes. Methods of enzyme electrode construction are discussed in the section on biosensors.

Substrate determinations (Table 8.10) using enzyme electrodes must be performed under controlled conditions of temperature and pH and if standard solutions are used to calibrate the electrode they must be analysed at the same time. The response time of some enzyme electrodes may be several minutes and although for many the time is shorter this factor must be considered in the design of an assay method.

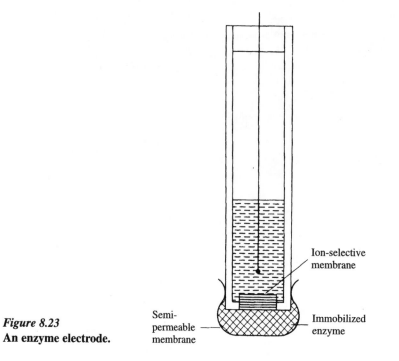

Figure 8.23
An enzyme electrode.

Table 8.10   Substrate assays using enzyme electrodes

Substrate	Enzyme	Substrate or product measured	Electrode system
Amino acids	D- and L-amino acid oxidases	Hydrogen peroxide	Platinum
		Ammonia	Cation
Ethanol	Alcohol oxidase	Hydrogen peroxide	Platinum
		Oxygen	Oxygen
Glucose	Glucose oxidase	Hydrogen peroxide	Platinum
		Oxygen	Oxygen
L-Glutamate	Glutamate dehydrogenase	Ammonia	Cation
Lactic acid	Lactate dehydrogenase	Ferricyanide ions	Platinum
Urea	Urease	Ammonia	Cation

Self test questions

Sections 8.4/5/6/7

1. Which of the following substances may be useful in stabilizing enzyme solutions?
   (a) Proteins.
   (b) Bile salts.
   (c) Nucleic acids.
   (d) Amino acids.
2. What are the enzymes associated with sub-cellular organelles known as?
   (a) Secreted enzymes.
   (b) Soluble enzymes.
   (c) Marker enzymes.
   (d) Indicator enzymes.

Self test questions

3. End-point methods for the quantitation of substrates are not appropriate when the assay involves coupled reaction systems
   BECAUSE
   automated instruments are available that can perform kinetic assays of substrates.
4. Covalent linking of protein to insoluble polymers is unsuitable as a method of enzyme immobilization
   BECAUSE
   any chemical modification of an enyzme always inactivates it.

8.8   Further reading

Bergmeyer, H.U. (ed.) (1993) *Methods of enzymatic analysis*, 3rd edition, VCH Publishers, USA.

Palmer, T. (1995) *Understanding enzymes*, 4th edition, Horwood (Ellis) Ltd, UK.

Eisenthal, R. and Danson, M.J. (eds) (1992) *Enzyme assays – a practical approach*, IRL Press, UK.

Rothe, G.M. (1994) *Electrophoresis of enzymes*, Springer-Verlag, UK.

Davis, J.M. (1995) *Basic cell culture: a practical approach*, Oxford University Press, UK.

9 Carbohydrates

Key topics

- General structure and function
- Chemical methods of carbohydrate analysis
- Enzymic methods of carbohydrate analysis
- Separation and identification of carbohydrate mixtures

Carbohydrates are aldehydes or ketones of higher polyhydric alcohols or components that yield these derivatives on hydrolysis. They occur naturally in plants (where they are produced photosynthetically), animals and microorganisms and fulfil various structural and metabolic roles. Monosaccharides are the simplest carbohydrates and they often occur naturally as one of their chemical derivatives, usually as components of disaccharides or polysaccharides.

Difficulties are encountered in the qualitative and quantitative analysis of carbohydrate mixtures because of the structural and chemical similarity of many of these compounds, particularly with respect to the stereoisomers of a particular carbohydrate. As a consequence, many chemical methods of analysis are unable to differentiate between different carbohydrates. Analytical specificity may be improved by the preliminary separation of the components of the mixture using a chromatographic technique prior to quantitation and techniques such as gas–liquid and liquid chromatography are particularly useful. However, the availability of purified preparations of many enzymes primarily involved in carbohydrate metabolism has resulted in the development of many relatively simple methods of analysis which have the required specificity and high sensitivity and use less toxic reagents.

The preparation of the sample prior to its analysis will depend upon the nature of both the sample and the analytical method chosen and may involve the disruption of cells, homogenization and extraction procedures as well as the removal of protein or other interfering substances. It may be necessary to prevent the decomposition and degradation of the carbohydrate content during such treatments or during storage by the addition of antibacterial agents such as thymol or merthiolate, or substances such as fluoride ions, which will inhibit the enzymic transformation of the carbohydrates.

➤ Cell disruption techniques – see Section 8.4.3.

9.1   General structure and function

Monosaccharides are composed of carbon, hydrogen and oxygen and have the general formula of $C_nH_{2n}O_n$. They are usually classified according to the number of carbon atoms which they contain and the nature of the reactive carbonyl group, i.e. an aldehyde or ketone (Table 9.1). The number of carbon atoms in a monosaccharide can range from three to nine although those most frequently occurring are composed of six or less, of which the hexoses are particularly significant. The monosaccharides are often represented as a linear structure but owing to the reactive nature of the carbonyl group, those composed of five or more carbon atoms usually exist in cyclic forms and are present as such in the more complex carbohydrate molecules.

Table 9.1   Examples of commonly occurring monosaccharides

General class	Number of carbon atoms	Formula	Trivial name	
			Aldose	Ketose
Trioses	3	$C_3H_6O_3$	Glycerose (glyceraldehyde)	Dihydroxyacetone
Tetroses	4	$C_4H_8O_4$	Erythrose	Erythrulose
Pentoses	5	$C_5H_{10}O_5$	Ribose	Ribulose
Hexoses	6	$C_6H_{12}O_6$	Glucose	Fructose

9.1.1   Stereoisomerism

▶ An asymmetric carbon is a carbon atom that is bonded to four different atoms or groups.

Although almost all compounds of biological origin contain some asymmetric carbon atoms, the phenomenon of stereoisomerism is particularly important in the case of carbohydrates. The simplest aldose, glyceraldehyde, has only one asymmetric carbon atom or chiral centre (carbon 2), and therefore only two possible arrangements of the OH and H of the CH_2OH unit are possible. If the OH group is shown as projecting to the left on the penultimate carbon atom in the straight chain representation of the molecule, then by convention the molecule is called the L isomer, while if the OH group is represented as projecting to the right it is known as the D isomer. These two stereoisomers of the same carbohydrates are enantiomers or mirror images of one another.

All carbohydrates can exist in either of these two forms and the prefix of D or L only refers to the configuration around the highest numbered asymmetric carbon atom. Enantiomers have the same name (e.g. D-glucose and L-glucose) and are chemically similar compounds but have different optical properties. The majority of naturally occurring monosaccharides, whether they be aldoses or ketoses, are of the D configuration.

▶ Enantiomers are compounds that are mirror images of each other.

The two isomers of glyceraldehyde (Figure 9.1) are the parent compound of what are known as the D and L series of the aldoses. A similar series of carbohydrates can be derived for the ketoses but because the parent compound

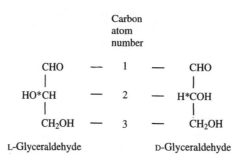

Figure 9.1   Enantiomers of glyceraldehyde. Enantiomers are mirror images of each other and are chemically similar but will cause a beam of plane polarized light to rotate in opposite directions. Glyceraldehyde has only one asymmetric centre (*) and the designation of D or L is determined by the orientation of the O and OH groups about this carbon atom. For carbohydrates with more than one asymmetric carbon atom, the prefix D or L refers only to the configuration about the highest numbered asymmetric carbon atom, although in such enantiomers the configuration about all the asymmetric centres will also be reversed.

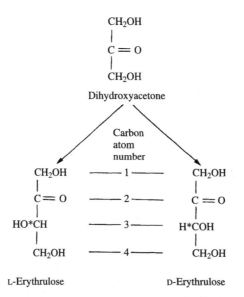

Figure 9.2   Dihydroxyacetone and enantiomers of erythrulose. The triose, dihydroxyacetone, is the parent compound of the ketoses because it has the lowest number of carbon atoms but it does not contain an asymmetric centre. Therefore the D and L series of the ketoses are built up from the two enantiomers of the tetrose, erythrulose, which has one asymmetric carbon atom (*).

dihydroxyacetone does not contain an asymmetric carbon atom, the series is built up from the ketotetrose, erythrulose (Figure 9.2), which is formed when a CHOH group is added to dihydroxyacetone.

Each time the number of carbon atoms is increased by one (with the addition of a CHOH group) a new carbohydrate is produced which has a chemically distinct form and a completely different name. Because this group may be added as either $HO-C-H$ or $H-C-OH$, an extra asymmetric (chiral) centre is created within the molecule, allowing it to exist in two isomeric forms. These two new isomers which differ at only one asymmetric centre are called epimers, and each epimer has its own name (Figure 9.3).

Each series is built up by adding additional CHOH units between carbon 1 and carbon 2 to produce increasing numbers of isomers, the actual number being 2^n, where n is the number of additional chiral centres. Thus there are eight stereoisomers of the D-aldohexoses (Figure 9.3), eight stereoisomers of

> ➤ Epimers are compounds that differ from each other only at one asymmetric centre.

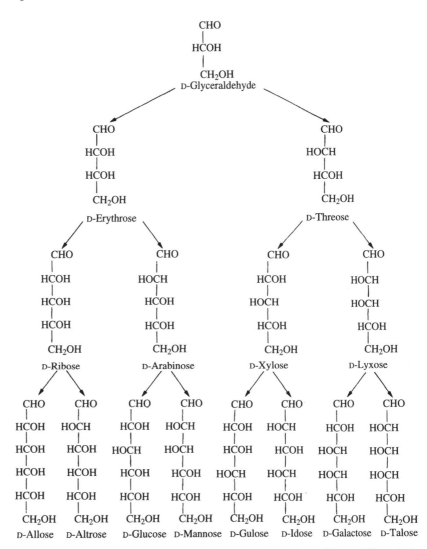

Figure 9.3 **Stereoisomers of the D-aldoses.** D-Ribose and D-arabinose differ only in their configuration about a single carbon atom (carbon 2) and are examples of epimers. Diastereoisomers are stereoisomers which are not enantiomers of each other but are chemically distinct forms, the eight D-hexoses being examples. Some, however, are also epimers of each other, for example D-allose and D-altrose. The number of aldoses in the L series is equal to that of the D series and each compound is an enantiomer of one in the other series.

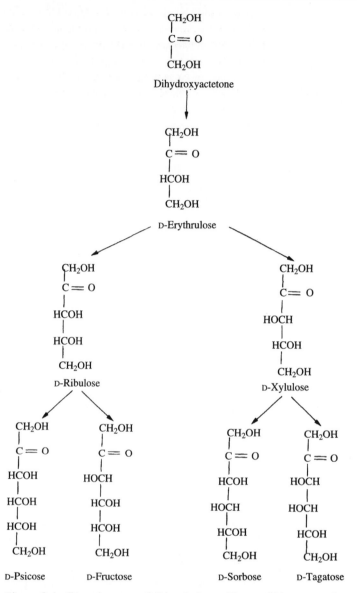

Figure 9.4   Stereoisomers of the D-ketoses. There will be an equal number of isomers in the L series of ketoses.

the L-aldohexoses, and four stereoisomers of both the D-ketohexoses (Figure 9.4) and the L-ketohexoses.

The presence of an asymmetric carbon atom confers the property of optical activity on the molecule, enabling it to cause the rotation of a beam of plane polarized light in either a clockwise or an anticlockwise direction. Thus all naturally occurring carbohydrates containing asymmetric carbon atoms are optically active and can be designated (+) for clockwise (dextro) rotation or (−) for anticlockwise (laevo) rotation. The designation of D or L to glyceraldehyde, the simplest monosaccharide, which has only one asymmetric centre, refers to the capability of these two optical isomers for dextro (+) or laevo (−) rotation of light, respectively. It does not follow, however, that a member of a D series of monosaccharides will automatically be dextrorotatory, although the D and L forms of the same compound will always show opposite directions of rotation. When equal amounts of dextro- and laevorotatory isomers are present, as they are in a synthetically prepared compound, the resulting mixture has no optical activity, is designated DL and is called a racemic mixture.

➤ A racemic mixture contains both enantiomers of the compound.

9.1.2 Structure

An aldehyde and alcohol group can react together to form a hemiacetal, and when this occurs within an aldohexose, a five- or six-membered ring structure is formed. Both pentoses and ketohexoses form similar cyclic structures and it is as such that hexoses and pentoses exist predominantly in nature, with only traces of the aldehyde or ketone forms being present at equilibrium. The cyclic hemiacetal of an aldohexose is formed by reaction of the aldehyde at carbon 1 with the hydroxyl group at carbon 4 or 5 producing an oxygen bridge (Figure 9.5). Similarly the cyclic form of the ketohexose is produced by reaction of the ketone group at carbon 2 with the hydroxyl at carbon 5.

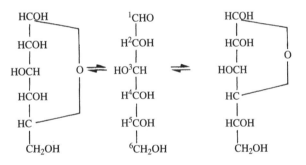

Figure 9.5 **Cyclic, hemiacetal structures of D-glucose.** The reaction between an alcohol and aldehyde group within an aldohexose results in the formation of a hemiacetal. The only stable ring structures are five- or six-membered. Ketohexoses and pentoses also exist as ring structures due to similar internal reactions.

This introduces a new asymmetric centre at carbon 1 and increases the possible number of isomers by a factor of two. The configuration of this new hydroxyl group (the glycosidic hydroxyl group at carbon 1) relative to the oxygen bridge results in two additional isomers of the original carbohydrate (Figure 9.6) which are designated as either the α form (OH group represented on the same side of the carbon atom as the oxygen bridge) or the β form (OH group represented on the opposite side of the carbon atom to the oxygen bridge). Such monosaccharides, differing only in this configuration around the carbon atom to which the carbonyl group is attached (the anomeric carbon), are called anomers.

➤ Anomers are enantiomers formed when the symmetric carbonyl carbon is made asymmetrical by the formation of the cyclical form of a carbohydrate.

These two possible orientations of the hydroxyl group confer differences in optical properties when the substance is present in the crystalline anhydrous

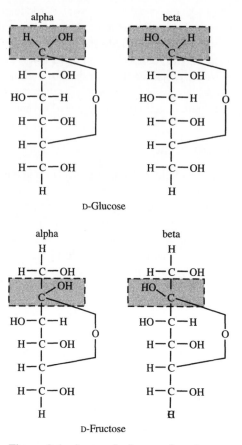

Figure 9.6 **Anomeric forms of D-glucose and D-fructose.** The alpha and beta anomers are named with reference to the configuration of the glycosidic hydroxyl group associated with the oxygen bridge.

form and α-D-glucose has a specific rotation of $+113°$ whereas that of β-D-glucose is $+19.7°$. However, either form in aqueous solution gives rise to an equilibrium mixture which has a specific rotation of $+52.5°$, with approximately 36% being in the α form and 64% in the β form, with only a trace present as the free aldehyde. Because it takes several hours for this equilibrium to be established at room temperature, any standard glucose solution for use with a specific enzyme assay (e.g. glucose by the glucose oxidase which is specific for β-D-glucose) should be allowed to achieve equilibrium before use, so that the proportions of each isomer will be the same in the standard and test solutions. Enzymes that accelerate the attainment of this equilibrium are called mutarotases and can be incorporated in assay reagents in order to speed up the equilibrium formation.

The cyclic forms adopted by the hexoses and pentoses can be depicted as symmetrical ring structures called Haworth projection formulae, which give a better representation of the spatial arrangement of the functional groups with respect to one another. The nomenclature is based on the simplest organic compounds exhibiting a similar five- or six-membered ring

➤ A mutarotase is an enzyme that catalyses the inter-conversion of alpha and beta forms of a carbohydrate.

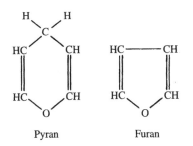

Pyran   Furan

Figure 9.7 **Six- and five-membered cyclic ethers.** The stable ring structures which are adopted by hexoses and pentoses are five- or six-membered and contain an oxygen atom. They are named as derivatives of furan or pyran, which are the simplest organic compounds with similar ring structures, e.g. glucofuranose or glucopyranose for five- or six-membered ring structures of glucose respectively.

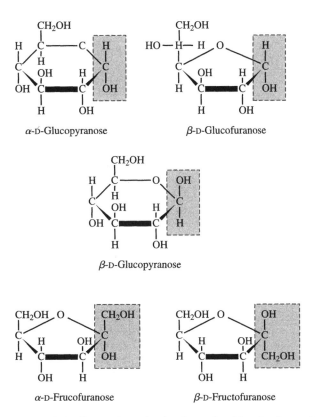

α-D-Glucopyranose   β-D-Glucofuranose

β-D-Glucopyranose

α-D-Frucofuranose   β-D-Fructofuranose

Figure 9.8 **Haworth projection formulae.** The ring is considered to be planar with the substituent groups projecting above or below the plane. The thickened lines represent the portion of the ring that is directed out of the paper towards the reader. The alpha and beta anomeric forms are shown with the hydroxyl group at carbon 1 below or above the plane of the ring respectively.

structure, namely furan and pyran (Figure 9.7). The cyclic five-membered structures of sugars are called furanoses and the six-membered rings are called pyranoses. Thus the name given to the five-membered ring formation adopted by D-glucose is D-glucofuranose and the six-membered ring structure is D-glucopyranose (Figure 9.8). Only traces of glucose occur in the furanose form in solution, the pyranose form being more stable. Galactose, however, as well as some other aldohexoses, does exist in appreciable amounts in the furanose form in solution and as a constituent of polysaccharides. Ketoses may also show ring configuration and the ketohexose, fructose, exists as a mixture of furanose and pyranose forms with the latter form predominating in the equilibrium mixture. However, the configuration of fructose when it occurs as a constituent of a disaccharide or a polysaccharide is usually the furanose form.

In the symmetrical ring structures depicted by the Haworth formulae, the ring is considered to be planar and the configuration of each of the substituent groups represented as being either above or below the plane. Those groups that were shown to the right of the carbon backbone in the linear representation are drawn below the plane of the ring and those that were originally on the left are drawn above the plane. An exception is the hydrogen attached to the carbon 5, the hydroxyl group of which is involved in the formation of the oxygen bridge, which is drawn below the plane of the ring, despite the fact that in the linear representation of the D form it is shown to the left. The orientation of the hydroxyl group at carbon 1 determines the α or β designation, and is shown below the plane for the α form and above the plane for the β form (Figure 9.8).

The planar Haworth projection formulae bear little resemblance to the shape of the six-membered pyranoses that actually adopt a non-planar ring conformation comparable to that of cyclohexane. The chair form is the most

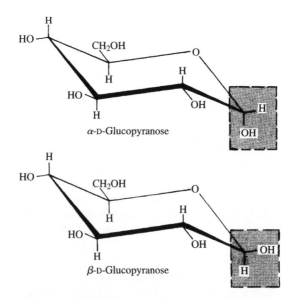

Figure 9.9 **Conformation of D-glucose.** The pyranoses adopt a non-planar ring conformation and the chair form with the highest number of equatorial rather than axial hydroxyl groups is favoured. It should be noted that α-D-glucopyranose, in contrast to β-D-glucopyranose, has an axial hydroxyl group.

accurate representation of the molecule (Figure 9.9) and the configuration of the ring substituents may be described as either equatorial (in the plane of the ring) or axial (perpendicular to the ring). Although two conformations are possible for each simple sugar, in general the form with the least axial constituents is the more stable. It should be noted that with the conversion of β-D-glucopyranose to α-D-glucopyranose, an axial hydroxyl group becomes evident and the number of equatorial and axial ring substituents of the stereoisomers of the aldohexoses will vary. The adopted conformation is of importance in carbohydrate chemistry because the axial or equatorial configuration influences the reactivity of the molecule.

9.1.3   Naturally occurring derivatives of carbohydrates

Monosaccharides are often encountered biochemically as components of complex molecules, in which they are linked to other residues that may or may not be monosaccharides. Such naturally occurring carbohydrates vary in size and complexity ranging from disaccharides, which as the name implies are composed of two monosaccharide units, to the large complex polysaccharides. Homopolysaccharides are polymers of one monosaccharide whereas heteropolysaccharides are composed of more than one type of monosaccharide and may also involve non-carbohydrate residues. Glycoproteins are mainly protein but contain between 5 and 10% carbohydrate by weight. Related molecules which are predominantly polysaccharide are normally referred to as proteoglycans. A whole range of different structures are found that vary in size, proportion of carbohydrate and the nature of the carbohydrate residue. They are prevalent in mammalian tissues, especially mucous secretions, and many important proteins, including some enzymes, serum proteins, pituitary hormones and blood group substances, are glycoproteins.

> ➤ Glycoproteins are proteins to which carbohydrate chains are covalently bound.
> ➤ Proteoglycans are predominantly carbohydrate but also contain protein.

> ➤ Cerebrosides and gangliosides – see Section 12.3.2.

Among the important glycolipids, which are combinations of carbohydrate and lipid, are the cerebrosides and gangliosides. These are constituents of brain and nervous tissue and are usually considered with lipids because they are water-insoluble. Water-soluble polymers of high relative molecular mass containing lipid and carbohydrate, known as lipopolysaccharides, are found in bacterial cell walls.

In many cases the monosaccharides found in these complex structures are present as one of their chemical derivatives, which may be an oxidation or reduction product, a phosphate or sulphate ester or an amino derivative, etc. However, these modified forms of monosaccharides may themselves have important biochemical roles and are not always found incorporated in polysaccharides.

Glycosamines

Monosaccharides that contain an amino group, usually replacing the hydroxyl group on carbon 2, are called glycosamines or amino sugars. They commonly occur as *N*-acetyl derivatives (Figure 9.10) and frequently appear in this form in a variety of heteropolysaccharides. Among these are the mucopolysaccharides (more correctly called glycosaminoglycans, the

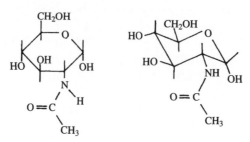

α-D-N-acetylglucosamine (GlcNAc)

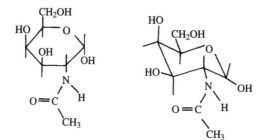

Figure 9.10   N-Acetyl derivatives of monosaccharides.

α-D-N-acetylgalactosamine (GalNAc)

accepted term for polysaccharides that contain amino sugars) and peptidogly-cans. Glycosaminoglycans are found in animal connective tissues and body fluids and are complex structures that also contain uronic acids. Peptidoglycans are components of bacterial cell walls and structures vary between species. They are composed of N-acetyl derivatives linked to short chains of amino acids. Several antibiotics contain glycosamines, including erythromycin, which is an N-methyl derivative and carbomycin, in which the uncommon 3-amino derivative of ribose is found.

Esters

The phosphate esters and, to lesser extent, the sulphate esters of monosaccha-rides are very important naturally occurring derivatives. Metabolism of carbo-hydrates involves the formation and interconversion of a succession of mono-saccharides and their phosphate esters of which glucose-1-phosphate and fruc-tose-6-phosphate are important examples. The sulphate esters of monosaccha-rides or their derivatives (usually esterified at carbon 6) are found in several polysaccharides, notably chondroitin sulphate, which is a constituent of con-nective tissues.

Oxidation products

➤ An aldonic acid is formed when the aldehyde group of a carbohydrate is oxidized to a carboxylic group.

There are three possible classes of sugar acids which may be produced by the oxidation of monosaccharides (Figure 9.11). The aldonic acids are produced from aldoses when the aldehyde group at carbon 1 is oxidised to a carboxylic acid. If, however, the aldehyde group remains intact and only a primary alcohol group (usually at carbon 6 in the case of hexoses) is oxidised then a uronic acid is formed. Both aldonic and uronic acids occur in nature as intermediates in

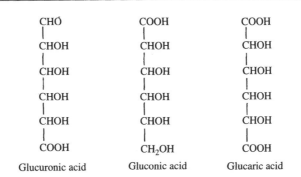

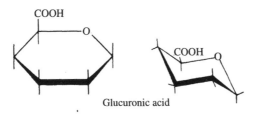

Figure 9.11 **Oxidation products of glucose.** Gluconic acid is an aldonic acid formed when the aldehyde group is oxidized. Glucuronic acid, a uronic acid, is a result of oxidation of the primary alcohol group. When both the aldehyde and the primary alcohol groups are oxidized, glucaric acid is formed, which is an aldaric acid.

> ➤ A uronic acid is formed when a carbohydrate is oxidized to a carboxylic acid but still retains the aldehyde group.

carbohydrate metabolism and the uronic acids are important constituents of the glycosaminoglycans. Glucuronic acid, formed from glucose, plays a particularly useful role in the detoxification of many compounds before their excretion by the kidneys. It is capable of assisting their removal by forming esters with an alcohol or phenolic group and many drugs and acids are excreted in the urine conjugated with glucuronic acid in this way.

Aldaric acids are formed when stronger oxidizing conditions are employed, such as nitric acid, when both the aldehyde and primary alcohol groups are oxidized.

Glycosylamines

Members of this group of compounds, which includes the extremely important nucleosides, are formed when the carbon atom of a monosaccharide, or often its deoxy derivative (Figure 9.12), is linked directly to the nitrogen of a nitrogenous base, including the amino group of an amino acid in a peptide chain (Figure 9.13), with the loss of the hydroxyl group. Such *N*-glycosidic linkages must not be confused with the bonding in a glycoside, which is through an oxygen atom.

> ➤ Nucleic acids – see Section 13.1.

Those nucleosides found in the nucleic acids DNA and RNA involve the joining of ribose of deoxyribose to a purine or a pyrimidine base. One such nucleoside is adenosine, in which a nitrogen of adenine is linked to carbon 1 of the pentose, ribose. In this form it is a component of RNA but as a phosphorylated derivative of adenosine (e.g. ATP), which is a high energy compound, it fulfils an important role in metabolism. The dinucleotides NAD and NADP are two cofactors necessary for many enzymic transformations and these also contain *N*-glycosides of ribose phosphate. Other important nucleosides are found

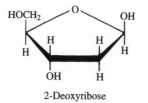

2-Deoxyribose

***Figure 9.12*   Deoxy derivatives.** These contain one less oxygen atom than the monosaccharide from which they are derived. 2-Deoxyribose is a most important deoxy pentose and is a major constituent of deoxyribonucleic acid (DNA). Deoxy hexoses are widely distributed among plants, animals and microorganisms especially as components of complex polysaccharides. Examples are rhamnose (6-deoxymannose), a component of bacterial cell walls, and fucose (6-deoxygalactose), which is often found in glycoproteins and is an important constituent of human blood group substances.

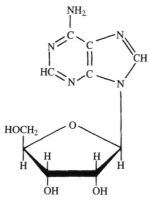

Adenosine

***Figure 9.13*   A glycosylamine.** Adenosine is a nucleoside and is an example of a glycosylamine in which the nitrogen atom of the purine, adenine, is linked directly to carbon 1 of β-D-ribofuranose.

in some antibiotics, puromycin, for example, which is an antibiotic of fungal origin whose structure includes *N*-acetyl ribose linked to a purine base.

Glycosides

Glycosides are an important group of carbohydrate derivatives and are composed of a carbohydrate linked via the hydroxyl group on the anomeric carbon atom (the glycosidic hydroxyl group) to another group, which may or may not be another carbohydrate (Figure 9.14). If this group is not a carbohydrate it is known as the aglycone group and may be a methyl, steroid, glycerol or more complex group. Such a bond is known as a glycosidic bond and involves a condensation reaction between the hydroxyl group at the anomeric carbon of the cyclic form of the carbohydrate and an alcohol group of the other molecule. It is in this way that the monosaccharide units are linked in disaccharides and polysaccharides and also how carbohydrates may be linked to proteins via the hydroxyl group of an amino acid, e.g. serine. Glycosides can vary considerably in their composition and may be found as constituents of

> ➤ A glycosidic bond is formed by a condensation reaction between an alcohol and the hydroxyl group of the anomeric carbon of a carbohydrate.

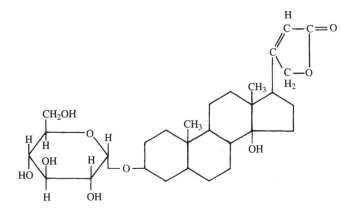

Digitoxenin

Figure 9.14   A steroid glycoside. Digitoxenin is a cardiac-stimulating drug. It is a steroid glucoside in which a compound with a steroid structure is linked to α-D-glucopyranose by condensation of the anomeric hydroxyl group on the carbohydrate and the alcohol group at position 5 on the steroid nucleus.

plant and animal tissues, where they fulfil a variety of roles. Various extracts containing specific glycosides are used as drugs, dyes and flavourings.

The glycosidic bond may be classified as either an alpha or beta linkage depending upon the orientation of the glycosidic hydroxyl group, but a complete description of the linkage should also state the carbon atom through which the oxygen bridge is formed and the ring type of the monosaccharide, i.e. pyranose or furanose. Thus the complete specification of maltose (a disaccharide composed of two glucose units with an alpha linkage) is α-D-glucopyranosyl-(1$\rightarrow$4)-D-glucopyranose (Figure 9.15). The structure of cellobiose, which is also composed of glucopyranose units but involves a beta linkage, is also shown for comparison.

The nature of the glycosidic linkage between two monosaccharides, or the modified forms in which they naturally occur, is an important feature of the structure of di-, oligo- and polysaccharides, especially with respect to their capacity to act as substrates for enzymes. The action of those hydrolases which break glycosidic bonds depends on the nature of the bond as well as the two substances that it links. An example of such stereospecificity is shown by the capacity of human salivary pancreatic amylase to split the α glycosidic linkages between the glucose units in starch but not the β linkages of cellulose. However, the glycosidic hydrolases, although being generally dependent upon the composition of the glycoside, may act on several similar glycosides. Thus testicular hyaluronate glycanohydrolase (EC 3.2.1.35) will cleave $\beta(1\rightarrow4)$ linkages between N-acetyl-D-glucosamine and D-glucuronic acid residues which make up the glycosaminoglycan, hyaluronic acid, but will also degrade chondroitin sulphates, which are composed of D-glucuronic acid linked to sulphated N-acetyl-D-galactosamine.

Disaccharides and polysaccharides form a high proportion of the diet, and are present in natural and processed foods; starch, the polysaccharide

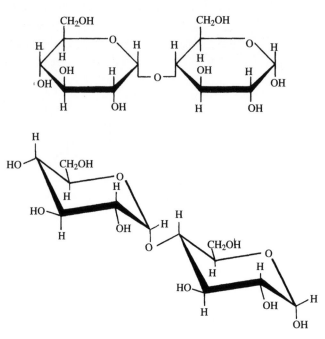

Maltose α (1→ 4) glycosidic linkage

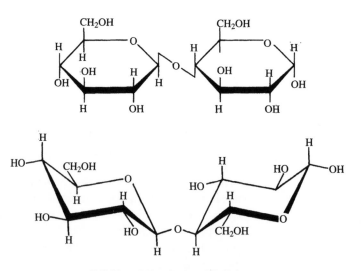

Cellobiose β (1→ 4) glycosidic linkage

Figure 9.15 **Structure of disaccharides.** Maltose and cellobiose are both disaccharides which contain only glucose units, the difference in their structures lying in the way the glucose units are joined. The glycosidic linkage in maltose is α(1→4), whereas that in cellobiose is β(1→4).

> ▶ Homopolysaccharides are carbohydrate polymers consisting of only one monosaccharide.

storage form of glucose (Figure 9.16) found in potatoes and grain, forms a major proportion of the total. The digestion of these dietary carbohydrates relies on the capability of certain enzymes, present in the saliva and intestinal secretions, to hydrolyse the α glycosidic linkage between the monosaccharide units. Thus the polysaccharide, is enzymically degraded with the eventual

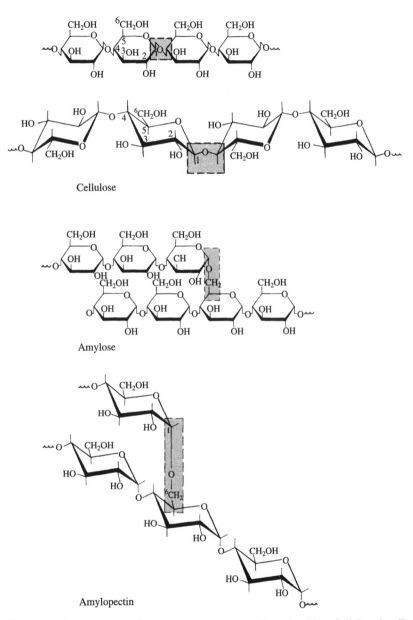

Figure 9.16   Structures of glucopyranose homopolysaccharides. Cellulose is a linear structure of the glucopyranose units linked $\beta(1\rightarrow4)$. Starch consists of amylose, which has a linear $\alpha(1\rightarrow4)$ structure, and amylopectin, which has $\alpha(1\rightarrow6)$ branch points on the linear $\alpha(1\rightarrow4)$ chains. Glycogen has a similar structure to amylopectin, but with a greater degree of $\alpha(1\rightarrow6)$ branching.

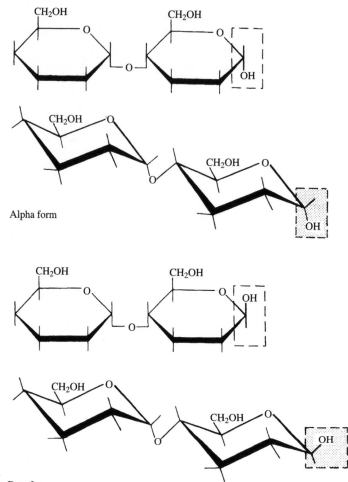

Figure 9.17 **Alpha and beta anomers of maltose.** When only the hydroxyl group on the anomeric carbon of one of the monosaccharide residues is involved in the glycosidic linkage, the anomeric hydroxyl group on the other residue is still free. This permits two possible orientations, which are described as either alpha or beta forms of the monosaccharide residue.

release of the monosaccharides, which are then absorbed from the gastrointestinal tract and subsequently metabolized to produce energy. Such hydrolytic enzymes are a specific for the α linkages and are ineffective towards the β linkages. As a consequence of this, those polysaccharides that are β-linked glucose polymers are of limited dietary value in humans, the structural polysaccharide cellulose being an example. It is, however, a substantial and effective nutritional component of the diet of ruminants, which have a high bacterial population in the rumen capable of metabolizing cellulose with the release of glucose. Polysaccharides also serve as energy reservoirs and the catabolism of glycogen, which is an α-linked polymer of glucose similar to amylopectin and is found in animal muscle and liver tissue, fulfils this function.

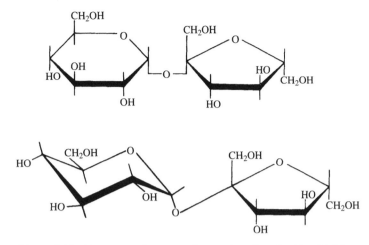

Figure 9.18   **Structure of sucrose.** Sucrose is a disaccharide which has only one possible structural form, there being no free anomeric hydroxyl group.

When the glycosidic linkage involves the anomeric carbon of only one of the monosaccharides, the free aldehyde group on the other monosaccharide can have two possible orientations. Thus, α and β forms of such a compound are possible (Figure 9.17). If the two monosaccharides are linked by an oxygen bridge between the anomeric carbon of each, as is the case with sucrose (Figure 9.18), the possibility of a free aldehyde or ketone group remaining is eliminated and only one structural form is possible which does not have reducing properties.

Self test questions

Section 9.1
1. Identify each of the following carbohydrates as either monosaccharide (M), disaccharide (D) or polysaccharide (P).
   (a) Maltose.
   (b) Cellulose.
   (c) Sucrose.
   (d) Fructose.
2. Which of the following have reducing properties?
   (a) Fructose.
   (b) Sucrose.
   (c) Lactose.
   (d) Glucose-1-phosphate.
3. Starch is of nutritional value to humans
   BECAUSE
   starch is composed of glucose units linked by a $\beta(1\rightarrow4)$ bond.
4. α-D-Glucose and β-D-glucose are anomeric forms of each other
   BECAUSE
   anomers are mirror images of each other.

9.2   Chemical methods of carbohydrate analysis

➤ Hazardous
substances – see
Section 1.2.4.

Many of the earliest methods available for the measurement of carbohydrates
were based on their chemical reactivity and involved the addition of a particular reagent with the subsequent formation of a coloured product. Although
the cheapness and technical simplicity of these procedures contributed to their
previous popularity, they are inherently non-specific and often involve the use
of substances that are now recognized as hazardous. It is essential that the hazard assessment is undertaken before using such techniques.

The lack of specificity is largely attributed to the fact that many monosaccharides have similar chemical composition and properties, although their
biological significance varies widely. The numerous isomeric forms in which
the hexoses exist, all having identical chemical composition, with some even
showing interconversion to an entirely common structure in alkaline solution
(Figure 9.19), demonstrate the difficulties encountered in attempting to measure an individual member of this group in a sample containing others.
Chemical methods are not even capable of differentiating effectively between
classes of monosaccharides because the reactions of the carbonyl group are
used. This function is common to all monosaccharides and thus they may give
similar or sometimes identical reaction products. Nevertheless, providing that
their limitations are appreciated, the use of chemical methods is justified in a
variety of situations.

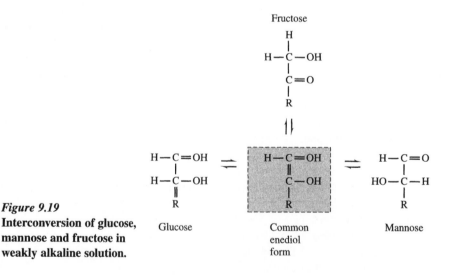

Figure 9.19
**Interconversion of glucose,
mannose and fructose in
weakly alkaline solution.**

9.2.1   Reactions of the carbonyl group

The reactions of the carbonyl group form the basis of many qualitative methods for the detection of carbohydrates and several have been used quantitatively. These are general methods that often only measure the total amount present in the sample. However, in some cases, reagents or reaction conditions
have been modified to improve specificity.

Reduction methods

Carbohydrates that have a potentially free aldehyde or ketone group exist in solution at equilibrium with the enediol form. At a slightly alkaline pH this conversion is favoured and the resulting enediol is an active reducing agent (Figure 9.19). Reduction methods can be used for disaccharides provided that the aldehyde or ketone group of at least one of the monosaccharides has not been eliminated in the glycosidic bond. Sucrose is an example of a disaccharide in which the anomeric carbon atoms of both monosaccharides are involved in the glycosidic bond and the reducing power is lost. However, this distinction between reducing and non-reducing disaccharides can sometimes be used to advantage in qualitative tests.

One of the commonest methods involves the reduction of **cupric ions** (Cu^{2+}) to cuprous ions (Cu^+), which in alkaline solution form yellow cuprous hydroxide, which is in turn converted by the heat of the reaction to insoluble red cuprous oxide (Cu_2O). In the qualitative tests based on this reaction, the production of a yellow or orange–red precipitate indicates the presence of a reducing carbohydrate. It is necessary to keep the cupric salts in solution and to this end Benedict's reagent incorporates sodium citrate while Fehling's reagent uses sodium potassium tartrate. Under carefully controlled reaction conditions, the amount of cuprous oxide formed may be used as a quantitative indication of the amount of reducing carbohydrate present, although different carbohydrates will result in the formation of different amounts of cuprous oxide.

The methods of measuring the amount of cuprous oxide formed are numerous but the one most frequently used involves the reduction of either phosphomolybdic acid or arsenomolybdic acid by the cuprous oxide to lower oxides of molybdenum. The intensity of the coloured complex produced is related to the concentration of the reducing substances in the original sample. The colour produced with arsenomolybdic acid is more stable and the method is more sensitive than with phosphomolybdic acid.

The **neocuproine** method for the measurement of the cuprous oxide is more sensitive than the phosphomolybdic acid reagent and uses 2,9-dimethyl-1, 10-phenanthroline hydrochloride (neocuproine), which produces a stable colour and is specific for cuprous ions.

Although a variety of oxidants other than copper salts have been used **ferricyanide** is the only other one of note. Ferricyanide ions (yellow solution) are reduced to ferrocyanide ions (colourless solution) by reducing carbohydrates when heated in an alkaline solution. The concentration of the carbohydrate can be related to the decrease in absorbance at 420 nm.

$$Fe(CN_6)^{3-} \rightarrow Fe(CN_6)^{4-}$$

The precision of this type of method in which quantitation involves inverse colorimetry (i.e. the absorbance decreases with increasing concentration of the analyte) is questionable, especially at low concentrations of the analyte, because of the difficulty of measuring slight absorbance differences from the high blank reading.

In addition to the lack of specificity of such reduction methods already mentioned, non-carbohydrate reducing substances present in the sample will also react similarly resulting in positive error. Over the years, workers have modified the reagent composition, sample preparation and even the shape of the test-tubes in attempts to reduce interference. Such names as Fehling, Benedict, Nelson, Somogy, Folin and Wu are still associated with these reduction methods. Such considerations are of little consequence nowadays, when reduction methods are less frequently used and the more specific and precise enzymic or chromatographic methods are preferred.

Reactions with aromatic amines

Various aromatic amines will condense with aldoses and ketoses in glacial acetic acid to form coloured products whose absorbance maxima are often characteristic of an individual carbohydrate or group. The use of different aromatic amines or absorbance measurements at alternative wavelengths gives a degree of specificity for individual sugars. The carcinogenic nature of some aromatic amines has ruled out their use as laboratory reagents.

Aromatic amines that have been used include o-toluidine, p-aminosalicylic acid, p-aminobenzoic acid, diphenylamine and p-aminophenol. Their ability to react preferentially with a particular carbohydrate or class of carbohydrate is often useful, e.g. p-aminophenol, which shows some specificity for ketoses compared with aldoses and is useful for measuring fructose. These reagents have proved particularly useful for the visualization and identification of carbohydrates after separation of mixtures by paper or thin-layer chromatography, when colour variations and the presence or absence of a reaction aid the interpretation of the chromatogram.

Reactions with strong acids and a phenol

When heated with a strong acid, pentoses and hexoses are dehydrated to form furfural and hydroxymethylfurfural derivatives respectively (Figure 9.20), the aldehyde groups of which will then condense with a phenolic compound to form a coloured product. This reaction forms the basis of some of the oldest qualitative tests for the detection of carbohydrates, e.g. the Molisch test using concentrated sulphuric acid and α-naphthol.

By careful choice of both the reaction conditions and the phenolic compound used, it may be possible to produce a colour that is characteristic of a particular carbohydrate or related group, so giving some degree of specificity to the method. Thus, Seliwanoff's test uses hydrochloric acid and either resorcinol or 3-indolylacetic acid to measure fructose with minimal interference from glucose. The colour produced by pentoses with orcinol (Bial's reagent) or p-bromoaniline is sufficiently different from that produced by hexoses to permit their quantitation in the presence of hexoses. However, none of the methods based on the formation of furfural or its derivatives can be considered to be entirely specific.

There may be non-carbohydrate substances present in a biological sample that will decompose on heating under the acidic conditions and will react in a similar manner to a carbohydrate. Glucuronic acid is an example and is often present in abundance in urine and may give a false positive reaction for carbohydrates.

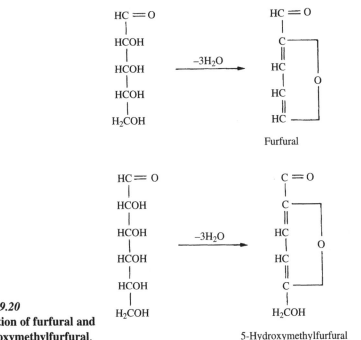

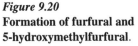

Figure 9.20
**Formation of furfural and
5-hydroxymethylfurfural**.

9.2.2 Structural studies of polysaccharides

Although the reaction with iodine may be used as a qualitative test for the presence of some homopolysaccharides (starch – blue; glycogen and dextrins – red; cellulose and inulin – no colour), the quantitation of polysaccharides is usually achieved by hydrolysis of the glycosidic linkages with the release of the individual components. This can be accomplished by heating at 60 °C with concentrated hydrochloric acid for 30 minutes followed by the quantitation of the monosaccharide using a suitable method. However, this acid hydrolysis procedure may result in the destruction of some carbohydrates and milder conditions using acetic acid or lower concentrations of hydrochloric acid (0.1–2.0 mol l^{-1}) are often preferable. Methanolysis under mild conditions will yield the methyl glycosides, which are especially suitable for analysis by gas–liquid chromatography. Enzymic hydrolysis of specific linkages is also of great value in the investigation of the structure and composition of hetero- and homopolysaccharides, and in such studies it may also be important to measure any non-carbohydrate moieties or monosaccharide derivatives that have been released by chemical or enzymic hydrolysis. Thus the uronic acid residues may be determined in the analysis of certain glycosaminoglycans as may the lipid component of a glycolipid and the sialic acid of a glycoprotein.

Structural investigations into the degree of branching and into the position and nature of glycosidic bonds and of non-carbohydrate residues in polysaccharides may include periodate oxidation and other procedures such as exhaustive methylation. X-ray diffraction and spectroscopic techniques such as nuclear magnetic resonance and optical rotatory dispersion also give valuable information especially relating to the three-dimensional structures of these polymers.

Periodate oxidation

Periodate oxidation involves the simultaneous oxidation and cleavage of carbon to carbon bonds that have adjacent free hydroxyl groups or an aldehyde or ketone group adjacent to a hydroxyl group. The action of periodic acid is represented by the following equation:

$$\begin{array}{c} R \\ | \\ H-C-OH \\ | \\ H-C-OH \\ | \\ R \end{array} + IO_4^- \rightarrow 2R-CHO + H_2O + IO_3^-$$

The reaction with aldoses and ketoses is different and the procedure may be used to distinguish between them. Free monosaccharides and glycosides do not react in an identical manner with periodic acid and this reveals information about the structure, sequence and linkage of monosaccharides or their derivatives in a polysaccharide.

The oxidation may be performed at pH 5.0 by addition of an aqueous solution of sodium metaperiodate. The reaction is allowed to proceed in the dark at 4 °C and after 24 hours the excess periodate is destroyed by the addition of ethylene glycol. The aims in periodate oxidation are to elucidate the number of neighbouring hydroxyl groups by estimating the number of moles of periodate consumed and to determine the structure of the moiety remaining after the reaction. The amount of periodate used in the reaction may be determined in several ways, including titrimetric and spectrophotometric methods.

If the products of periodate oxidation are reduced with sodium borohydride, polyalcohols are produced which may be readily hydrolysed under mildly acidic conditions and the reaction products can be determined. The identification of these products gives considerable information about the linkages between the polysaccharide components and also their sequence in the overall structure.

The laboratory procedures and the interpretation of the results of periodate oxidation studies are complicated exercises and this brief account is intended only as a general introduction to the topic and further details, if required, should be sought from a more specific text.

9.3   Enzymic methods of carbohydrate analysis

There are a large number of enzymes that are capable of modifying carbohydrates or carbohydrate derivatives, and that may be used in various analytical methods. The hydrolytic enzymes, which break glycosidic linkages, are useful in the study of disaccharide or polysaccharide structure and in methods for quantitation (Table 9.2). Such enzymes will hydrolyse the glycosidic linkages between the monosaccharide residues and release the individual components for further analysis. The enzyme is chosen bearing in mind the nature of the glycosidic linkage involved, which may not be unique to one particular disaccharide or polysaccharide. Thus α-glucosidase will hydrolyse both the α(1→4) linkage of maltose and the α(1→2) linkage of sucrose, resulting in the release of glucose in both cases.

Table 9.2 Examples of glycosidases

Hydrolysing enzyme	Glycosidic linkage	Trivial name of substrate	Hydrolysis products
β-Galactosidase (EC 3.2.1.23)	$\beta(1{\rightarrow}4)$	Lactose	D-Galactose. D-glucose
α-Glucosidase (EC 3.2.1.20)	$\alpha(1{\rightarrow}4)$	Maltose	D-Glucose. D-glucose
β-Fructofuranosidase (EC 3.2.1.26)	$\beta(1{\rightarrow}2)$	Sucrose	D-Glucose. D-fructose
α-Galactosidase (EC 3.2.1.22)	$\alpha(1{\rightarrow}6)$	Melibiose	D-Galactose. D-glucose
Amyloglucosidase (EC 3.2.1.3)	$\alpha(1{\rightarrow}4)$ $\alpha(1{\rightarrow}6)$	Glycogen	D-Glucose
Cellulase (EC 3.2.1.4)	$\beta(1{\rightarrow}4)$	Cellulose	D-Glucose

Enzymic methods for the quantitation of monosaccharides are employed when a higher degree of specificity is required than can be achieved by the majority of the chemical methods. They often enable the quantitation of one stereoisomer in the presence of others and can often differentiate between the α and β anomeric forms.

Absolute specificity of an enzyme for only one substrate is rare and there may be several monosaccharides present in a sample that can be acted on to varying degrees by the same enzyme. Hexokinase (EC 2.7.1.1), although used in a method for the measurement of glucose, is not specific for that substrate and will catalyse the phosphorylation of other hexoses. The specificity of a particular enzyme may also vary according to the source from which it was prepared (e.g. microbial, fungal, animal or plant origin) and the method of commercial preparation can affect the purity and thus may influence the results obtained when the enzyme is used. Although it is now possible to purchase many enzymes in a highly purified form, the possibility of the presence of small amounts of an unwanted enzyme, whose substrate is also present in the sample being analysed, should not be discounted.

9.3.1 Assay of glucose using glucose oxidase

The flavoprotein enzyme, glucose oxidase (EC 1.1.3.4), which may be obtained from *Aspergillus niger*, catalyses the oxidation of β-D-glucose by atmospheric oxygen to produce D-gluconolactone, which is converted to gluconic acid with the production of hydrogen peroxide:

$$\beta\text{-D-glucose} + O_2 + H_2O \rightarrow \text{D-gluconic acid} + H_2O_2$$

The oxidation of α-D-glucose occurs at less than 1% of the rate of oxidation of the β anomer. Because these two forms exist in solution in equilibrium in the proportion of 36% (α) and 64% (β), mutarotation of the α to the β form must be allowed to reach equilibrium in the sample and standards for consistent results. The inclusion of aldose-1-epimerase (glucomutarotase) (EC 5.1.3.3) in the glucose oxidase reagent will permit rapid restoration of the α–β equilibrium, effectively enabling the reaction to go to completion.

The rate of oxidation of other monosaccharides (e.g. galactose, mannose, xylose, arabinose and fructose) by glucose oxidase has been shown to be negligible or zero but some derivatives of glucose do react slightly, e.g. 2-deoxy-D-glucose shows a reaction rate of less than 5% of that with β-D-glucose.

Quantitation of glucose using glucose oxidase is achieved by measurement of either the hydrogen peroxide formed or the oxygen consumed during the reaction, both of which are proportional to the β-D-glucose content of the sample.

Measurement of the hydrogen peroxide formed

Spectrophotometric methods

The methods that were originally developed for routine use were colorimetric procedures in which the hydrogen peroxide formed was measured by monitoring the change in colour of a chromogenic oxygen acceptor in the presence of the enzyme peroxidase (EC 1.11.1.7). Such chromogens are colourless in their reduced form but exhibit characteristic colours when oxidized, and *o*-tolidine and *o*-dianisidine were among the substances originally used. However, owing to their potential carcinogenic nature, they have been superseded by other, less toxic, chemicals displaying similar chromogenic properties and offering the methodological advantages of faster reaction time and the production of a more stable coloured product.

> ➤ Oxidized chromogens – see Section 2.1.1.

While glucose oxidase is highly specific for β-D-glucose, the colorimetric determination of hydrogen peroxide is far less specific and significant errors may be introduced in this second stage of the assay reaction. Difficulties will arise if the glucose oxidase preparation is contaminated with catalase, which destroys the hydrogen peroxide by converting it to oxygen and water. In addition some substances, such as ascorbic acid, glutathione and haemoglobin, interfere with the reaction by competing with the chromogen as hydrogen donors. However, some of the more recently introduced chromogens are said to minimize these effects and the reaction involving the peroxidase-catalysed oxidative coupling of 4-amino-phenazone and phenol to produce a coloured complex is widely used. Another commonly used chromogen is 2,2′-azino-di-(3-ethyl-benzthiazolone sulphonic acid), which provides a simple and sensitive assay method.

Procedure 9.1: Quantitation of glucose using glucose oxidase

Reagents

Glucose oxidase reagent

Glucose oxidase (EC 1.1.3.4) 3000 units (50 μkatal)

Peroxidase (EC 1.11.1.7) 5000 units (85 μkatal)

2,2′-Azino-di-(3-ethyl-benzthiazolone) sulphonic acid (ABTS) 1.0 g

Phosphate buffer (0.1 mol l^{-1}) pH 7.0, 1 litre

Method

To 4.0 ml glucose oxidase reagent add 0.1 ml sample.
Mix and allow to react at 30°C for exactly 30 min.
Measure the absorbance at 560 nm.

Standard

A series of standard solutions of glucose (0–20 mmol l^{-1}) should be treated in exactly the same manner as the sample and a calibration graph drawn using the results.

Calculation

The concentration of glucose in the sample is read off the calibration graph using the absorbance value obtained for the sample.

There are numerous commercially produced kits and dry reagent test devices which are available for the determination of glucose using glucose oxidase. They contain all the required reagents although their composition may vary between manufacturers, especially with respect to the chromogenic oxygen acceptor that is used.

➤ Assay kits – see Section 1.1.3.

Electrochemical methods

The electrochemical measurement of the hydrogen peroxide produced forms the basis of instruments often referred to as glucose analysers. Several are commercially available and although the design varies from one manufacturer to another, a common feature of those that amperometrically measure the hydrogen peroxide produced is the use of glucose oxidase in an immobilized form. This is often incorporated in an enzyme electrode which is surrounded by a small chamber of buffered reagents into which the sample is introduced. Other similar biosensor devices have more recently been developed which demonstrate improved specificity and linear range.

➤ Immobilized enzymes – see Section 8.7.
➤ Enzyme electrodes – see Section 8.7.
➤ Biosensors – see Section 4.5.

Alternatively, the immobilized enzyme may be packed in a bed permitting continuous sample analysis, for example of a process stream. The sample is passed through the bed of immobilized enzyme and the hydrogen peroxide produced is monitored amperometrically. Dual channels permit simultaneous analysis of glucose and lactose or glucose and sucrose (Figure 9.21), the disaccharides being hydrolysed enzymically and the glucose content measured. Thus in the analysis of sucrose, an immobilized sucrase breaks the glycosidic linkage to yield glucose and fructose and an immobilized mutarotase ensures the conversion of α-D-glucose to β-D-glucose for its measurement by the glucose oxidase method. The measurement of lactose is achieved similarly using β-galactosidase as the hydrolytic enzyme.

Measurement of oxygen consumed

Alternatively, the initial oxidation of glucose can be monitored and this is most easily achieved by measuring the amount of oxygen consumed during the reaction. An electrochemical method using a polarographic Clark oxygen electrode has been used and the first oxygen electrode to be described for the measurement of glucose in 1962 contained soluble glucose oxidase held between cuprophane membranes. Recent modifications have resulted in the incorporation of a variety of forms of immobilized glucose oxidase into the electrode.

➤ Oxygen electrode – see Section 4.4.3.

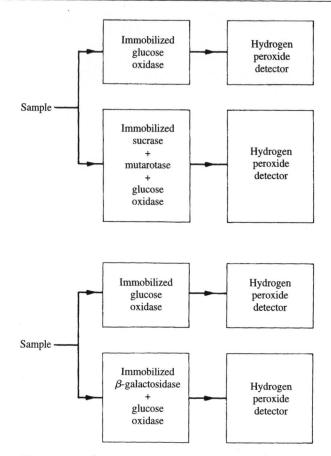

Figure 9.21 **Schematic diagram of the simultaneous continuous automated analysis of glucose and sucrose or glucose and lactose mixtures.**

> ► Catalase is an enzyme that catalyses the conversion of hydrogen peroxide to water and oxygen.

Catalase is a common contaminant of glucose oxidase preparations and will result in erroneous measurement of oxygen consumption owing to the regeneration of oxygen from the hydrogen peroxide produced in the reaction. This error can be prevented by destroying the hydrogen peroxide as it is produced, by adding ethanol, which results in the formation of acetaldehyde:

$$H_2O_2 + ethanol \rightarrow acetaldehyde + H_2O$$

Iodide and ammonium molybdate are also used and rapidly reduce the hydrogen peroxide to water:

$$H_2O_2 + 2I^- + 2H^+ \xrightarrow[molybdate]{ammonium} I_2 + 2H_2O$$

Several types of glucose analysers are commercially available and although the purchase of such an instrument involves considerable initial financial outlay, it will have advantages of speed of analysis (seconds), low reagent volumes and very small sample size (e.g. 10 μl). Glucose analysers are usually simple to operate, not requiring highly trained staff to carry out the analysis, and for these reasons they are becoming increasingly popular.

9.3.2 Assay of glucose using glucose dehydrogenase

Glucose can be measured using bacterial glucose dehydrogenase (EC 1.1.1.47), which catalyses the dehydrogenation of β-D-glucose to D-gluconolactone, the hydrogen being transferred to NAD^+ or $NADP^+$. The method involves measuring the increase in absorbance at 340 nm caused by the production of NADH using either a kinetic or fixed time assay technique. The molar absorption coefficient for NADH is used in the final calculation. A mutarotase (EC 5.1.3.3) should be included in the assay to accelerate the conversion of the α to the β form.

A colorimetric version of the method uses a tetrazolium salt. The oxidation of the NADH is coupled to the reduction of 3-(4,5-dimethyl-2-thiazolyl)-2,5-diphenyltetrazolium bromide by a diaphorase (EC 1.6.4.3) and a deep blue formazan develops. The absorbance is read at a wavelength between 540 and 600 nm. This is more sensitive than the ultraviolet method but is more prone to interference, especially from any reducing agents present in the sample, which will result in a positive error.

Procedure 9.2: Quantitation of glucose using glucose dehydrogenase

Reagents

Glucose dehydrogenase reagent
   Glucose dehydrogenase (EC 1.1.1.47) 3000 units (60 μkatal)
   Mutarotase (EC 5.1.3.3) 100 units (1.5 μkatal)
   Phosphate buffer (0.1 mol l^{-1}) pH 7.6, 1 litre
Coenzyme solution
   Nicotinamide adenine dinucleotide (NAD^+) 30 mmol l^{-1}

Method

Mix in a cuvette:
   2.6 ml glucose dehydrogenase reagent
   0.2 ml sample
   Record the 'baseline' absorbance at 340 nm
Initiate the reaction by adding:
   0.2 ml NAD^+ solution
   Allow to react until the absorbance at 340 nm is maximal and stable (5–10 min).

Calculation

a) Calculate the concentration of NADH formed using the molar absorption coefficient for NADH at 340 nm (6.22×10^3 l mol^{-1} cm^{-1}).

$$\text{Concentration of NADH (mol l}^{-1}) = \frac{\text{Final absorbance} - \text{Baseline absorbance}}{6.22 \times 10^3}$$

b) Calculate the concentration of glucose in the original sample.

$$\text{Glucose concentration (mol l}^{-1}) = \frac{\text{concentration of NADH formed} \times 3.0}{0.2}$$

9.3.3   Assay of glucose using hexokinase

➤ NADP assay – see Section 8.2.

Another enzyme used for the measurement of glucose is hexokinase (EC 2.7.1.1) which catalyses the phosphorylation of glucose to produce glucose-6-phosphate with adenosine triphosphate as the phosphate donor and magnesium ions as an activator. The rate of formation of glucose-6-phosphate can be linked to the reduction of NADP by the enzyme glucose-6-phosphate dehydrogenase (EC 1.1.1.49). This indicator reaction can be monitored spectrophotometrically at 340 nm or fluorimetrically:

$$\text{glucose} + \text{ATP} \xrightarrow{\text{hexokinase}} \text{glucose-6-phosphate} + \text{ADP}$$

$$\begin{array}{l}\text{glucose-6-phosphate} \\ + \text{NADP}^+ \end{array} \xrightarrow{\text{glucose-6-phosphate dehydrogenase}} \begin{array}{l}\text{6-phosphogluconate} \\ + \text{NADPH} + \text{H}^+ \end{array}$$

The enzyme hexokinase is, however, not specific for glucose and is capable of converting some other hexoses to their corresponding 6-phosphate derivatives. Additionally, the specificity of the enzyme may vary slightly depending on its source. Yeast hexokinase will catalyse the phosphorylation of a number of other hexoses as well as glucose including D-mannose, D-fructose, 2-deoxy-D-glucose and D-glucosamine. This provides an assay system for mannose or fructose (Figure 9.22) if the initial reaction is linked to a suitable second reaction.

In the determination of glucose, although there is a lack of specificity of the hexokinase, the overall assay is very specific for glucose because the linking enzyme is specific for glucose-6-phosphate and will react with neither fructose-6-phosphate nor mannose-6-phosphate without the incorporation of phosphoglucose isomerase (EC 5.3.1.9) to convert it to glucose-6-phosphate. Thus it is a requirement for the specific glucose assay that the hexokinase preparation is not contaminated with phosphoglucose isomerase.

9.3.4   Miscellaneous methods for the measurement of hexoses

Two other commonly occurring hexoses which are usually found as components of polysaccharides or combined with other molecules in complex structures are galactose and fructose and, in a similar manner to other monosaccharides, enzymic methods are available for their measurement. An enzymic method for the measurement of fructose using hexokinase was described earlier, together with the method for mannose and glucose (Figure 9.22).

Galactose oxidase (EC 1.1.3.9) catalyses the oxidation of β-D-galactose in a similar manner to the oxidation of β-D-glucose by glucose oxidase and forms the basis of an identical quantitative method. The specificity is less than that of glucose oxidase and although glucose is not a substrate, other monosaccharides and glycosides, including L-altrose, D-talose, D-galactosamine, melibiose and raffinose, may also be oxidized depending upon the source of the enzyme. An assay using galactose dehydrogenase (EC 1.1.1.48), which catalyses the conversion of

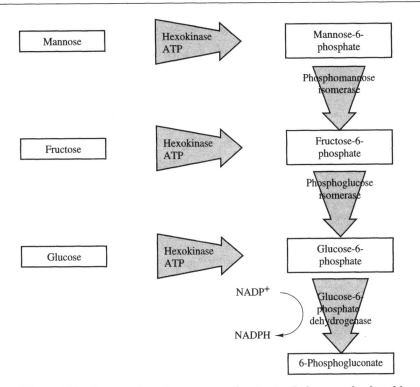

Figure 9.22   Assay systems for mannose, fructose and glucose using hexokinase.
Mannose, fructose and glucose may all be assayed independently using hexokinase and the appropriate additional enzymes. All the reactions can be monitored by the increase in absorbance at 340 nm as NADP+ is reduced to NADPH.

D-galactose to D-galactonolactone in the presence of the coenzyme NAD^+, is more specific and enzyme preparations are available for which the only other substrates are α-L-arabinose and β-D-fucose. The generation of NADH is conveniently monitored at 340 nm and permits quantitation of the galactose:

$$\text{D-galactose} + NAD^+ \xrightarrow{\substack{\text{galactose} \\ \text{dehydrogenase}}} \text{D-galactonolactone} + NADH + H^+$$

9.4   Separation and identification of carbohydrate mixtures

Historically, techniques such as the formation of osazones and the demonstration of fermentation have contributed significantly to the separation and identification of carbohydrates. Observation of the characteristic crystalline structure and melting point of the osazone derivative, prepared by reaction of the monosaccharide with phenylhydrazine, was used in identification. This method is not completely specific, however, because the reaction involves both carbon atoms 1 and 2 with the result that the three hexoses, glucose, fructose and mannose (Figure 9.19), will yield identical osazones owing to their common enediol form.

Fermentation tests are based on the ability of yeast to oxidize the sugar to yield ethanol and carbon dioxide, although only the D-isomers are fermentable and only relatively few of these. Modern chromatographic techniques are, however, much more acceptable and paper and thin-layer techniques are useful for routine separation and semi-quantitation of carbohydrate mixtures, although GLC or HPLC techniques may be necessary for the more complex samples or for quantitative analysis.

9.4.1 Paper and thin-layer chromatography

> ➤ Thin-layer
> chromatography – see
> Section 3.2.1.

Both ascending and descending paper chromatographic techniques have been used and, when thin-layer supports are employed, the use of either silica gel or cellulose is applicable. As the number of carbohydrates present in the sample is often small, the careful choice of solvent will generally make it unnecessary to perform the two-dimensional separations that are often needed when large numbers of substances, such as amino acids, are present. Reference solutions of each carbohydrate can be made up in concentrations of approximately 2 g l^{-1} dissolved in an isopropanol solvent (10% v/v in water) and samples of about 10 μl should give discernible spots after separation.

Solvent systems
Several monophasic solvent systems are useful for the separation of carbohydrate mixtures, and in all those listed in Table 9.3 the smallest solute molecules have the fastest mobility. Thus pentoses have higher R_F values than hexoses, followed by disaccharides and oligosaccharides.

The distance moved by different oligosaccharides is a reflection of the number of monosaccharide units of which they are composed, with the smallest molecules again moving the furthest. However, in general, the distances moved by all classes of carbohydrates are small and although modification of solvent composition may result in greater overall mobility, the relative differences between the components is still low and it may be necessary to run the solvent off the end of the support to achieve a satisfactory separation. In these circumstances the distance moved by glucose is used as a reference and is given the value $R_F = 100$ in any solvent system. The migration of another carbohydrate is reported as its R_g value:

$$R_g = \frac{\text{distance moved by substance}}{\text{distance moved by glucose}} \times 100$$

Locating reagents
A variety of reagents can be used for visualization of the separated components and it may be useful to run duplicate chromatograms and use a different stain on each one to assist in identification of unknown spots. The most commonly used reagents make use of the chemical reactions of carbohydrates already described in the section on quantitative methods and appropriate safety precautions must be taken when using the various locating reagents. The actual reagent composition may be modified either in terms of concentration of components or by substitution of one chemical for another very similar one, although the principle of

Table 9.3   Some monophasic solvents for thin-layer chromatography of carbohydrates

Solvent composition	Proportions*	Comments
Ethyl acetate Pyridine Water	60 30 20	Commonly used. Gives good separation of pentoses and hexoses. Will resolve glucose and galactose (C)
n-Butanol Pyridine Water	60 40 30	Many variations in composition may be used to increase or decrease overall mobilities (C)
Formic acid Methyl ethyl ketone Tertiary butanol Water	15 30 40 15	Gives good separation of monosaccharides and disaccharides (C)
Ethyl acetate Ethanol Pyridine Acetic acid Water	70 10 10 10 10	Useful for separation of pentoses and hexoses (SG)
n-Butanol Acetone Acetic acid Water	35 35 10 20	May be useful if amino acid separations are also performed. R_F values tend to be low, and glucose and galactose are not resolved. Use second after solvents containing pyridine to remove it (C)
n-Butanol Acetic acid Water	2 1 1	Separates monosaccharides and disaccharides. Especially useful for sugar acids. Triple development useful for oligosaccharides (SG)
n-Butanol Acetic acid Diethyl ether Water	9 6 3 1	Separates monosaccharides and disaccharides. Especially useful for methyl glycosides and sugar alcohols. Gives good separation with mono-, di-, tri- and oligosaccharides. Used as second solvent after n-butanol/acetic acid/water (SG)

* Proportions of constituents can be varied.
Recommended supported media: (C), cellulose; (SG), silica gel.

the reaction with a carbohydrate remains unaltered. Such variations in reagent composition may be advantageous in promoting the production of characteristic colours for different carbohydrates (Table 9.4) either within a class (e.g. hexoses with aniline diphenylamine phosphate reagent) or by differentiating on a broader basis (e.g. pentoses from hexoses, or aldoses from ketoses). In certain situations the use of a reagent that incorporates a specific enzyme may be advocated and it will be necessary to be aware of any apparent lack of enzyme specificity and to have a knowledge of all the substrates on which the enzyme will act.

Table 9.4 Colour reactions of common mono- and disaccharides

Carbohydrate	Locating reagents		
	Naphthoresorcinol	4-Aminobenzoic acid	Aniline diphenylamine phosphate
Ribose	—	Red/brown	Blue/green
Xylose	—	Violet	Green/blue
Arabinose	—	Red/brown	Blue/green
Xylulose	Green/brown	Red/brown	Blue/green
Glucose	—	Brown	Grey/blue
Galactose	—	Brown	Grey/blue
Fructose	Red	Pink*	Orange/brown
Sucrose	Red	Brown*	Brown
Lactose	—	Brown	Blue

* Indicates that the reagent shows poor sensitivity for that carbohydrate.

Table 9.5 gives examples of some useful locating reagents, the composition of which may be modified for dipping or spraying techniques, the colours usually appearing after heating at 100°C for 5–10 min. The reaction of acid locating reagents may be impaired if any pyridine present in the chromatographic solvent is not completely removed. This is particularly important when paper and cellulose thin layers are used, although with silica it is not so critical. The pyridine is absorbed by cellulose and cannot be removed completely by oven drying, although the problem can be overcome by dipping the dried chromatogram in *n*-butanol and re-drying before applying the locating reagent or by employing a second solvent system which includes an alcohol, usually *n*-butanol.

9.4.2 Gas–liquid chromatography

> ➤ Gas–liquid chromatography – see Section 3.2.3.

Gas–liquid chromatography may be the method of choice when it is necessary to identify or to quantitate one or more carbohydrates especially when they are present in small amounts. Although this technique is often used because it is possible to resolve carbohydrates with very similar structures, the fact that α and *β* anomers and pyranose and furanose forms of the same carbohydrate all give separate peaks is sometimes a disadvantage.

Gas chromatography must be carried out using volatile derivatives of the carbohydrates and although many have been studied (e.g. *O*-methyl ethers, *O*-acetyl ethers, trimethylsilyl-*O*-methyl oximes and *O*-trimethylsilyl ethers) using a variety of stationary phases, the *O*-trimethylsilyl (TMS) derivatives are probably used most frequently, although circumstances may dictate the use of an alternative. Such TMS derivatives are simple to prepare and have been successfully applied to a wide variety of compounds. It should be noted that, in general, carbohydrates require only weak silylating conditions, otherwise random isomerization will occur with the production of spurious peaks making interpretation of the chromatographic trace very difficult. The use of a mixture of HMDS (hexamethyldisilazane), TMCS (trimethylchlorosilane) and pyridine

Table 9.5 Examples of locating reagents suitable for the TLC of carbohydrates

Reagent	Composition and use	Reaction principle	Comments
Concentrated sulphuric acid	Concentrated acid **Spray**	Dehydration of carbohydrates	General locating reagent for all classes of carbohydrate. Some slight colour differences for different classes (yellow/brown/black). Not suitable for use on cellulose plates.
Orcinol–sulphuric acid	200 mg orcinol in 100 ml 10% sulphuric acid **Spray**	Formation of furfural and derivatives with heat and acid which condense with a phenol	Detects mono-, di-, tri- and oligosaccharides. Colour variations of green/purple/brown
Naphthoresorcinol	200 mg naphthoresorcinol in 3.2 ml 90% orthophosphoric acid in 100 ml methanol **Dip**	Formation of furfural and derivatives with heat and acid which condense with a phenol	Reacts with ketoses. Colours (red/brown) fade at room temperature but not at −20°C. Residual pyridine from solvent will interfere
4-Aminobenzoic acid	1.4 g 4-aminobenzoic acid + 3.2 ml 90% orthophosphoric acid in 100 ml methanol **Dip**	Reaction with aromatic amine in hot acid	Very useful for many different carbohydrates. Very sensitive for pentoses. Colours (red/brown) can be preserved by coating with vinyl from aerosol
Anilinediphenylamine phosphate	1 ml aniline + 1g diphenylamine in 100 ml acetone. Add 10 ml 85% orthophosphoric acid **Dip**	Reaction with aromatic amine in hot acid	Not as sensitive as some reagents but colour variations are useful in interpretation (grey/blue/green/brown)

(2:1:10) is recommended although carbohydrates combined with or containing amino, phosphate, carboxylic groups or nucleic acids will require a more powerful silylating agent and BSA (*N,O*-bis(trimethylsilyl) acetamide) or BSTFA (*N,O*-bis(trimethylsilyl) trifluoroacetamide) in conjunction with TMCS (trimethylchlorosilane) and pyridine are widely used.

The choice of a stationary phase will depend upon the nature of the carbohydrates to be separated and whereas an OV-17 column (phenylmethyl polysiloxy gum) may give satisfactory isomeric separations, a non-polar phase such as OV-1 (methylpolysiloxy gum) may be more useful for a wider range of carbohydrates (Figure 9.23).

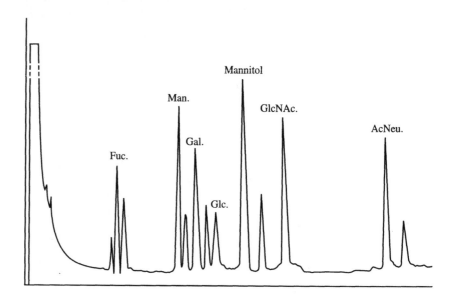

Figure 9.23 **Separation of equimolar concentrations of methylglycosides by gas–liquid chromatography.** The analysis was performed on an OV-1 stationary phase using a temperature gradient from 120 to 220 °C.

L-Fucose and D-galactose each gave three separate peaks corresponding to the furanoside and the alpha and beta methylpyranosides. D-Mannose, D-glucose and *N*-acetylglucosamine (GlcNAc) each gave two peaks due to their alpha and beta pyranosides. D-Mannitol, which was used as an internal standard, and *N*-acetylneuraminic acid (AcNeu) gave single peaks. Under these chromatographic conditions, complete resolution of all the components was not achieved and some had identical retention times, e.g. β-D-methylgalactopyranoside and α-D-methylglucopyranoside.

9.4.3   High performance liquid chromatography

► HPLC – see Section 3.2.2.

Separation and quantitation of carbohydrate mixtures may be achieved using HPLC, a method that does not necessitate the formation of a volatile derivative as in GLC. Both partition and ion-exchange techniques have been used with either ultraviolet or refractive index detectors. Partition chromatography is usually performed in the reverse phase mode using a chemically bonded stationary phase and acetonitrile (80:20) in 0.1 mol l^{-1} acetic acid as the mobile phase. Anion- and cation-exchange resins have both been used. Carbohydrates form anionic complexes in alkaline borate buffers and quaternary ammonium anion-exchange resins in the hydroxyl form can be used for their separation. Problems caused by rearrangement reactions in alkaline solution have been minimized by the use of boric acid/glycerol buffers at pH 6.7. It is possible to separate monosaccharides at this pH by elution with such a buffer to which sodium chloride has been added. This is omitted for di- and trisaccharides and a weaker boric acid/glycerol buffer is used which extends their elution times and permits good resolution. Sulphonated cationic exchange resins with metal counter-ions are also useful for carbohydrate analysis. Water or mildly acidic eluents are normally used under temperature-controlled conditions, e.g. 85 °C. Resins are available in a variety of forms. The calcium or lead form is generally recommended for monosaccharides and disaccharides whereas, for oligosaccharides, the silver or sodium form is preferred.

Self test questions

Sections 9.2/3/4

1 Which of the following enzymes act on a carbohydrate?
   (a) Catalase.
   (b) Cellulase.
   (c) Amylase.
   (d) Peroxidase.

2 In the glucose oxidase assay of glucose which of the following are used to measure the hydrogen peroxide produced?
   (a) A diaphorase and a tetrazolium salt.
   (b) A peroxidase and a hydrogen acceptor.
   (c) A peroxidase and an oxygen acceptor.
   (d) A mutarotase and an aromatic amine.

3 Glucose-6-phosphate dehydrogenase is used as an indicator enzyme in the hexokinase assay of glucose
   BECAUSE
   hexokinase is specific for the phosphorylation of glucose.

4 Refractive index detectors are not suitable for use in the HPLC of carbohydrates
   BECAUSE
   carbohydrates are readily detected by their natural fluorescence.

9.5   Further reading

Chaplin, M.F. and Kennedy, J.F. (eds) (1994) *Carbohydrate analysis*, 2nd edition, IRL Press, UK.

Birch, G.G. (ed.) (1985) *Analysis of food carbohydrate*, Elsevier Applied Science Publishers, UK.

Bochkov, A.F., Zaikov, G.E. and Afanasiev, V.A. (1991) *Carbohydrates*, VSP, Netherlands.

10 Amino acids

- General structure and properties
- General reactions
- N-terminal analysis
- Reactions of specific amino acids
- Separation of amino acid mixtures
- Amino acid analyser

Amino acids are organic molecules of low relative molecular mass (approximately 100–200) which contain at least one carboxyl (COOH) and one amino (NH$_2$) group and are essential constituents of plant and animal tissues. The variations that occur between the different amino acids lie in the nature of their R groups (side chains) (Figure 10.1), a feature that is of fundamental importance and confers individuality upon each amino acid. From a knowledge of the chemical nature of the R group, predictions can be made regarding the properties of a particular amino acid and, conversely, a knowledge of the properties of an amino acid under investigation will assist in the identification of the R group and hence of the amino acid.

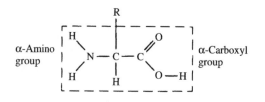

Figure 10.1   General structure of an α-amino acid. The part of the molecule shown inside the box is common to all α-amino acids while the R represents the side chain, which is different for each amino acid.

There are several methods available for the quantitation of amino acids which only give information about the total amino acid content of the sample regardless of whether one or several amino acids are present and cannot differentiate between the individual components. When it is necessary to detect a particular amino acid in the presence of others, in theory a method is chosen which makes use of a chemical or physical characteristic that is specific to the amino acid in question. In practice this is usually difficult to achieve and many such methods show varying degrees of specificity. Thus, some of the most useful and widely employed methods involve the separation of the various amino acids in the sample by a chromatographic or electrophoretic technique followed by the qualitative or quantitative determination of each component using one of the general colorimetric or fluorimetric methods. However, if the aim of the analysis is to detect and identify the various amino acid components without prior separation, a more specific reagent may be used where the reaction depends on the presence of a particular type of amino acid. Such reagents are also often used as visualization agents with paper and thin-layer chromatographic or electrophoretic techniques.

10.1   General structure and properties

> ➤ Proteins are composed of α-amino acids and α-imino acids.

There are approximately 20 amino acids found in proteins, all of which are α-amino acids with the exception of the two α-imino acids proline and hydroxyproline (Figure 10.2), which for the purpose of this discussion will be considered with the amino acids because of their similarity. The α-amino acids are so called because the amino group is attached to the α-carbon of the chain which is, by convention, the carbon atom adjacent to the carboxyl group. Succeeding carbon atoms are designated β, γ, δ and ε (Figure 10.3). Hence in

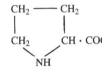

Figure 10.2
α-Imino acids.

Proline

Hydroxyproline

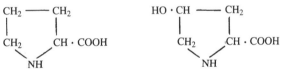

Figure 10.3
Structure of alternative forms of amino acids.

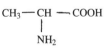

$$\overset{\epsilon}{C} - \overset{\delta}{C} - \overset{\gamma}{C} - \overset{\beta}{C} - \overset{\alpha}{C} - COOH$$

α-Alanine

β-Alanine

$$CH_3 - CH - COOH$$
$$\quad\quad\; | $$
$$\quad\quad NH_2$$

$$CH_2 - CH_2 - COOH$$
$$\; | $$
$$NH_2$$

α-Aminobutyric acid

γ-Aminobutyric acid

$$CH_3 - CH_2 - CH - COOH$$
$$\quad\quad\quad\quad\; | $$
$$\quad\quad\quad\; NH_2$$

$$CH_2 - CH_2 - CH_2 - COOH$$
$$\; | $$
$$NH_2$$

α-amino acids both the amino group and the carboxyl group are attached to the same carbon atom. Many naturally occurring amino acids not found in protein, but which are of importance either in metabolism or as constituents of plants and antibiotics, have structures that differ from the α-amino acids. In these compounds the amino group is attached to a carbon atom other than the α-carbon atom and they are called β, γ, δ or ϵ amino acids accordingly (Table 10.1).

Table 10.1   Some naturally occurring amino acids not found in proteins

Amino acid	Metabolic significance or tissue source	Formula
α-Aminobutyric acid	Plant and animal tissues	CH_3—CH_2—$\underset{\underset{NH_2}{\vert}}{CH}$—$COOH$
α,γ-Diaminobutyric acid	Antibiotics	$\underset{\underset{NH_2}{\vert}}{CH_2}$—$CH_2$—$\underset{\underset{NH_2}{\vert}}{CH}$—$COOH$
β-Alanine	Coenzyme A	$\underset{\underset{NH_2}{\vert}}{CH_2}$—$CH_2$—$COOH$
γ-Aminobutyric acid	Brain tissue	$\underset{\underset{NH_2}{\vert}}{CH_2}$—$CH_2$—$CH_2$—$COOH$
α,ε-Diaminopimelic acid	Bacterial cell wall	$COOH$—$\underset{\underset{NH_2}{\vert}}{CH}$—$(CH_2)_3$—$\underset{\underset{NH_2}{\vert}}{CH}$—$COOH$

10.1.1   Classification

It is helpful when considering the principles and applications of methods for the determination of amino acids to be able to appreciate the characteristics of these compounds. Although it is not always essential to know the exact structural formula of individual amino acids it is useful to be able to remember particular properties or the presence of functional groups.

Table 10.2   Classification of amino acids based on chemical structure

Chemical nature of R group	Examples
Aliphatic	Gly, Ala, Val, Leu
Aromatic	Phe, Tyr, Trp
Hydroxylic	Ser, Thr
Carboxylic	Asp, Glu
Sulphur containing	Cys, Met
Imino	Pro, Hyp
Amino	Lys, Arg
Amide	Asn, Gln

It is convenient to group together amino acids with similar R groups that show common chemical or physical characteristics. The classification may be based on the chemical nature of the R group and such a system (Table 10.2) makes it easier to remember the general properties of each amino acid. However, a classification system based on the polarity of the R group may also be useful and this, together with the structural formula and abbreviated forms of the α-amino acids, is given in Table 10.3.

Table 10.3 Amino acids

Common name	Abbreviation	Structure	pI	Comments
Uncharged polar R groups				
Glycine	Gly	$H—\overset{\displaystyle NH_2}{\underset{\displaystyle COOH}{C}}—H$	6.0	Hydrogen atom on the side chain – not optically active
Serine	Ser	$H—\overset{\displaystyle NH_2}{\underset{\displaystyle COOH}{C}}—CH_2OH$	5.7	Hydroxyl group in side chain
Threonine	Thr	$H—\overset{\displaystyle NH_2}{\underset{\displaystyle COOH}{C}}—\overset{\displaystyle H}{\underset{\displaystyle OH}{C}}—CH_3$	6.5	Hydroxyl group in side chain – has two asymmetric carbon atoms
Hydroxyproline	Hyp	$H—\overset{\displaystyle HN}{\underset{\displaystyle COOH}{C}}\langle\overset{CHOH}{\underset{CH_2}{}}$ ring with CH₂	5.8	The hydroxyl group is added to proline after synthesis into protein and is only found in collagen and gelatine – has two asymmetric carbon atoms
Cysteine	Cys	$H—\overset{\displaystyle NH_2}{\underset{\displaystyle COOH}{C}}—CH_2—SH$	5.0	Contains sulphur – forms disulphide bonds
Tyrosine	Tyr	$H—\overset{\displaystyle NH_2}{\underset{\displaystyle COOH}{C}}—CH_2—$ (benzene ring) $—OH$	5.7	Phenolic aromatic side chain
Asparagine	Asn	$H—\overset{\displaystyle NH_2}{\underset{\displaystyle COOH}{C}}—CH_2—CONH_2$	5.4	Amide of aspartic acid

Table 10.3 Amino acids (*cont.*)

Common name	Abbreviation	Structure	p*I*	Comments
Glutamine	Gln	$\begin{array}{c}NH_2\\|\\H—C—CH_2—CH_2—CONH_2\\|\\COOH\end{array}$	5.7	Amide of glutamic acid

Non-polar R groups

Common name	Abbreviation	Structure	p*I*	Comments
Alanine	Ala	$\begin{array}{c}NH_2\\|\\H—C—CH_3\\|\\COOH\end{array}$	6.0	Aliphatic side chain
Valine	Val	$\begin{array}{c}NH_2 \quad CH_3\\|\qquad|\\H—C——CH\\|\qquad|\\COOH \quad CH_3\end{array}$	6.0	Branched aliphatic side chain
Leucine	Leu	$\begin{array}{c}NH_2 \qquad CH_3\\|\qquad\quad|\\H—C—CH_2—CH\\|\qquad\quad|\\COOH \qquad CH_3\end{array}$	6.0	Branched aliphatic side chain
Isoleucine	Ile	$\begin{array}{c}NH_2 \quad CH_3\\|\qquad|\\H—C---CH\\|\qquad|\\COOH \quad CH_2\\\qquad\quad|\\\qquad\quad CH_3\end{array}$	6.0	Branched aliphatic side chain – has two asymmetric carbon atoms
Proline	Pro	$\begin{array}{c}\qquad CH_2\\HN\diagdown\quad|\\\qquad CH_2\\H—C\diagdown\quad|\\\qquad CH_2\\|\\COOH\end{array}$	6.1	Imino acid – distorts the regular ∝-helix structure
Phenylalanine	Phe	$\begin{array}{c}NH_2\\|\\H—C—CH_2—\bigcirc\\|\\COOH\end{array}$	6.0	Aromatic side chain
Tryptophan	Typ	$\begin{array}{c}NH_2\\|\\H—C—CH_2\\|\\COOH \quad NH\end{array}$	5.9	Heterocyclic aromatic side chain
Methionine	Met	$\begin{array}{c}NH_2\\|\\H—C—CH_2—CH_2—S—CH_3\\|\\COOH\end{array}$	5.8	Aliphatic side chain contains sulphur

Table 10.3 Amino acids (*cont.*)

Common name	Abbreviation	Structure	pI	Comments
Polar with an extra carboxyl group				
Aspartic acid	Asp	$H{-}\overset{\displaystyle NH_2}{\underset{\displaystyle COOH}{C}}{-}CH_2{-}COOH$	2.9	Dicarboxylic
Glutamic acid	Glu	$H{-}\overset{\displaystyle NH_2}{\underset{\displaystyle COOH}{C}}{-}CH_2{-}CH_2{-}COOH$	3.2	Dicarboxylic
Polar with an extra ionizable N-containing group				
Lysine	Lys	$H{-}\overset{\displaystyle NH_2}{\underset{\displaystyle COOH}{C}}{-}CH_2{-}CH_2{-}CH_2{-}CH_2{-}NH_2$	9.7	Diamino
Hydroxylysine	Hyl	$H{-}\overset{\displaystyle NH_2}{\underset{\displaystyle COOH}{C}}{-}CH_2{-}CH_2{-}CHOH{-}CH_2{-}NH_2$	9.2	Hydroxy group is added after synthesis into protein – only found in collagen and gelatine
Arginine	Arg	$H{-}\overset{\displaystyle NH_2}{\underset{\displaystyle COOH}{C}}{-}CH_2{-}CH_2{-}CH_2{-}NH{-}\underset{\displaystyle NH}{\overset{\displaystyle}{C}}{-}NH_2$	10.8	Guanidino group – important in the urea cycle
Histidine	His	$H{-}\overset{\displaystyle NH_2}{\underset{\displaystyle COOH}{C}}{-}CH_2{-}C{=}CH$ (heterocyclic imidazole ring: $HN{-}\underset{\displaystyle C-H}{}{-}NH$)	7.6	Heterocyclic imidazole group in side chain

10.1.2 Isomerism

The α-carbon of all amino acids, with the exception of glycine, has four different substituent groups and is therefore an asymmetric carbon atom. Such an atom can exist in two different spatial arrangements which are mirror images of each other. These structural forms of molecules are known as stereoisomers and the common notation of D and L forms is used, a nomenclature that refers to their absolute spatial configuration when compared with that of glyceraldehyde (Figure 10.4).

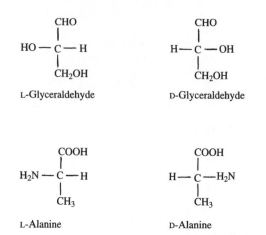

Figure 10.4   **The stereochemical relationship between amino acids and glyceraldehyde.** The designation of D or L to an amino acid refers to its absolute configuration relative to the structure of D- or L-glyceraldehyde respectively. The D and L forms of a particular compound are called enantiomers.

➤ D and L isomers – see Section 9.1.1.

All amino acids except glycine exist in these two different isomeric forms but only the L isomers of the α-amino acids are found in proteins, although many D amino acids do occur naturally, for example in certain bacterial cell walls and polypeptide antibiotics. It is difficult to differentiate between the D and the L isomers by chemical methods and when it is necessary to resolve a racemic mixture, an isomer-specific enzyme provides a convenient way to degrade the unwanted isomer, leaving the other isomer intact. Similarly in a particular sample, one isomer may be determined in the presence of the other using an enzyme with a specificity for the isomer under investigation. The other isomer present will not act as a substrate for the enzyme and no enzymic activity will be demonstrated. The enzyme L-amino acid oxidase (EC 1.4.3.2), for example, is an enzyme that shows activity only with L amino acids and will not react with the D amino acids.

Although the D and L notation is still commonly used in amino acid and carbohydrate terminology, modern nomenclature employs a system that permits the configuration of an asymmetric atom to be specified. This is called the *R–S* convention or the sequence rule and involves assigning a priority ($a>b>c>d$) to the four different substituent atoms attached to the asymmetric atom on the basis of the atomic number of each. The sequence of substituent groups relative to the axis between the asymmetric carbon atom and the substituent group with the lowest priority (d) is used to designate the atom as either *R* (*rectus*, right) or *S* (*sinister*, left) (Figure 10.5).

➤ *R–S* convention is based upon the relative positions of the four different groups attached to the asymmetric carbon atom.

10.1.3   Ionic properties

Amino acids contain both acidic (COOH) and basic (NH_2) groups. As a result they can act as both weak acids and weak bases and are therefore called

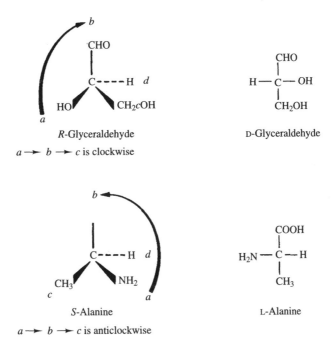

R-Glyceraldehyde

D-Glyceraldehyde

$a \longrightarrow b \longrightarrow c$ is clockwise

S-Alanine

L-Alanine

$a \longrightarrow b \longrightarrow c$ is anticlockwise

Figure 10.5 **Systematic naming of an amino acid using the *R–S* convention.** The compound is named by designating the configuration of the asymmetric carbon atom as either *R* or *S*. For amino acids containing more than one asymmetric centre, e.g. threonine and isoleucine, the configuration about each asymmetric atom is specified.

➤ An ampholyte, depending upon the pH, can act as either a weak acid or a weak base.

ampholytes. Their behaviour is termed amphiprotic because they can either accept or donate a proton, a reaction that can be represented by the following equation:

$$R.NH_3^+.COOH \rightleftharpoons R.NH_2COOH \rightleftharpoons R.NH_2.COO^- + H^+$$

➤ In the zwitterion form, amino acids carry no net charge.

Even this representation is not completely true, because it implies that an amino acid exists in an uncharged form ($R.NH_2.COOH$), whereas the molecule in this state carries one negative and one positive charge and as a result shows no net charge. This is known as the dipolar form or **'zwitterion'** of the amino acid (Figure 10.6).

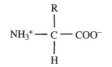

Figure 10.6 **Dipolar or zwitterionic form of an amino acid.** Amino acids exist in a charged form in aqueous solution, the carboxyl group being dissociated and the amino group associated. Some amino acids also have an extra ionizable group present in their side chain (R group). The ionization of each group is pH-dependent and for each amino acid there is a pH at which the charges are equal and opposite and the molecule bears no net charge. This is called the iso-ionic pH (p*I*).

In solution the dissociation of each ionizable group in the molecule may be represented as follows:

$$COOH \rightleftharpoons COO^- + H^+$$
$$NH_3^+ \rightleftharpoons NH_2 + H^+$$

The dissociated and undissociated forms of each group exist in equilibrium with each other and the position of the equilibrium (or the tendency of each of the groups to dissociate) may be expressed in terms of the equilibrium (dissociation constant) K, often termed K_a because it refers to the dissociation of groups that liberate protons, i.e. acids. The actual values for K_a are often very small and are conventionally expressed as the negative logarithm of the value, a term known as the pK_a value:

> ➤ The pK_a value expresses the tendency of an acid to dissociate and become charged.

$$pK_a = -\log K_a$$

This function results in a numerical value which is less cumbersome to use and is comparable with the method of expressing the hydrogen ion concentration of a solution, the pH value:

$$pH = -\log [H^+]$$

The concentration of hydrogen ions liberated by the dissociation of an acid is related to the dissociation constant for that acid and this relationship can be expressed by the Henderson–Hasselbalch equation:

$$pH = pK_a + \log \frac{[salt]}{[acid]}$$

where the square brackets indicate the molar concentration of the named substance.

An examination of this equation reveals the fact that when the concentrations of salt and undissociated acid are equal, then the pH of the solution is numerically equal to the pK_a for that acid. The lower the value of pK_a for an acid, the greater is the ability of the acid to dissociate, yielding hydrogen ions, a characteristic known as the strength of the acid. Amino acids with two ionizable groups, an α-carboxyl and an α-amino group, will be characterized by a pK_a value for each group and the actual value will give an indication of the strength of the acidic or basic group concerned.

The ionization of an amino acid is most easily demonstrated in a titration curve, which can be prepared by titrating a solution of the amino acid in the fully protonated form with a solution of sodium hydroxide (Figure 10.7) and plotting the amount of alkali added against the resulting pH of the solution. The titration curve for a simple amino acid will show two regions where the addition of alkali results in only a small change in the pH value of the mixture. The buffering action of an amino acid is most significant over these pH ranges.

The first end-point in such a titration is due to the carboxyl group and the pK_a value for this is called pK_{a1} while the second pK value is for the amino group and is called pK_{a2}. In practice each acid and its salt will act as a buffer over a pH range of approximately one unit on either side of its pK_a value. For the amino acid alanine, where pK_{a1} is 2.4 and pK_{a2} is 9.6, the most effective buffering action occurs over the pH ranges 2.4 ± 1.0 and 9.6 ± 1.0.

In addition to the α-amino and α-carboxyl groups those amino acids with an extra ionizable group will also have a pK_a value. Glutamic acid is an example of an amino acid with an extra acidic group (COOH) on the γ-carbon, and lysine is an example of an amino acid with an extra amino group on the ϵ-carbon atom. As a result they each have three ionizable groups and three pK_a

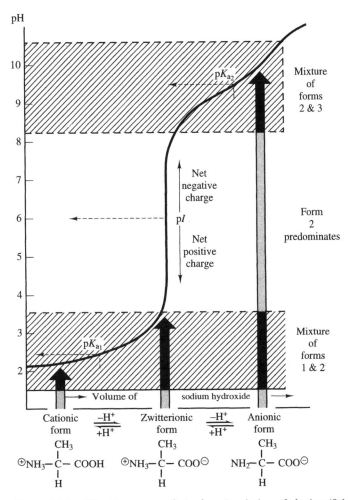

Figure 10.7 **Titration curve of alanine.** A solution of alanine (0.1 mol l^{-1}) in the fully protonated form at pH 2.0 is titrated with 0.1 mol l^{-1} sodium hydroxide. The volumes of sodium hydroxide added are recorded and plotted against the resulting pH values to give a titration curve which is typical of an amino acid with only two ionizable groups (one carboxyl and one amino). The two shaded areas show the pH range over which the addition of alkali results in only a very small change in pH and where the amino acid exhibits its most significant buffering action. At a pH equivalent to pK_{a1}, there are equal amounts of forms 1 and 2, while at a pH equivalent to pK_{a2}, forms 2 and 3 are in equal concentrations. The pI value for alanine is 6.0 and is the mean of pKa_1 (2.4) and pKa_2 (9.6). At a pH below its pI value, an amino acid will carry a net positive charge but it will carry a net negative charge at pH values greater than its pI.

Table 10.4

Amino acid	Formula	pK_a values
Glutamic acid (extra COOH group)	COOH*	4.1
	CH$_2$	
	CH$_2$	
	H — C — NH$_2$*	9.5
	COOH*	2.1
Lysine (extra NH$_2$ group)	NH$_2$*	10.8
	CH$_2$	
	CH$_2$	
	CH$_2$	
	CH$_2$	
	H — C — NH$_2$*	9.2
	COOH*	2.2

The asterisks (*) show where proton gain or loss can occur.

Table 10.5   pK_a values for some amino acids

Amino acid	Extra ionizable group	α-Carboxyl pK_{a1}	α-Amino pK_{a2}	Extra group pK_{a3}
Arginine	Guanidinium	1.8	8.9	12.5
Cysteine	Sulphydryl	1.7	10.8	8.3
Tyrosine	Phenolic	2.0	9.1	10.1
Histidine	Imidazole	1.8	9.2	6.0

values can be demonstrated (Table 10.4), the pK_{a3} value being for the extra group. Other functional groups present in an amino acid may also be ionizable and will have characteristic pK_a values (Table 10.5). Such amino acids result in complex titration curves.

The overall charge carried by an amino acid depends upon the pH of the solution and the pK_a values of the ionizable groups present. If the pH is greater than the pK_a value for a group, a proton will be lost and the molecule will carry a negative charge (Figure 10.7), but if the pH is less than the pK_a value, a positive charge will predominate. The fact that at different pH values different amino acids will be present in different ionic forms and will carry different net charges is utilized in many analytical methods, e.g. electrophoresis and ion-exchange chromatography.

The **iso-ionic point** of a molecule is the pH at which the number of negative charges due to proton loss equal the number of positive charges due to

▶ The iso-electric point (p*I*) of an amino acid is the pH at which it will show no migration in an electric field.

proton gain and the zwitterionic form predominates. The iso-electric point (p*I*) is the pH of the solution at which the molecules show no migration in an electric field and can be determined experimentally by electrophoresis: for amino acids it is equal to the iso-ionic point. The iso-ionic point of an amino acid with one carboxylic and one amino group is the mean of the two pK_a values (Figure 10.7). However, when three ionizable groups are present, the effect of an extra acid group will be to reduce the ionic character of the other acid group and hence the p*I* value will not be the mean of the three separate pK_a values but will more closely approximate to the mean of the closest pK_a values.

10.1.4 Peptides

When two amino acids are linked together by the condensation of the α-amino group from one amino acid and the α-carboxyl group from another to form a peptide bond, the resulting compound is called a dipeptide (Figure 10.8). The ionic character of the constituent amino acids will be modified due to the loss of either an amino or a carboxyl group and the properties of the dipeptide will depend not only on the terminal amino and carboxyl groups but also on any ionizable R groups. For peptides containing increasing numbers of amino acids the significance of the two terminal groups (COOH and NH₂) becomes less important and the ionic nature of the R groups becomes more important. Molecules containing many amino acids linked in such a manner are known as polypeptides and are generally only classed as proteins when they are composed of more than 50 amino acids and their relative molecular mass exceeds 5000.

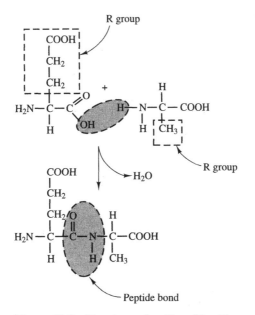

Figure 10.8 **Structure of a dipeptide.** The peptide bond joins glutamic acid and alanine by condensation of the α-carboxyl group of glutamic acid and the α-amino group of alanine. The resulting dipeptide is called glutamylalanine, which can be abbreviated to NH₂-Glu-Ala-COOH or Glu-Ala.

The peptide bonds in proteins are between the α-amino and the α-carboxyl groups but peptides do occur naturally where the peptide linkage involves a carboxyl or amino group which is attached to a carbon atom other than the α-carbon. The amino acid glutamic acid contains two carboxyl groups attached to the α- and γ-carbon atoms and either may be involved in peptide linkages. A dipeptide formed between the γ-carboxyl group of glutamic acid and the amino group of alanine is called γ-glutamylalanine (Figure 10.9). Linkages may also be formed between the ϵ-amino group of lysine and other amino acids. Peptides are also found whose constituents are amino acids other than the α-amino acids; carnosine, for instance, is a dipeptide found in muscle and consists of β-alanine and histidine.

> The term amino acid residue is applied to that part of an amino acid that is not involved in the peptide bond when incorporated in a polypeptide.
> The residues at the ends of a polypeptide chain are designated N-terminal and C-terminal by virtue of the free α-amino and the free α-carboxylic groups respectively.

The only residual evidence of the amino acids that make up polypeptides is their R group and the individual components are now termed amino acid residues. It is conventional when representing peptides to show the terminal amino acid residue with the free amino group to the left of any diagram and to designate it Residue 1 (the N-terminal residue) and the one with the free carboxyl group (the C-terminal residue) on the right of any diagram (Figure 10.10).

Many naturally occurring hormones and antibiotics are polypeptides and investigation into both the amino acid constituents and their sequence in the polypeptide chain are important areas of research. These investigations may reveal information regarding the biologically active part of the molecule, a fact that may then be used in the commercial production of a synthetic peptide

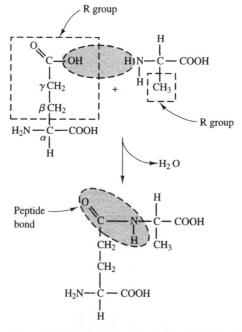

Figure 10.9 **Structure of γ-glutamylalanine.** The dipeptide consists of glutamic acid linked by a peptide bond which involves the carboxyl group attached to the γ-carbon atom and the α-amino group of alanine.

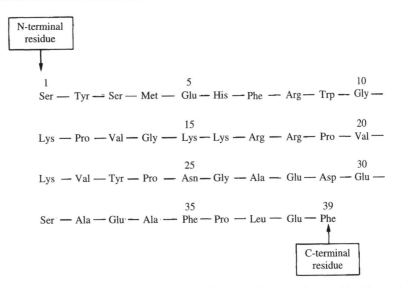

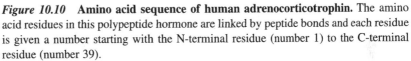

Figure 10.10 **Amino acid sequence of human adrenocorticotrophin.** The amino acid residues in this polypeptide hormone are linked by peptide bonds and each residue is given a number starting with the N-terminal residue (number 1) to the C-terminal residue (number 39).

containing only that small part of the original polypeptide but showing a physiological activity comparable with the whole molecule. This is true for several hormones, a good example being the anterior pituitary hormone, adrenocorticotrophic hormone (ACTH), which is naturally composed of 39 amino acid residues (Figure 10.10), but a synthetic peptide containing only residues 1 to 23 of the original hormone shows comparable physiological activity.

Self test questions

Section 10.1

1. What do all α-amino acids contain?
   (a) At least two amino (NH_2) groups.
   (b) At least two carboxyl (COOH) groups.
   (c) At least one amino (NH_2) group.
   (d) One amino (NH_2) group and one carboxyl (COOH) group.
2. Which of the following amino acids are aromatic (A) and which are basic (B)?
   (a) Glycine.
   (b) Lysine.
   (c) Tyrosine.
   (d) Cysteine.
3. The bond linking amino acids together in proteins is known as the peptide bond
   BECAUSE
   a condensation reaction between an amino group and a carboxyl group produces an amide.

4. At a pH greater than its p*I* value, an amino acid will carry a negative charge
BECAUSE
under appropriate pH conditions, an amino acid can either accept or donate a proton.

10.2   General reactions

There are several compounds that will react with amino acids to give coloured or fluorescent products and as a result can be used in qualitative or quantitative methods. Fluorimetric methods are gaining in popularity and offer some important advantages over absorption spectrophotometry for amino acid analysis.

10.2.1   Ninhydrin

Ninhydrin (triketohydrindene hydrate) reacts with an amino acid when heated under acidic conditions (pH 3–4) to produce ammonia, carbon dioxide and a blue–purple complex. This reaction forms the basis of many widely used methods (Figure 10.11). One mole of carbon dioxide is liberated from each mole of amino acid, exceptions being the dicarboxylic amino acids, which produce two moles of carbon dioxide, and the α-imino acids, proline and hydroxyproline, which do not produce carbon dioxide. Although this formed the basis of a gasometric technique, colorimetric methods are now the most common.

> Molar absorption coefficient – see Section 2.2.3.

The molar absorption coefficient of the coloured product can be used in the quantitation of individual amino acids but this value varies from one amino acid to another and must be determined under the conditions of the assay. An accepted value, however, must be used in the quantitation of the total amino acids in a mixture when absorbance readings are normally taken at 570 nm.

Amines other than α-amino acids will also give a colour reaction with ninhydrin but without the production of carbon dioxide. Thus β-, γ-, δ- and ϵ-amino acids and peptides react more slowly than α-amino acids, to give the blue complex, while imino acids result in the formation of a yellow-coloured product which can be measured at 440 nm. Removal of substances such as protein, ammonia and urea from biological samples may be necessary in quantitative work because they also react in a similar manner.

The ninhydrin colour reaction has proved very useful in qualitative work and is widely used in the visualization of amino acid bands after electrophoretic or chromatographic separation of mixtures. The reagent used in such circumstances is usually prepared in ethanol and, if 2,4,6-collidine is added, the variations in colour produced by different amino acids will aid their identification (Table 10.6).

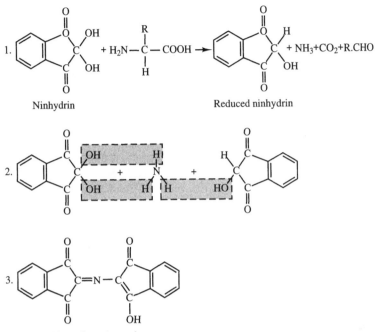

Figure 10.11 The ninhydrin reaction. The overall reaction of amino acids with ninhydrin is:
1. oxidative decarboxylation of the amino acid and the production of reduced ninhydrin, ammonia and carbon dioxide;
2. reduced ninhydrin reacts with more ninhydrin and the liberated ammonia;
3. a blue-coloured complex is formed.

Table 10.6 Modified ninhydrin reagent

Amino acid	Colour produced
Histidine	Brown
Phenylalanine	Brown/grey
Glycine	Blue
Glutamic acid	Bright blue
Lysine	Grey/blue
Tyrosine	Grey
Proline	Orange
Hydroxyproline	Orange
Aspartic acid	Orange/yellow

Reagent composition	
Ninhydrin	2.5 g
2,4,6-Collidine	73 ml
Ethanol	1750 ml
Glacial acetic acid	73 ml

10.2.2 *o*-Phthalaldehyde

➤ Fluorescence – see Section 2.4.

Primary amino acids will react with *o*-phthalaldehyde in the presence of the strongly reducing 2-mercaptoethanol (pH 9–11) to yield a fluorescent product (emission maximum, 455 nm; excitation maximum, 340 nm). Peptides are less reactive than α-amino acids and secondary amines do not react at all. As a result, proline and hydroxyproline must first be treated with a suitable oxidizing agent such as chloramine T (sodium *N*-chloro-*p*-toluene-sulphonamide) or sodium hypochlorite, to convert them into compounds which will react. Similarly cystine and cysteine should also be first oxidized to cysteic acid.

The aqueous reagent is stable at room temperature and the reaction proceeds quickly without requiring heat. The method is approximately ten times more sensitive than the ninhydrin method and is particularly useful when the quantitation of many amino acids is being carried out using amino acid analysers or HPLC. However, the fluorescent yield of individual amino acids varies and fluorescence values must be determined for quantitative work in the same manner as the colour values for ninhydrin.

➤ HPLC – see Section 3.2.2.

10.2.3 Fluorescamine

All primary amines react with fluorescamine under alkaline conditions (pH 9–11) to form a fluorescent product (Figure 10.12) (excitation maximum, 390 nm; emission maximum, 475 nm). The fluorescence is unstable in aqueous solution and the reagent must be prepared in acetone. The secondary amines, proline and hydroxyproline, do not react unless they are first converted to primary amines, which can be done using *N*-chlorosuccinimide. Although the reagent is of interest because of its fast reaction rate with amino acids at room temperature, it does not offer any greater sensitivity than the ninhydrin reaction.

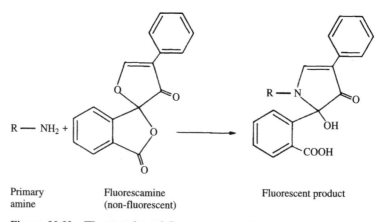

Primary amine

Fluorescamine (non-fluorescent)

Fluorescent product

Figure 10.12 **The reaction of fluorescamine with a primary amino group.** The reaction of an amino acid containing a primary amino group with a solution of fluorescamine in acetone at pH 9.0 results in the conversion of the non-fluorescent fluorescamine to a fluorescent product.

▶ R_F values – see
Section 3.2.1.

10.3   N-terminal analysis

10.3.1   1-Fluoro-2,4-dinitrobenzene

1-Fluoro-2,4-dinitrobenzene (FDNB) reacts in alkaline solution (pH 9.5) with the free amino group of an amino acid or a peptide to form a yellow dinitrophenyl (DNP) derivative (Figure 10.13). This reaction cannot be used for the accurate quantitation of mixtures of amino acids because the molar absorption coefficients of the DNP derivatives of different amino acids vary, but it is very useful in qualitative methods of analysis. The yellow DNP derivatives of free amino acids in a sample can be clearly seen after separation by paper or thin-layer chromatography and, for identification purposes, comparison of the R_F values is made with those of known amino acids treated in an identical manner. FDNB is a hazardous substance and should be used only in exceptional circumstances and under strict safety conditions.

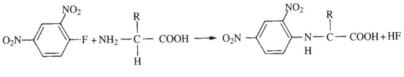

FDNB   Yellow-coloured derivative

Figure 10.13   **The reaction of FDNB with compounds containing a free amino group.** The reaction at pH 9.5 between FDNB and amino acids or peptides results in the formation of yellow-coloured dinitrophenyl derivatives.

The amino group of the N-terminal amino acid residue of a peptide will react with the FDNB reagent to form the characteristic yellow DNP derivative, which may be released from the peptide by either acid or enzymic hydrolysis of the peptide bond and subsequently identified. This is of historic interest because Dr F. Sanger first used this reaction in his work on the determination of the primary structure of the polypeptide hormone insulin and the reagent is often referred to as Sanger's reagent.

10.3.2   Dansyl chloride

Dansyl chloride (dimethylaminonaphthalene-5-sulphonyl chloride) will react with free amino groups in alkaline solution (pH 9.5–10.5) to form strongly fluorescent derivatives (Figure 10.14). This method can also be used in combination with chromatographic procedures for amino acid identification in a similar manner to the FDNB reagent but shows an approximately 100-fold increase in sensitivity. This makes it applicable to less than 1 nmol of material and more amenable for use with very small amounts of amino acids liberated after hydrolysis of peptides. The dansyl amino acids are also very resistant to hydrolysis and they can be located easily after chromatographic separation by viewing under an ultraviolet lamp; see Procedure 10.1.

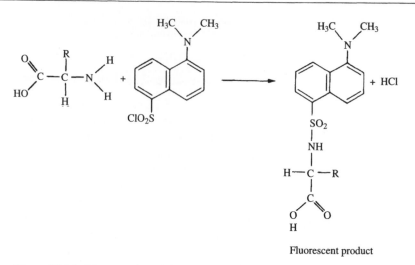

Fluorescent product

Figure 10.14 **The reaction of dansyl chloride with compounds containing a free amino group.** At an alkaline pH, the reaction results in the formation of fluorescent derivatives of free amino acids and the N-terminal amino acid residue of peptides.

Procedure 10.1: Preparation of dansyl derivatives of amino acids for subsequent separation by thin layer chromatography

Reagent
Dansyl chloride solution (in acetone) 25 mg l^{-1}
NB. This reagent is hazardous and must be handled in accordance with approved safety procedures.

Method
1. In small tubes, mix
     10 μl amino acid solution (approximately 1 mmol l^{-1})
     10 μl sodium bicarbonate solution (0.4 mol l^{-1})
     20 μl dansyl chloride reagent
2. Cover the tubes and heat at 37°C for 1 h.
3. Apply about 5 μl of the reaction mixture to a TLC plate and develop the chromatogram using the solvents of choice.
4. Air dry the plates and view under an ultraviolet lamp.

10.3.3   Peptide sequencing

The elucidation of the sequence of amino acids in a polypeptide chain is a complex process but it has been considerably simplified with the introduction of fully automated instruments. These are normally capable of sequencing peptides containing up to about 20 amino acid residues with certainty. Larger molecules must first be split into manageable fragments by chemical or enzymic digestion followed by further cleavage in different positions using residue-specific enzymes in order to produce the overlapping fragments required for full analysis of a protein or long polypeptide.

Peptide sequencers automatically carry out all the reactions of the Edman degradation procedure under controlled conditions, and a typical scheme is described below. The released N-terminal derivatives are then analysed by reverse-phase HPLC.

The sample to be sequenced is first applied to a solid support within an enclosed chamber. A variety of support media are available but glass fibre discs offer the advantage of high reaction efficiencies owing to their large surface area. Pre-treatment of the disc with polybrene confers a slight charge to the surface, which attracts the proteins but does not produce the problems which were experienced when peptides were covalently bound to a glass support.

The programmed cycle (Figure 10.15) begins with the removal of the oxygen in the chamber by purging with argon. Trimethylamine is added to give the alkaline conditions required for the coupling of the phenylisothiocyanate to the N-terminus of the peptide. Trifluoroacetic acid (100%) is then used to cleave the N-terminal derivative, leaving the N-terminus of the

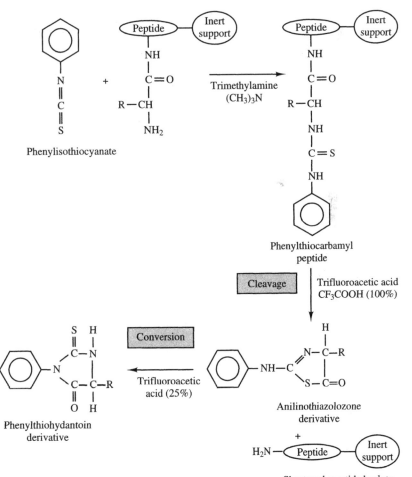

Figure 10.15
Solid-phase peptide sequencing procedure based on Edman degradation technique.

remaining peptide ready for the next cycle. The anilinothiazolozone (ATZ) derivative released is eluted from the disc with butyl chloride into a separate chamber where it is converted to the more stable phenylthiohydantoin (PTH) derivative by the addition of trifluoroacetic acid (25%). After drying with argon and reconstitution in approximately 100 μl of solvent, the sample is ready for analysis by reverse-phase HPLC.

10.4   Reactions of specific amino acids

➤ Absorption maxima of proteins – see Section 11.2.1.

It is often difficult to quantitate one particular amino acid in the presence of others because of chemical similarities. Interference from substances other than amino acids is also a problem in many reputedly specific methods. Ultraviolet spectroscopy is of little value in the detection of aromatic amino acids because they have similar absorbance maxima and considerably different molar absorption coefficients.

10.4.1   Colorimetric methods

There are several colour reagents which are of little quantitative value without the prior removal of interfering substances, although in some cases it may be possible to increase the specificity of the reagent for the determination of a particular amino acid by modifying its composition or altering the reaction conditions. Pauly's reagent (diazotized sulphanilic acid reagent), for instance, reacts with histidine and tyrosine to give a red-coloured product but other phenolic compounds also give this reaction. Erhlich's reagent (p-aminobenzaldehyde in HCl) gives a purple–red product with tryptophan and other indoles and a yellow-coloured product with aromatic amines and ureides, of which urea is the most widely distributed in biological fluids. They are, however, useful qualitatively, especially as locating reagents after the separation of amino acids by electrophoresis or chromatography and their use in multiple-dip sequences aids identification.

10.4.2   Fluorimetric methods

The fluorimetric methods often offer improved specificity and sensitivity over colorimetric procedures and the quantitative assays for the aromatic amino acids tyrosine and phenylalanine illustrate this point.

Tyrosine
1-Nitroso-2-naphthol reacts with tyrosine in the presence of sodium nitrite to form an unstable red compound which is converted, by heating with nitric acid, to a stable yellow fluorescent product. After removal of the excess unreacted nitroso-naphthol, the fluorescence is measured at 570 nm with excitation at 460 nm. This reagent is very hazardous and must be treated accordingly.

Phenylalanine

Phenylalanine reacts with ninhydrin in the presence of a dipeptide (usually glycyl-L-leucine or L-leucyl-L-alanine) to form a fluorescent product. The fluorescence is enhanced and stabilized by the addition of an alkaline copper reagent to adjust the pH to 5.8 and the resulting fluorescence is measured at 515 nm after excitation at 365 nm; see Procedure 10.2.

10.4.3 Microbiological methods

Microbiological assays have been widely used for the quantitation of amino acids because they were, until recently, the most reliable, sensitive and specific tests available. They are applicable to any type of biological material although the presence of activators or inhibitors sometimes causes problems. Such methods lend themselves to the analysis of large batches of samples and are inexpensive, but they are time consuming and not suitable for all amino acids. They have now been largely superseded by ion-exchange chromatography for the quantitation of specific amino acids and will only be discussed briefly.

Certain microorganisms require amino acids for growth and without

---

Procedure 10.2: Fluorimetric determination of phenylalanine

Reagents

Copper reagent
    Sodium carbonate 1.6 g l^{-1}
    Sodium potassium tartrate 65.0 g l^{-1}
    Copper sulphate 60 mg l^{-1}
Buffer pH 5.8
    Sodium succinate (0.3 mol l^{-1})
Dipeptide reagent
    L-Leucyl-L-alanine (5 mmol l^{-1})
or Glycyl-L-leucine (5 mmol l^{-1})
Ninhydrin reagent
    Triketohydrindene hydrate (30 mmol l^{-1})
NB. This reagent is hazardous and must be handled in accordance with approved procedures.

Method

1. Mix
    20 μl sample
    20 μl succinate buffer
    80 μl ninhydrin reagent
    40 μl dipeptide reagent
2. Heat at 60 °C for 2 h and then cool to 20 °C.
3. Add 2 ml copper reagent.
4. Measure the fluorescence at 515 nm after excitation at 365 nm.

Calculation

The fluorescence resulting from an identical treatment of several standard solutions of phenylalanine ($0.1–1.0 \text{ mmol l}^{-1}$) is used to calculate the test concentration.

them they cannot replicate. Many strains of such microorganisms have been produced which show dependence on a particular amino acid. Hence attempts to culture such a microorganism in the presence of only a small amount of that amino acid will result in a limited degree of growth. This can be assessed using turbidimetry or by measuring the increase in lactic acid production by either microtitration or pH change.

A modification of the microbiological assay which utilizes diffusion in gels has been successfully introduced into clinical biochemistry laboratories for the mass screening of blood samples for the raised phenylalanine levels found in phenylketonuria (PKU), which is an inherited disorder of amino acid metabolism. It is often called the 'Guthrie test' after its originators, Guthrie and Susi, and is the most extensively used microbiological assay for the measurement of an amino acid. It is a bacterial inhibition assay and is based on the ability of phenylalanine to counteract the effects of a competitive metabolic antagonist β-2-thienylalanine (Figure 10.16) on the growth of a special strain of *Bacillus subtilis* which requires phenylalanine as a growth factor.

> ➤ The Guthrie test is a bacterial inhibition assay for the measurement of phenylalanine.

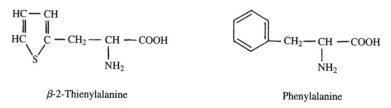

β-2-Thienylalanine   Phenylalanine

Figure 10.16   **The Guthrie test.** The similarity in structure between phenylalanine and its metabolic antagonist, β-2-thienylalanine, provides the basis for a microbiological assay for phenylalanine.

The assay is performed on a layer of agar in which is incorporated a mixture of the suspension of *Bacillus subtilis* spores, the minimum amount of growth nutrients and a fixed amount of the metabolic antagonist β-2-thienylalanine. Blood-soaked filter paper discs of identical diameter (approximately 4 mm) are placed on the surface of the agar together with a range of phenylalanine standards also in the form of blood discs and the agar plates are incubated overnight at 37 °C. Bacterial growth will occur only when the concentration of phenylalanine in the blood discs is sufficient to overcome the effects of the metabolic antagonist, resulting in zones of growth around each disc. The following day the diameter of bacterial growth around each disc is measured and is related to phenylalanine concentration.

10.4.4   Enzymic methods

There are several enzymes that, in theory, may be used for quantitation but because they react with more than one amino acid cannot be used to measure an individual amino acid in a mixture.

The specificity of an enzyme for a particular isomer may be used for measurement of D amino acids in the presence of the L isomers, or vice versa, and the amino acid oxidases are useful in this respect. They catalyse the oxidative deamination of amino acids:

$$\text{amino acid} + O_2 \xrightarrow{\text{amino acid oxidase}} \text{oxo acid} + H_2O_2 + NH_3$$

D-Amino acid oxidase (EC 1.4.3.3) extracted from sheep kidney possesses low selectivity and at pH 8–9 will oxidise many D amino acids, whereas L-amino acid oxidase (EC 1.4.3.2) from snake venom (*Crotalus adamanteus*) at pH 8–9 catalyses the oxidation of many L amino acids. However, as these enzymes show different reactivity towards different amino acids, the results for a sample that contains several D and L amino acids may be difficult to interpret. The use of these enzymes is therefore only recommended for the measurement of one isomer of an isolated amino acid. They may also be used to remove an unwanted isomer from a sample containing both to allow subsequent measurement of the other.

One approach to the measurement of amino acids using an amino acid oxidase is to measure the amount of ammonia formed during the reaction either using an ion-selective electrode or by linking it to the oxidation of NADH by the enzyme glutamate dehydrogenase (EC 1.4.1.3). This reaction is described in Procedure 8.8. In an alternative method the amount of hydrogen peroxide formed is measured either using an ion-selective electrode or by the oxidation of a suitable chromogen using a peroxidase. These procedures are common to other assays employing oxidases (for example, glucose oxidase).

> Ion-selective electrode – see Section 4.1.3.
> Oxidase assays – see Section 9.3.1.

L-Amino acid oxidase has been used to measure L-phenylalanine and involves the addition of a sodium arsenate–borate buffer, which promotes the conversion of the oxidation product, phenylpyruvic acid, to its enol form, which then forms a borate complex having an absorption maximum at 308 nm. Tyrosine and tryptophan react similarly but their enol–borate complexes have different absorption maxima at 330 and 350 nm respectively. Thus by taking absorbance readings at these wavelengths the specificity of the assay is improved. The assay for L-alanine may also be made almost completely specific by converting the L-pyruvate formed in the oxidation reaction to L-lactate by the addition of lactate dehydrogenase (EC 1.1.1.27) and monitoring the oxidation of NADH at 340 nm.

A group of enzymes which may be employed in the measurement of L amino acids are the L-amino acid decarboxylases (EC 4.1.1) of bacterial origin, many of which are substrate specific. They catalyse reactions of the type:

$$\underset{\displaystyle NH_2\!\!-\!\!\overset{\displaystyle \overset{R}{|}}{CH}\!\!-\!\!COOH}{} \rightarrow R\!\!-\!\!CH_2\!\!-\!\!NH_2 + CO_2$$

Some enzymes with improved single amino acid specificity are commercially available. An example is phenylalanine dehydrogenase (EC 1.4.1.1), derived from bacterial sources, which acts on phenylalanine with the simultaneous conversion of NAD to NADH. Quantitation of the phenylalanine is based on determining the amount of NADH produced using standard procedures. In the direct methods, the absorbance at 340 nm is measured, whereas in the colorimetric methods, the reaction is coupled to an electron acceptor

> NADH reactions – see Section 8.3.2.

such as 1-methoxy-5-methyl-phenazium methylsulphate (1-MPMS) or alternatively to a tetrazolium salt.

► Substrate specificity
– see Section 8.1.3.

Variations in the substrate specificity of enzymes derived from different sources does occur and cross-reactivity should always be checked when developing an enzymic assay. This includes an investigation of the interference from a variety of substances that may be present in the sample in addition to studies on amino acid specificity.

Self test questions

> **Sections 10.2/3/4**
> 1. The ninhydrin reaction:
>   (a) is specific for α-amino acids;
>   (b) yields hydrogen peroxide with amino acids;
>   (c) produces yellow-coloured products with imino acids;
>   (d) can be monitored at 570 nm?
> 2. Which of the following reagents (when used under appropriate conditions) react with an N-terminal amino acid residue?
>   (a) Ninhydrin.
>   (b) Dansyl chloride.
>   (c) 1-Fluoro-2,4-dinitrobenzene.
>   (d) Phenylisothiocynate.
> 3. Dansyl chloride reacts with the carboxyl group of an amino acid
>   BECAUSE
>   dansyl derivatives of amino acids are fluorescent.
> 4. Ultraviolet spectroscopy can be used to identify individual amino acids
>   BECAUSE
>   aromatic amino acids show significantly different absorption maxima from each other.

10.5   Separation of amino acid mixtures

The identification and quantitation of the individual amino acids in a mixture is often required in metabolic studies and investigations of protein structure. The use of thin-layer chromatography or electrophoresis may be adequate to indicate the relative amounts and number of different amino acids in a sample but the use of gas–liquid chromatography or an amino acid analyser is essential for quantitative analysis.

10.5.1   Paper and thin-layer chromatography

Paper chromatography has been used successfully for many years and is still a useful tool despite the fact that thin-layer techniques, especially with readily available commercially prepared plastic or foil-backed plates, offer advantages of speed, resolution and easier handling. Larger volumes of sample can be applied to paper, permitting the subsequent elution of a particular amino acid for further purification and analysis, and this may be of particular importance in the identification of an unknown sample constituent.

Prior to chromatography, it may be necessary to remove interfering substances such as protein, carbohydrates and salts and this may be done using an ion-exchange resin. A small column containing a cation-exchange resin, e.g. Zeo-Karb 225, is prepared in the acid form by treating it with hydrochloric acid (2 mol l^{-1}). After the resin column has been washed with water, the acidified sample is applied and the interfering substances washed through with water and discarded. The amino acids are retained on the resin and may be subsequently eluted by adding a small volume of ammonia solution (2 mol l^{-1}) to the column and washing through with distilled water. The alkaline eluent is collected and reduced in volume using a rotary evaporator. Other techniques such as solvent extraction, dialysis or protein precipitation may also be used to separate the amino acids from the other components of the sample.

> ➤ Ion-exchange chromatography – see Section 3.3.1.

Table 10.7 Solvents for both paper and thin-layer chromatography

Solvent	Proportions	Comments
n-Butanol	12	Very widely used for one-way runs or as a first
Glacial acetic acid	3	solvent in two-way chromatography
Water	5	
n-Butanol	7	Very widely used for one-way runs or as a first
Acetone	7	solvent in two-way chromatography
Glacial acetic acid	2	
Water	4	
Phenol	160 g	Gives a large spread of R_F values for a range of
Water	40 ml	amino acids. Must always be used second in any two-way runs. The removal of the phenol is time consuming. Precautions must be exercised in its use because of the toxic and corrosive nature of phenol
Phenol	160 g	Useful for the separation of basic amino acids and
Water	40 ml	must always be used second in any two-way
Ammonia	1 ml	runs. The addition of ammonia causes discoloration of the solvent and so should be prepared frequently. Other comments as above for phenol
n-Butanol	10	This can often replace phenol as the solvent in
Acetone	10	some two-way runs on paper. For thin-layer
Diethylamine	2	work it should be used as the first solvent. The
Water	5	diethylamine must be removed from the paper before locating agents are used
Isopropanol	20	This gives very good separation if used second for
Formic acid	1	thin-layer chromatography
Water	5	

The identification of an amino acid is achieved by comparison of R_F values with those of reference solutions and the use of at least three different solvent systems is recommended before its identity can be established with any degree of certainty.

The nature of the amino acids is an important factor in the choice of a solvent and different solvents will permit better resolution of acidic, basic or neutral components (Table 10.7). In general, increasing the proportion of water in the solvent will increase all R_F values and the introduction of small amounts of ammonia will increase the R_F of the basic amino acids. Some solvents contain noxious chemicals, e.g. phenol, and this may restrict their routine use. The chemical composition may also limit the range of locating reagents which can be satisfactorily applied. For example, sulphanilic acid reagent cannot be used with phenolic solvents.

The resolving power of paper or thin-layer chromatography can be increased by the use of two-dimensional techniques, which involve the use of two different solvent systems. A larger volume ($\times$3) of sample than is normally used for one-dimensional chromatography separation is applied to one corner of the support medium and separation in the first dimension is carried out. The chromatogram is then air dried thoroughly, turned through an angle of 90° and run in the second solvent. After further drying, it is dipped in the chosen locating reagent. In two-dimensional chromatography the composition of the two solvents will determine the order in which they are used (Table 10.7).

Two-dimensional separations permit the resolution of large numbers of amino acids present in a sample and those having a similar mobility in one dimension will usually be separated from each other in the second. This is especially useful in the detection of components that are present only in low concentrations and might be obscured in one dimension by other amino acids that are present in higher concentrations.

Locating reagents

Reagents used for the visualisation of amino acids on the dried chromatogram may be applied either by spraying or dipping. Those commonly used produce intensely coloured bands with approximately 20 nmol of each amino acid for paper chromatography and 5 nmol for thin-layer separations, although smaller amounts can be detected.

Ninhydrin is the most commonly used reagent. If the composition is modified from the original 2.0 gl^{-1} in acetone by the addition of acetic acid and 2,4,6-collidine, the colours produced vary from different amino acids and this greatly aids interpretation and identification (Table 10.6). All the α-amino acids will react in the cold within a few hours and if heat is applied for 10 min at about 100 °C, all compounds containing a primary or secondary amino group attached to an aliphatic carbon atom will react with the formation of various colours. Thus, if a compound yields a colour on heating but not in the cold, it is almost certainly not an α-amino acid. The ninhydrin colours will slowly fade, especially in the presence of strong acid fumes, but are more stable if the chromatogram is stored in the dark at 4 °C. Preservation of the spots can also be achieved by treatment with a solution of a copper or nickel salt but the colours are altered to deep pink. Alternatively cadmium acetate (1.0 g l^{-1}) may be incorporated into the ninhydrin reagent.

Table 10.8 Reagents for the colorimetric detection of amino acids

Common name	Composition and reaction conditions*	Amino acids detected**	Colour produced	Comments
Isatin	2.0 g l⁻¹ in acetone. Heat at 105 °C for 2–3 min	**Pro** **Hyp** Asp Glu Ala Phe Tyr β-Ala Gln	Dark blue Blue/green Light blue Light brown Grey/blue Blue Blue/grey Light blue Light red	Non-specific, mainly for proline but certain sulphur-containing amino acids and some aromatic acids react. Use first in multiple-dip procedures.
Ehrlich	p-Dimethylaminobenzaldehyde (100 g l⁻¹) in conc. HCl. Mix 1 volume with 4 volumes of acetone. No heat required. Reacts within 20 min	**Trp** Citrulline	Pink/red Yellow	Some indoles, aromatic amines and ureides react. Use after ninhydrin in multiple-dip sequences
Pauly	Sulphanilic acid (9 g l⁻¹) in conc. HCl. Mix 1 volume with 10 volumes of water, 1 volume sodium nitrite (50 g l⁻¹) and 1 volume sodium carbonate (100 g l⁻¹)	**His** Tyr	Red Light orange	Some imidazoles and phenolic compounds and ammonium salts react. Colours vary – red, brown, yellow. Use after ninhydrin, isatin or Ehrlich in multiple-dip sequences
Diacetyl	10 g l⁻¹ α-naphthol in 80 g l⁻¹ NaOH plus an equal volume of diacetyl (1 ml per litre water). Mix before use. Heat at 100 °C for 2–3 min	**Arg**	Purple/red	Mono- and di-substituted guanidines, e.g. creatine, creatinine, also react
Cyanide nitro prusside	100 g l⁻¹ of each of the following: sodium hydroxide, sodium nitroprusside, potassium ferricyanide. Mix 1 volume of each solution with 3 volumes of water and stand for 30 min before use. No heat required	**Cys** **Cystine** Homocys Homocystine	Purple Purple Purple Purple	Some sulphur-containing amino acids react. Colours fade in about 30 min

* Details given are for chromatographic or electrophoretic location and will differ for use with liquid samples.

** Amino acids customarily detected are in bold type.

Other reagents that are more specific for particular amino acids have been described (Table 10.8) and their use significantly assists in the identification process. The different locating reagents may be applied either to separate chromatograms or as part of a multi-dip sequence, when they should be used in the recommended order to prevent interference of one reagent by another. Many reagents are hazardous and must be handled in accordance with approved safety procedures.

The formation of **DNP or dansyl amino acid derivatives** followed by chromatography or electrophoresis is a useful technique in certain circumstances. The preparation of DNP derivatives may be indicated when the sample for analysis contains a variety of other substances, removal of which would be complicated, leading possibly to considerable analytical errors. However, the derivative formation and extraction is time consuming and itself can introduce inaccuracies into the analysis and should be used only when it offers an advantage over the separation of untreated amino acids.

The use of dansyl derivatives is not recommended for routine analysis of free amino acids but is very suitable in the identification of an unknown amino acid that has been selectively extracted from the original sample and is present in small quantities. Both kinds of derivative can be easily separated by chromatography or electrophesis and no locating reagent is required for either because the DNP derivatives are themselves yellow in colour and the dansyl derivatives are fluorescent.

10.5.2   Electrophoresis

➤ Electrophoresis – see Section 3.3.2.

The fact that different amino acids carry different net charges at any particular pH permits mixtures to be separated using low or high voltage electrophoresis. The most frequently used supporting media are paper or thin-layer sheets (cellulose or silica gel) and the locating reagents already described for chromatography may be used for visualization of the spots. Separations at high voltages can be achieved more quickly than at low voltages and one of the principal advantages of the former is that salts and other substances that may be present in the sample affect the quality of electrophoretogram to a lesser extent. This permits the separation of amino acids in relatively crude extracts and untreated fluids, whereas prior to low voltage electrophoresis it is necessary to remove interfering substances such as proteins, carbohydrates and salts using the same methods as described for chromatography.

Although electrophoretic separations can be achieved using buffers over a wide range of pH values, in practice the pH values chosen are either pH 2.0 or pH 5.3. At pH 2.0 all amino acids will carry a positive charge and the basic amino acids, having the highest positive charge, will migrate furthest towards the cathode whereas at pH 5.3 migration will occur towards both electrodes depending on the charge carried. Separations at pH 5.3 are particularly useful to determine the acidic or basic nature of an unknown amino acid or dipeptide.

A two-dimensional technique involving initial separation by high voltage electrophoresis at pH 2.0 followed by chromatography is a useful means of separating similar amino acids and short peptides and does not require desalting or excessive purification of the sample (Figure 10.17).

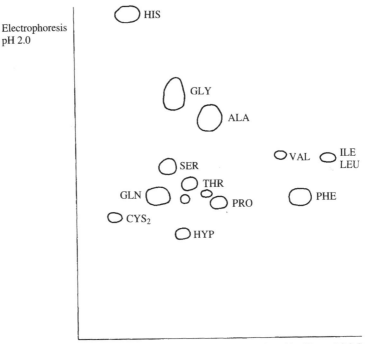

Butanol – Acetic acid – Water (12:3:5)

Figure 10.17   **Two-dimensional high voltage electrophoresis and chromatography of amino acids.** Paper high voltage electrophoresis (4000 V) in an acetic acid–formic acid buffer at pH 2.0 in the first dimension followed by descending chromatography in the second dimension in an *n*-butanol–acetic acid–water solvent (12 : 3 : 5). The spots were visualized with a ninhydrin–collidine reagent.

10.5.3   Gas–liquid chromatography

Derivatives of amino acids (Table 9.10) are required because amino acids are not themselves sufficiently volatile for gas–liquid chromatography and difficulties may be encountered in the choice and method of derivatization. In the past no single column was normally capable of resolving the derivatives of such a diverse group of compounds but the introduction of fused silica capillary columns has resulted in considerably improved resolution.

The advantage of trimethylsilyl (TMS) derivatives lies in the simplicity of the derivatization procedure, which is carried out by the addition of *N,O*-bis(trimethylsilyl)trifluoroacetamide (BSTFA) in acetonitrile and heating for approximately 2 h at 150 °C under anhydrous conditions in a sealed tube. However, there may be problems owing to the formation of multiple derivatives of each amino acid. Another technique involves the formation of *n*-butyl esters of the amino acids and their subsequent trimethylsilylation by a similar procedure. The *n*-butyl esters are formed by heating the amino acids for 15 min in *n*-butanol and HC1 and these are then converted to the *N*–TMS–*n*-butyl ester derivatives. *N*-acyl amino acid alkyl esters are commonly used. Acetylation of the butyl, methyl or propyl esters of amino acids,

Table 10.9 Derivatives of amino acids suitable for gas–liquid chromatography

Derivative	Formula
N — TMS — TMS ester	$$\text{TMS} - \text{NH} - \overset{\displaystyle R}{\underset{\displaystyle H}{\overset{\displaystyle |}{\underset{\displaystyle |}{C}}}} - \text{COOTMS}$$
N — TMS — n-butyl ester	$$\text{TMS} - \text{NH} - \overset{\displaystyle R}{\underset{\displaystyle H}{\overset{\displaystyle |}{\underset{\displaystyle |}{C}}}} - \text{COOC}_4\text{H}_9$$
TFA — n-butyl ester	$$\text{CF}_3 - \text{CO} - \text{NH} - \overset{\displaystyle R}{\underset{\displaystyle H}{\overset{\displaystyle |}{\underset{\displaystyle |}{C}}}} - \text{COOC}_4\text{H}_9$$
HFB — n-propyl ester	$$\text{C}_3\text{H}_7 - \text{CO} - \text{NH} - \overset{\displaystyle R}{\underset{\displaystyle H}{\overset{\displaystyle |}{\underset{\displaystyle |}{C}}}} - \text{COOC}_3\text{H}_7$$

> ➤ Detectors for gas chromatography – see Section 3.2.3.

to give trifluorocetyl (TFA) or heptafluorobutyryl (HFB) derivatives can be performed by reacting them with either TFA or HFB anhydrides in methylene dichloride at 150 °C for 5 min. Electron capture detectors may be used with these derivatives, giving greater sensitivity than with flame ionization detectors.

Until the introduction of capillary columns, it was not possible to separate all the amino acids found in proteins on one column. The choice of stationary phase will depend upon the types of derivatives that have been prepared and in some situations it still may be preferable to use two different columns simultaneously.

Another difficulty in the gas chromatographic separation of amino acids is the choice of detector and it may be necessary to split the gas stream and use two different detectors. The flame ionization detector, which is commonly used, is non-specific and will detect any non-amino acid components of the sample unless purification has been performed prior to derivatization. In addition the relative molar response of the flame ionization detector varies for each amino acid, necessitating the production of separate standard curves. As a consequence, although gas chromatography offers theoretical advantages, its practical application is mainly reserved for special circumstances when a nitrogen detector may be useful to increase the specificity.

10.5.4 High performance liquid chromatography

> ➤ Reverse-phase chromatography – see Section 3.2.2.

The use of reverse-phase columns with pre-column derivatization of the amino acids offers an acceptable alternative to the dedicated instrumentation of an amino acid analyser or separation by HPLC followed by post-column derivatization.

Buffered mobile phases are used and the proportions of polar solvents (e.g. methanol, tetrahydrofuran) depend upon the type of derivative employed. Gradient elution is required for the resolution of complex mixtures and analysis times are less than 1 h. Ultraviolet, fluorescence or electrochemical detectors are used depending upon the nature of the amino acid derivatives.

Derivatization of primary amino acids with *o*-phthalaldehyde (OPA) is simple and the poor reproducibility due to the instability of the reaction product can be improved by automation and the use of alternative thiols, e.g. ethanthiol in place of the 2-mercaptoethanol originally used. An alternative fluorimetric method using 9-fluoroenylmethylchloroformate (FMOC-CL) requires the removal of excess unreacted reagent prior to column chromatography. This procedure is more difficult to automate fully and results are less reproducible. However, sensitivity is comparable with the OPA method with detection at the low picomole or femtomole level, and it has the added advantage that both primary and secondary amino acids can be determined.

Dansyl chloride (5-dimethylaminonaphthlene-1-sulphonyl chloride) and the related dabsyl chloride (4-dimethyl-aminoazobenzene-4′-sulphonyl chloride) have also been used. Sensitivity of the latter method is poor and both suffer problems from interference with reagent excess.

Reaction with phenylisothiocyanate (PITC) in alkaline conditions produces stable phenylthiocarbamyl (PTC) adducts which can be detected either in the ultraviolet below 250 nm or electrochemically. However, this method involves a complex derivatization procedure and offers poorer sensitivity than the alternatives available for individual amino acids. It is useful, however, in conjunction with the automated analysis of peptides when single derivatized residues can be cleaved and analysed after conversion in acidic conditions to phenylthiohydantoins.

10.5.5   Ion-exchange chromatography

Ion-exchange resins may be used to isolate amino acids from other substances prior to analysis, but ion-exchange chromatography is also used to separate mixtures of amino acids. By using a cation-exchange resin and varying the pH of the eluting buffer, it is possible to separate a wide range of amino acids. The amino acid analyser, which also quantifies each amino acid, is based on this technique. Anion-exchange resins tend to be used only in special circumstances for the separation of strongly acidic amino acids.

10.6   Amino acid analyser

Although the chromatographic separation of amino acids on starch columns was introduced in 1941 it was more than 10 years later that Spackman, Stein and Moore developed the first amino acid analyser based on separation by ion-exchange chromatography and quantitation of each component in the column effluent using the ninhydrin reaction. The instruments currently available are based on their original design, permitting buffers of varying pH or ionic strength to be pumped through a thermostatically controlled resin column

(Figure 10.18), although many modifications resulting in improved performance are constantly being introduced. The most significant changes have been the advent of high quality resins, sophisticated automation and increased sensitivity of the detection systems. These have contributed to a reduction in analysis time from days to hours and extended the analytical range to below the nanomole (10^{-9} mol) level.

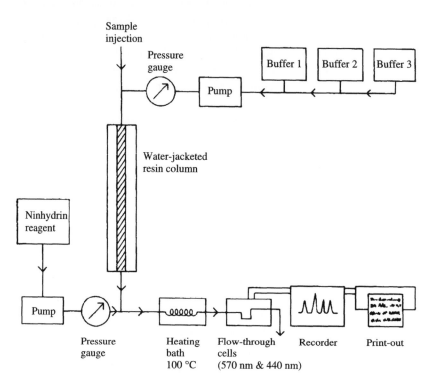

Figure 10.18 **Schematic diagram of an amino acid analyser using the ninhydrin reagent for quantitation.**

10.6.1   Principles of separation

➤ Ion-exchange chromatography – see Section 3.3.1.

The separation takes place in a column of sulphonated cross-linked poly-styrene resin, which is a strong cation exchanger. The matrix of the resin is strongly anionic in nature (SO_3^-) and at the low pH used initially, the amino acids will be positively charged and will be attracted to the negatively charged sulphonate groups.

As the pH of the buffer passing through the column is raised, the amino acids will be differentially eluted as their net positive charge diminishes and they are less strongly attracted to the sulphonate sites on the resin matrix. They will emerge from the column in a sequence relating to their individual p*I* values. At pH 3.25, the pH at which the analysis usually commences, the amino acids with an extra acidic group in their side chain will have very little affinity for the resin and will be the first to emerge from the column, whereas at the same pH value those amino acids whose side chain contains an extra ionizable group capable of carrying a positive charge, for example lysine and histidine, will be strongly held on the resin and will be eluted from the column only as the pH is raised substantially and their net positive charge reduced.

However, it is not only the pH of the eluting buffer that determines the relative elution position of the amino acids, but also the cation concentration of the buffer. Sodium citrate buffer solutions are commonly used and the positive sodium ions compete with the positively charged amino acids for the sulphonic acid sites on the resin:

$$\text{resin—SO}_3^-\ldots{}^+\text{AA} + \text{Na}^+ \rightleftharpoons \text{resin—SO}_3^- \ldots {}^+\text{Na} + \text{AA}^+$$

Although the amino acids have a considerable affinity for the resin, the sodium ions are constantly present in a much higher concentration and, as a result, the equilibrium of the above equation is shifted to the right and the amino acids are displaced from the resin. Thus the molarity of the eluting buffer affects elution and when the ionic concentration of the buffer is increased, the amino acids are eluted more rapidly from the column.

Non-ionic interactions between the amino acids and the resin also influence the elution sequence, allowing amino acids with identical pI values to be eluted separately. The polarity of the amino acid is important in this respect and those with a non-polar, hydrophobic R group interact with the strongly hydrophobic resin matrix. The degree of cross-linking also affects separation because it influences the rate of diffusion of the charged species to the exchange sites. Thus an accurate prediction of the positions of the amino acids on a chromatogram is difficult and may vary considerably from that determined in practice.

The quantitative resolution of amino acid mixtures is achieved by varying the composition of the buffer flowing through the column by either increasing the pH and maintaining a constant molarity (cation concentration) or by keeping the pH constant but varying the molarity, or by using a combination of both. Other factors such as column temperature, buffer flow rate and type of buffer cation play important, but less significant, parts in the quality of the separation.

10.6.2 Practical aspects

The column
In the prototype analysers, two columns were often needed to achieve complete separation of all the amino acids. A 50–100 cm column was used to separate the acidic and neutral amino acids and a 5–10 cm column for the basic amino acids, each with a diameter of 1 or 2 cm, but today's instruments use single columns with narrower diameters. As peak width is proportional to the square root of the column length, these glass or stainless steel columns give narrow peaks and improved separation of closely related amino acids.

Buffers
The composition and pH of the buffer should be accurate to 0.001 mol l^{-1} and 0.01 pH units. Most methods rely on the sequential application of a series of buffer solutions of increasing pH and molarity, with the initial pH around 3.2. Sodium citrate or, preferably, lithium citrate buffers are used, which incorporate a detergent (BRIJ 35), an antioxidant (thiodiglycol) and a preservative

(caprylic acid) and may be used to perform either stepwise or gradient elution. When stepwise elution is being carried out, each buffer is pumped through the column for varying lengths of time and these are chosen with reference to the types of amino acid to be separated and the column dimensions. Buffers often used are pH 3.25, pH 4.25, pH 5.25 and pH 10.0 with either constant or varying molarity if a separation of acidic, neutral and basic amino acids is to be achieved. However, when groups of amino acids with similar ionic characteristics are under investigation, for example those with an acidic R group, it is often necessary to use only one or two of the buffers.

When the technique of gradient elution is employed the buffers are mixed in a predetermined manner to give gradual changes in pH and ionic strength. A two-buffer gradient elution system is used in some models where an acidic buffer, pH 2.2, and an alkaline buffer, pH 11.5, are mixed in varying proportions to achieve increasing pH values. Gradient elution gives improved separation with decreased analysis time and eliminates dramatic baseline fluctuations due to sudden changes in the buffer. A less common practice is to use an 'Iso-pH' system which employs a stepwise or gradient elution using buffers with increasing cation concentration, which may be from 0.2 to 1.6 mol l^{-1} and an almost constant pH value between 3.25 and 3.65.

Temperature

The temperature of the resin column must be carefully maintained to avoid changes in both the pH of the buffers and ionization of the amino acids. Although increasing temperature usually results in faster elution, the effect may be variable for different amino acids and the relative elution positions can be altered, making interpretation of results difficult. The temperature often chosen is 60°C although lower temperatures are sometimes required to resolve two similar amino acids. Temperature programming, which entails an alteration in temperature at a specified time in the separation procedure, is widely used.

Column flow rate

Successful and reproducible separations require a steady buffer flow rate and this is achieved with either a constant pressure or a constant displacement pump. These pumps are designed to deliver a constant rate of fluid independent of the resistance to the flow and recent developments in pump design permit the production of a precise and pulseless flow; this has contributed towards the increased analytical precision and sensitivity that can now be achieved with amino acid analysers. The choice of flow rate is dependent upon the type of resin, the dimensions of the column and overall design of the instrument and this varies between models.

Sample preparation

The volume of sample which can be applied to the resin column will vary for the different instruments commercially available. With progressive refinements in instrumentation the tendency has been towards a decreased sample volume, of 50 μl or less. Correct sample preparation is very important if reproducible results are to be obtained along with trouble-free operation. This will vary according to the nature of the sample and its constituents but the fluid applied to the column must be clean and free from protein and other large molecules. Failure to comply with this requirement will result in a slow clogging of the resin column and may necessitate the removal, cleaning and re-packing of the resin, or purchase of a new column. In general, liquid samples need only filtration or centrifugation to remove artifacts but any protein present must be removed by precipitation with picric acid or salicylsulphonic acid or by dialysis or ultrafiltration. Solid samples, such as foodstuffs and animal or plant tissue, may require homogenization, extraction and deproteinization. Any peptides or proteins under investigation for their amino acid composition must first be hydrolysed to release the free amino acids.

The prepared sample may be applied in pH 2.2 buffer directly to the top of the column and the analysis sequence started, and after elution of all the amino acids and regeneration of the resin column, the next sample can be applied. However, many newer models incorporate an automatic loading device which enables several samples to be stored ready for analysis either in sample cups or in small Teflon coils. After the completion of an analysis the next stored sample is automatically applied to the resin and the buffer cycle restarted.

Detection

A second pump is required to deliver a constant flow of reagent to meet the column effluent. When the reaction has taken place the stream is monitored continuously using a flow-through cell in either a colorimeter or a fluorimeter. Ninhydrin reagent is the most widely used and after the solution has passed through a coil in the heating bath at 100°C, the absorbance is monitored at 570 and 440 nm to detect amino and imino acids respectively. Ninhydrin reagent should be prepared accurately using high quality chemicals. The ninhydrin is dissolved in peroxide-free ethyleneglycol, monomethyl ether (methylcellosolve) and buffered with acetate at pH 5.5. Nitrogen gas is bubbled through during the preparation of the reagent to exclude air and a small quantity of a reducing agent, stannous chloride or titanious chloride, is added to ensure the production of a limited but accurate amount of reduced ninhydrin. Some analysers introduce argon or nitrogen to the column effluent to produce a gas-segmented stream and in these instruments either sodium cyanide or hydrazine is used as the reducing agent. The prepared reagent should be a pale straw colour and must be stored in a dark bottle under nitrogen pressure because it is sensitive to light and oxygen.

The ninhydrin reaction requires heat and therefore the stream of column eluate plus ninhydrin reagent must pass through a coil of narrow-bore tubing (approximate diameter 1 mm) held in a 100°C heating bath. It is important to ensure that the flow is not restricted because excessive heating may cause the ninhydrin to precipitate in these micro-bore tubes, resulting in complete stoppage of the analyser.

The *o*-phthalaldehyde reaction compares favourably with the ninhydrin reaction in several respects. The reagent is stable and is in an aqueous form, which eliminates the use of potentially toxic chemicals and storage under nitrogen. Because the reaction proceeds quickly at room temperature there is no need for the 100°C heating bath with all its inherent problems, and the increased sensitivity permits detection at the picomole level.

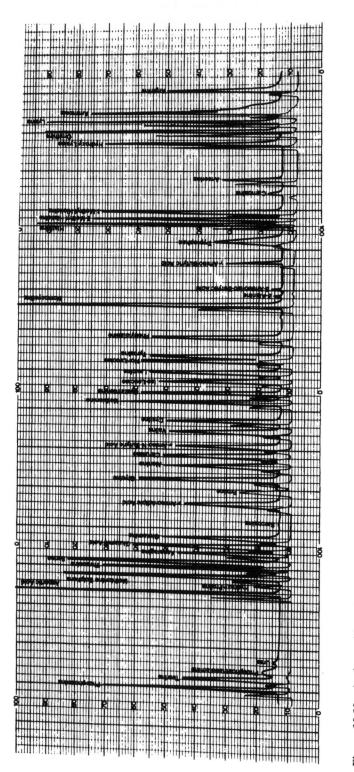

Figure 10.19 Amino acid analyser trace. Separation of a complex physiological standard mixture of amino acids in 3.5 hours using lithium citrate buffers and ninhydrin detection: 10 nmol of each amino acid, including the internal standard, nor-leucine, were applied to the column (0.3 × 35 cm) in a total volume of 50 ml.

(Photography by courtesy of Rank Hilger, Margate, UK.)

10.6.3 Quantitation

➤ Response factor – see Section 3.2.2.

The colour or fluorescence produced per mole of amino acid varies slightly for different amino acids and this must be determined for each one to be quantitated. This is done by loading a mixture of amino acids containing the same concentration of each amino acid including the chosen internal standard and from the areas of the peaks on the recorder trace calculating each response factor in the usual way (Figure 10.19). These values are noted and used in subsequent calculations of sample concentrations.

➤ Internal standard – see Section 3.2.2.

An internal standard should always be used for every analysis carried out. This is an amino acid that is known to be absent from the sample under investigation. For instance in blood plasma analysis either of the non-physiological amino acids, nor-leucine or α-amino-β-guanidinobutyric acid, may be used. This should be added in a known amount to the sample prior to any sample pre-treatment (for example, removal of protein).

If the amount of internal standard which was added to the sample is known, the concentration of the unknown amino acid can be determined using peak area relationships. These calculations must take the various response factors into account.

In this way losses due to sample preparation will be allowed for as will variation in the intensity of the colour or fluorescence produced with different preparations of reagent and changes in analysis conditions.

Self test questions

Sections 10.5/6

1. Formation of derivatives of amino acids is normally required prior to their separation by which of the following techniques?
   (a) Reverse-phase HPLC.
   (b) Gas–liquid chromatography.
   (c) High voltage electrophoresis.
   (d) Ion-exchange chromatography.
2. Elution from the ion-exchange column in an amino acid analyser is affected by which of the following?
   (a) The concentration of the eluting fluid.
   (b) The pH of the eluting fluid.
   (c) The amino acid derivative used.
   (d) The temperature of the column.
3. At pH 2, all amino acids migrate electrophoretically towards the cathode
   BECAUSE
   amino acids carry no net charge at their iso-electric pH.
4. A strong cation-exchange resin is used in the amino acid analyser
   BECAUSE
   amino acids are eluted from a strong cation-exchange resin by a buffer with a pH value below 4.

10.7   Further reading

Allen, G. (1989) Sequencing of proteins and peptides. In Burdon, R.M. and van Knippenberg, P.H. (eds) *Laboratory techniques in biochemistry and molecular biology*, Vol. 9, 2nd revised edition, Elsevier, USA.

Fini, C., Floridi, A. and Finelli, V.N. (1989) *Laboratory methodology in biochemistry: amino acid analysis and sequencing*, CRC Press, UK.

Bailey, P.D. (1992) *An introduction to peptide chemistry*, John Wiley, UK.

11 Proteins

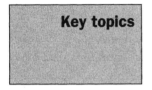

Key topics	• Protein structure
	• General methods of quantitation
	• Separation of proteins

The word protein describes only one type of polymer involving mainly α-amino acids and yet it includes many thousands of different molecules. It is possible to measure the total protein content of a sample despite the fact that relatively simple preparative techniques may be capable of demonstrating the presence of different proteins. However, if interest lies in only one of these proteins, then a measure of the total protein content would be completely inappropriate. Methods for the quantitation of proteins are either suitable for all proteins or designed to measure individual proteins. Such specific methods may depend on either a preparative stage in the analysis or the use of a specific characteristic of the protein in question.

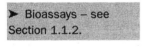
➤ Bioassays – see Section 1.1.2.

The quantitation of a protein that has a specific biological function, a hormone, for instance, may not give a true indication of its biological activity owing to the inactivation of some of the protein. For proteins that have definite biological functions the choice is between chemical quantitation and bioassays. For this reason the catalytic activity of an enzyme is more frequently measured than is its protein concentration.

11.1 Protein structure

➤ The primary structure of a protein is the sequence of amino acids in the chain.

➤ Amino acids – see Section 10.1.

The sequence of amino acids in a polypeptide chain is known as the **primary structure.** This feature is genetically determined and is responsible not only for the final shape of the protein but also for its physical characteristics and, ultimately, for its biological function. Proteins are made up of about 22 amino acids, which are linked by the peptide bond (Figure 11.1), an amide linkage involving the amino group of one amino acid and the carboxyl group of another.

The formation of a peptide bond results in the loss of an amino and a

➤ The peptide bond is formed by a condensation reaction between an amino group on one amino acid and the carboxyl group of another.
➤ Amino acid sequencing – see Section 10.3.3.

carboxyl group of each amino acid, the remaining portions of the molecules (the residues) being the major components in the structure of the protein. However, the amino acids at each end of a polypeptide chain retain either an amino group (the **N-terminal residue**) or a carboxyl group (the **C-terminal residue**). By convention the N-terminal residue is always designated as the number one residue in a numerical sequencing of the amino acids in the protein.

The three-dimensional shape of a polypeptide chain or a portion of a chain is known as the **secondary structure**. In its simplest form the fully extended polypeptide chain would show a structure similar to that indicated in Figure 11.2(a). However, it often assumes a helical structure similar to that shown in Figure 11.2(b) which is stabilized by intra-chain hydrogen

Figure 11.1
The peptide bond.

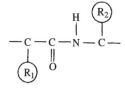

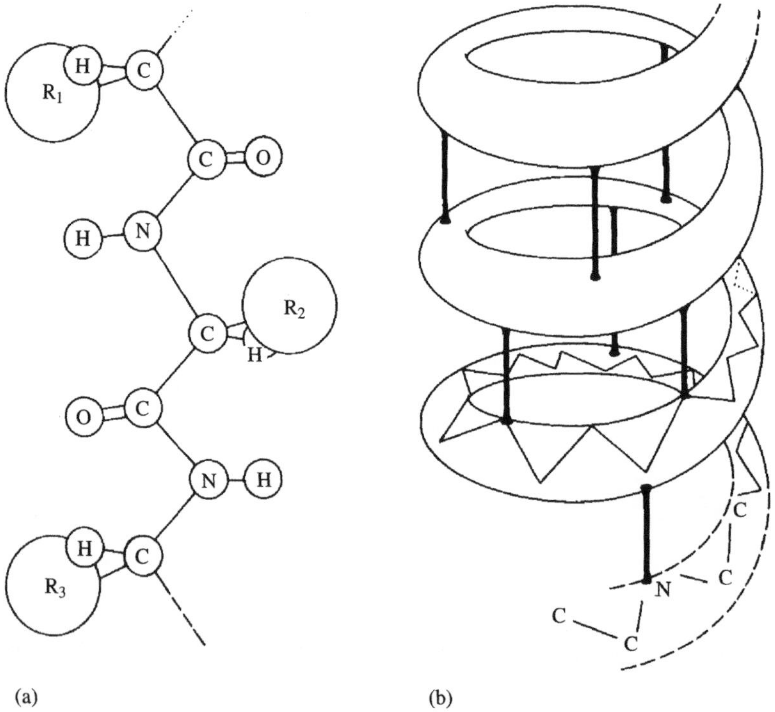

Figure 11.2 **The secondary structure of proteins.** The simplest spatial arrangement of amino acids in a polypeptide chain is as a fully extended chain (a) which has a regular backbone structure due to the bond angles involved and from which the additional atoms, H and O, and the amino acid residues, R, project at varying angles. The helical form (b) is stabilized by hydrogen bonds between the —NH group of one peptide bond and the —CO group of another peptide bond. The amino acid residues project from the helix rather than internally into the helix.

(a)   (b)

bonds formed between the amide hydrogen of one peptide bond and the oxygen of a carbonyl group of another. A hydrogen bond is a non-covalent bond between the slight positive charge induced in a hydrogen atom when it is covalently bonded to an electronegative atom (e.g. nitrogen) and the slight negative charge of another electronegative atom (e.g. oxygen).

Many proteins show not only these two structural features but also inter-mediary forms. The presence of the imino acids proline and hydroxyproline induces bends in the chain due to the variant of the resulting peptide bond (Figure 11.3). Glycine, which effectively has no side chain (H), permits greater flexibility at the peptide bond than do other amino acid residues. Inter-chain bonding can also occur between parallel extended chains to produce *pleated sheet structures* in which the hydrogen bonds are formed between hydrogen and oxygen atoms in different chains. Depending upon the arrangement of the chains these are known as either parallel pleated sheets or anti-parallel pleated sheets (Figure 11.4).

➤ The secondary structure of a protein is the three-dimensional shape of a polypeptide chain.
➤ The hydrogen bond is a weak electrostatic bond.

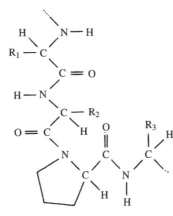

Figure 11.3 **The effect of imino acids on protein structure.** The presence of the imino acids proline and hydroxyproline introduces a constraint into the angles of the peptide bond which results in a bend in the previously regular chain structure.

➤ The tertiary structure of a protein is the overall shape of its polypeptide chain.

The **tertiary structure** of a protein is its overall shape and the level of organization at which the role of the protein becomes significant. Those proteins that have overall spherical or globular structure due to the internal folding of the chain are known as *globular proteins* (Figure 11.5). They are semi-soluble in water, forming colloidal solutions, and in the solid form often exhibit a crystalline structure. They are usually the functional proteins of the cell, enzymes and immunoglobulins being examples. The *fibrous proteins* are the structural proteins and are linear polypeptide chains which are associated with each other to form strands or sheets. In general those fibrous proteins that are helical in structure tend to be the elastic proteins, e.g. keratin, while those proteins with a pleated sheet structure tend to be non-elastic, e.g. silk. However, helical proteins become non-elastic if there is a high degree of bonding between the individual helices, collagen being a good example (Figure 11.6).

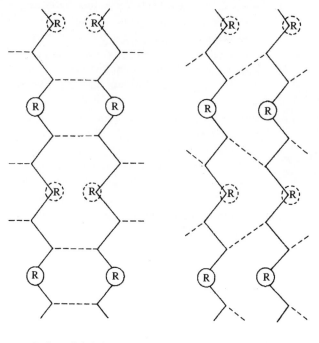

Anti-parallel chains   Parallel chains

Figure 11.4   **Pleated sheets of fibrous proteins.** Parallel pleated sheets are composed of polypeptide chains which all have their N-terminal amino acid at the same end whereas anti-parallel pleated sheets involve polypeptide chains which are alternately reversed in direction. Both forms of sheet show a high degree of hydrogen bonding between the chains.

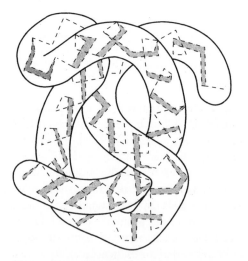

Figure 11.5   **Globular proteins.** The folding of a polypeptide chain in a globular form is stabilized by hydrophobic interactions and some covalent bonding, particularly the disulphide bond between cysteine residues. The polypeptide chain shows some sections which are regular and helical in nature and other sections, particularly at bends and folds, where the conformation of the chain is distorted.

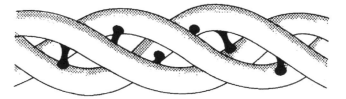

Figure 11.6 **Collagen.** The collagens are major constituents of cartilage and other connective tissue and contain large amounts of the imino acids proline and hydroxy-proline. The basic structure involves three polypeptide chains with a considerable degree of inter-chain hydrogen bonding.

The nature of the amino acid residues is of prime importance in the development and maintenance of protein structure. Polypeptide chains composed of simple aliphatic amino acids tend to form helices more readily than do those involving many different amino acids. Sections of a polypeptide chain which are mainly non-polar and hydrophobic tend to be buried in the interior of the molecule away from the interface with water, whereas the polar amino acid residues usually lie on the exterior of a globular protein. The folded polypeptide chain is further stabilized by the presence of disulphide bonds, which are produced by the oxidation of two cysteine residues. Such covalent bonds are extremely important in maintaining protein structure, both internally in the globular proteins and externally in the bonding between adjacent chains in the fibrous proteins.

➤ A disulphide bond is a covalent bond between the sulphur atoms of two cysteine residues, each in a different part of a polypeptide chain.

The residues of those amino acids that are classed as either acidic or basic are capable of accepting or donating a proton and will at any given pH carry a charge of characteristic sign and intensity. The presence of all these ionizable groups results in proteins showing not only acidic and basic features but also the characteristics of an electrolyte. Such substances are known as ampholytes. At low pH values the ionization of the basic groups will be dominant and the protein will carry a net positive charge, while at high pH values the ionization of the anionic groups will be most evident. At a pH peculiar to the molecule the anionic nature will exactly balance the cationic nature and the protein will carry no net charge. This is known as the *iso-ionic pH* of the protein.

➤ An ampholyte is a substance which, depending upon pH, can act as either an acid or a base and also is an electrolyte.
➤ The iso-ionic pH of a protein is that pH at which the acidic and basic characteristics balance each other.

At the iso-ionic pH the repulsive effect of like-charges, which is the major stabilizing force of colloidal suspensions, is lacking and the protein will show minimal solubility and may precipitate. Whether or not precipitation occurs will depend mainly upon the degree of hydration, a phenomenon in which molecules of water are bound to the polar exterior of the protein and act as a stabilizing factor in colloidal suspensions. The effective charge carried by a colloid is known as the zeta potential and is affected by not only the ionic nature of the amino acid residues but also external factors. In the presence of salts adsorption of ions occurs, resulting in a reduction in the charge carried by the colloid. At some point the concentration of salts will be such that the zeta potential is reduced to zero and the protein will again tend to precipitate, a process known as 'salting out'. This effect is enhanced by competition for the water molecules by these high concentrations of salts.

➤ Zeta potential – see Section 3.3.2.

➤ The quaternary structure of a protein is the aggregation of independent protein sub-units to produce a functional molecule.
➤ Iso-enzymes – see Section 8.1.7.
➤ Conjugated proteins consist of a protein part (apoprotein) and a non-protein part (prosthetic group).

Many globular proteins have a further level of organization known as the **quaternary structure**, which describes the association of protein units to produce an aggregate protein with a definite functional property. The complex catalytic functions of iso-enzymes is dependent upon their quaternary structure. The bonds involved are usually non-covalent and mainly hydrophobic between non-polar regions on the surfaces of the molecules concerned. Haemoglobin, for instance, is composed of four polypeptide chains, normally in two identical pairs, forming the haemoglobin tetramer, which is more effective in oxygen transport than is the monomer. In addition to the four protein chains, haemoglobin also incorporates an iron porphyrin which facilitates the binding of oxygen. Proteins such as haemoglobin that involve non-protein components are known as conjugated proteins.

Self test questions

Section 11.1

1. What is the main stabilizing force in the secondary structure of proteins?
   (a) Peptide bond.
   (b) Hydrogen bond.
   (c) Disulphide bond.
   (d) Hydrophobic bond.
2. What will a protein in solution at its iso-ionic pH show?
   (a) Minimal solubility in water.
   (b) No net charge.
   (c) Maximum enzymic activity.
   (d) Maximum stability.
3. The term globular structure in a protein is associated with the tertiary level of protein structure
   BECAUSE
   a globular protein is soluble in water.
4. The hydrophobic bond is important in the secondary level of protein structure
   BECAUSE
   the hydrophobic bond is produced by the affinity between non-polar regions of the polypeptide chain.

11.2   General methods of quantitation

The nature of the sample and the presence of any interfering substances are major considerations in the selection of a suitable method. Fluid samples are the most convenient to handle but some methods are appropriate for the analysis of solid material. The presence of interfering substances may necessitate an initial purification of the protein components. This may be achieved by precipitating the soluble proteins and, after washing, quantitating using a suitable method. The use of heat or strong acids results in irreversible

> ➤ Denaturation is the breaking down of protein structure, either by simple disruption of the structure or breaking up the polypeptide chain.

denaturation and a method such as the Kjeldahl will usually be necessary. However, the precipitate produced by salts, alcohol, etc., can be redissolved in alkali and a method such as the Biuret reaction could subsequently be used. Alternatively, the substances causing the interference, which are often small molecules such as amino acids and salts, may be removed by dialysis against a large volume of a suitable buffer. The sample may be contained in a sealed bag made from suitable Visking tubing or one of the commercial hollow fibre units may be used in which a long length of narrow bore dialysis tubing gives a very large surface area for rapid dialysis to take place.

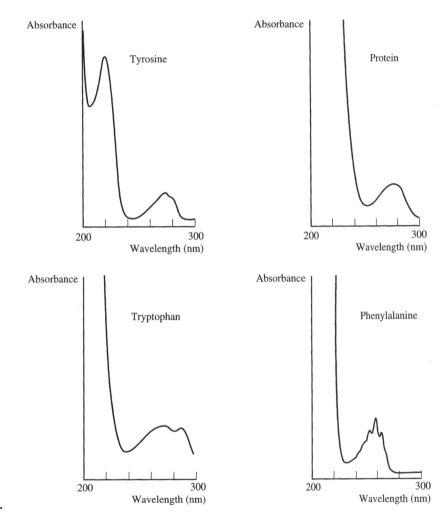

Figure 11.7
Absorption spectra of amino acids and protein.

11.2.1 Spectroscopic methods

The presence of aromatic amino acid residues results in proteins showing an absorption maximum at 280 nm and comparison of the absorbance value of a test solution with that of a standard solution provides a sensitive method of quantitation. This cannot be considered to be an absolute method of quantitation

➤ Molar absorption
coefficient – see Section
2.2.3.

because variations in the amino acid content mean that values for absorption coefficients vary from one protein to another. Figure 11.7 shows the absorption spectra of several amino acids and a protein and although the absorption maxima of the amino acids vary it is the cumulative effect that results in all proteins showing an absorption maximum at 280 nm. At this wavelength tryptophan has the highest molar absorption coefficient and contributes most to the absorbance value recorded. Absorption in the region of 220 nm due to the peptide bond has been used in the quantitation of proteins but many other compounds also absorb at this wavelength and as a result such methods suffer from a considerable degree of interference.

11.2.2   Chemical methods

Kjeldahl method
The Kjeldahl method measures the nitrogen content of a compound and may be used to determine the protein content of a sample provided that the proportion of nitrogen in the protein is known. Protein determinations are complicated by the presence of nitrogen from non-protein sources. The simplest way of eliminating this source of error is to precipitate the proteins using a suitable method and to determine the nitrogen content of the precipitate.

The nitrogen content of proteins is usually accepted as 16% of the total weight but this may not always be correct as values for individual proteins do differ. This is particularly true if the protein has either a high proportion of basic amino acids (additional nitrogen atoms) or is a conjugated protein with an appreciable non-protein component (Table 11.1).

Table 11.1   Nitrogen content of various proteins

Source of protein	Proportion of nitrogen % total weight	Conversion factor
Meat	16.0	6.25
Blood plasma	15.3	6.54
Milk	15.6	6.38
Flour	17.5	5.70
Egg	14.9	6.68

All the nitrogen-containing compounds are oxidized to ammonia by heating the sample with concentrated sulphuric acid together with a catalyst, usually cupric ions, although mercuric ions and metallic selenium have been used. Potassium or sodium sulphate is usually included to increase the boiling point of the mixture, although excessive heating must be avoided to prevent decomposition of the ammonium sulphate. This digestion stage is important and usually takes several hours during which the sample initially turns brown or black and sulphur dioxide fumes are given off. The solution eventually clears and heating should be continued for about a further 2 h.

The cooled mixture is transferred to a steam distillation flask, and after

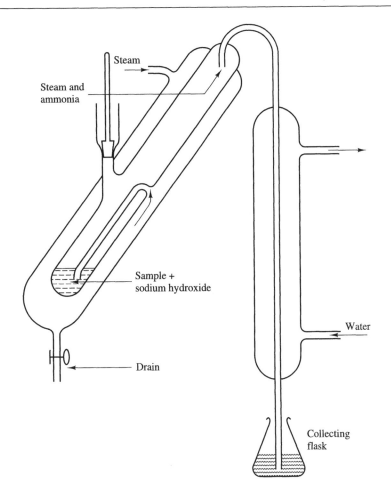

Figure 11.8
Markham still.

making the mixture alkaline with excess sodium hydroxide the ammonia is distilled into a receiver flask (Figure 11.8). The ammonia is trapped in a boric acid solution and the ammonium ions are titrated directly with a standard hydrochloric acid solution. Although boric acid is the most convenient, it is possible to trap the ammonia in a known volume of standard hydrochloric acid and back-titrate the residual acid with a sodium hydroxide solution. The advantage of the latter method is that the detection of the end-point of the titration (a strong acid and a strong base) is easier than that for the ammonium borate and hydrochloric acid titration.

The calculation is based on the fact that 1 atom of nitrogen will result in the formation of 1 mol of ammonia, which will subsequently require 1 mol of acid. Hence:

$$1.0 \text{ mol HCl} \equiv 14 \text{ g nitrogen}$$

If A ml of acid solution containing B mol l^{-1} is required to neutralize the ammonia formed from a given amount of sample, the amount of nitrogen present in the sample is:

$$\frac{14}{1000} \times A \times B \text{ grams}$$

and the amount of protein from which this nitrogen was derived:

$$\frac{14}{1000} \times A \times B \times \frac{100}{16} \text{ grams}$$

Although the Kjeldahl method is tedious and cumbersome it can be extremely accurate if the proportion of nitrogen in the protein sample is known and the complete recovery of nitrogen can be assured. This can be checked by recovery experiments particularly for the digestion stage of the process.

Biuret method

The name given to this method is in some ways unfortunate and misleading. The method was developed following the observation that biuret reacts with an alkaline solution of copper sulphate to give a purple-coloured complex. Protein and some amines react in a similar manner to biuret. In the absence of a more suitable but equally simple title, the name biuret was retained, although its chemical relevance to the quantitation of proteins is only vague. It was useful, however, in elucidating the nature of the copper complex formed (Figure 11.9). The cupric ions form a coordination complex with the four nucleophilic–NH groups, which in the reaction with proteins are provided by the peptide bonds linking the amino acids. The complex shows absorption maxima at 330 and 545 nm (Figure 11.10). The absorbance is usually measured at 545 nm and although the sensitivity at 330 nm is greater, measurements are more prone to interference.

Compounds that contain two of any of the following groups linked through a carbon or nitrogen atom give a similar reaction:

–CONH$_2$
–CH$_2$NH$_2$
–C(NH)NH$_2$
–CSNH$_2$

Some polyalcohols, particularly glycerol and ethyleneglycol, also form a similar complex although the absorption maxima do vary slightly from those

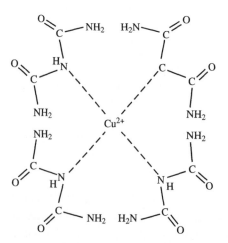

Figure 11.9 **Biuret reaction.** The coordination complex formed in alkaline solution between cupric ions and the nucleophilic nitrogen atoms in four molecules of biuret.

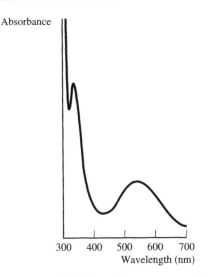

Figure 11.10   **Absorption spectrum of the protein–copper complex of the biuret reaction.**

of the protein–copper complex. The biuret reagent consists of an alkaline solution of copper sulphate with either sodium potassium tartrate or sodium citrate added to prevent the precipitation of cupric ions as the hydroxide.

The method is extremely robust and reliable and obeys the Beer–Lambert relationship to a final protein concentration of about 2 g l^{-1} (a sample concentration of about 20 g l^{-1}) with a lower limit of about 100 μg of protein (sample concentration of 1.0 g l^{-1}). The colour reaches maximum intensity in about 15 min and is stable for at least several hours. The method is simple and reliable and readily lends itself to automation but its lack of sensitivity is its greatest drawback. All proteins react in a similar manner and results show very little difference for different proteins.

Procedure 11.1: Quantitation of protein using the Biuret reaction

Reagent

Copper sulphate	1.5 g l^{-1}
Sodium potassium tartrate	6.0 g l^{-1}
Sodium hydroxide	30.0 g l^{-1}

Method

0.5 ml protein solution.
4.5 ml reagent.
Mix and allow to stand for 30 min.
Measure absorbance at 545 nm.

Calculation

A calibration graph should be prepared using standard protein solutions ranging in concentration from 0.5 to 20.0 g l^{-1}.

Lowry method

A reagent for the detection of phenolic groups known as the Folin and Ciocalteu reagent was used in the quantitation of proteins by Lowry (1951). In its simplest form the reagent detects tyrosine residues due to their phenolic nature but the sensitivity of the method was considerably improved by the incorporation of cupric ions. A copper–protein complex produced using a dilute version of the biuret reagent causes the reduction of the phosphotungstic and phosphomolybdic acids, the main constituents of the Folin and Ciocalteu reagent, to tungsten blue and molybdenum blue. The precise composition of the blue reaction products is not known but they show broad absorption peaks in the red portion of the visible spectrum (600–800 nm). Approximately 75% of the reduction that occurs is due to the copper–protein complex, and tyrosine (and to a lesser extent, tryptophan) residues are responsible for the remainder. Folin and Ciocalteu's reagent is of complex composition and is normally purchased ready for use. It is prepared by the reflux heating of sodium tungstate and sodium molybdate with orthophosphoric acid. The reagent is normally pale yellow and of limited shelf life.

The method is more sensitive than the biuret method and has an analytical range from 10 μg to 1.0 mg of protein. Using the method outlined below this is equivalent to sample concentrations of between 20 mg l^{-1} and 2.0 g l^{-1}. The relationship between absorbance and protein concentration deviates from a straight line and a calibration curve is necessary. The method is also subject to interference from simple ions, such as potassium and magnesium, as well as by various organic compounds, such as Tris buffer and EDTA (ethylenediaminetetraacetic acid). Phenolic compounds present in the sample will also react and this may be of particular significance in the analysis of plant extracts.

Procedure 11.2: Quantitation of protein using the Lowry method

Reagents

1. Alkaline copper reagent

Copper sulphate	20 mg l^{-1}
Sodium potassium tartrate	20 mg l^{-1}
Sodium carbonate	20 g l^{-1}
Sodium hydroxide	40 g l^{-1}

2. Folin and Ciocalteu reagent

Method

6.0 ml alkaline copper reagent.
0.5 ml protein sample.
Mix and allow to stand for 10 min.
Add 0.5 ml Folin and Ciocalteu reagent.
Mix and allow to stand for a further 30 min.
Measure the absorbance at 600 nm.

Calculation

A calibration graph should be prepared using standard protein solutions ranging in concentration from 0.1 to 2.0 g l^{-1}.

Bicinchoninic acid method

This is a modification of the Lowry method involving a dye-binding step. The copper–protein complex that forms the basis of the biuret and the Lowry methods can be chelated by bicinchoninic acid to produce a very stable complex with a strong absorption maximum at 562 nm.

It is said to suffer from less interference effects than the Lowry method and is capable of detecting protein levels as low as 10 μg. The presence of lipids does interfere with the assay and modification of the technique is required if detergents are present. The reagent is only stable for a short time but if the copper sulphate is only added prior to use the stock reagents keep indefinitely.

Procedure 11.3: Quantitation of protein using bicinchoninic acid

Reagent

Bicinchoninic acid	10 g l^{-1}
Sodium carbonate	20 g l^{-1}
Sodium hydroxide	4 g l^{-1}
Sodium bicarbonate	9 g l^{-1}
Sodium tartrate	2 g l^{-1}
Copper sulphate	8 g l^{-1}

Method

2.0 ml bicinchoninic acid reagent.

0.1 ml protein sample.

Mix and heat at 37°C for 30 min.

Cool to room temperature.

Measure absorbance at 562 nm.

Calculation

A calibration graph should be prepared using standard protein solutions ranging in concentration from 0.1 to 2.0 g l^{-1}.

11.2.3   Dye-binding methods

Observations that the presence of protein affects the colour change of some indicators used in acid–base titrations led to the development of methods for the quantitation of proteins based on these altered absorption characteristics of such dyes. As the presence of protein alters the colour produced by these indicators when measuring pH, so in the quantitation of proteins using dye-binding methods the control of pH is vital.

Methyl orange buffered at pH 3.5 binds to albumin with greater affinity than other proteins and the resulting complex shows reduced absorbance at 550 nm. The method is unsuitable as a general method for protein determination because of considerable variation in the binding of the dye with different proteins.

Coomassie brilliant blue has been used extensively in a general quantitative method for proteins, and when complexed with protein shows a shift in

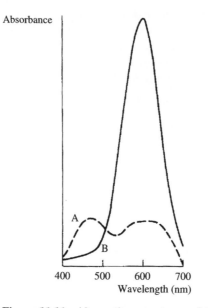

Figure 11.11 **Absorption spectrum of the protein–Coomassie brilliant blue G 250 complex:** A, Coomassie brilliant blue only; B, protein–dye complex.

its absorption maximum from 464 nm to 595 nm. The increase in absorbance at 595 nm can be used as a measure of the protein concentration (Figure 11.11). The maximum absorbance is developed very rapidly (2–5 min) and is stable for at least an hour.

The method is suitable for all proteins although the amount of dye bound does vary from one protein to another in a manner that seems to relate to the proportion of basic amino acid residues in the protein. Bovine serum albumin, for instance, gives absorbance values that are 60% greater than the same concentration of egg albumin. Consequently it is important that the standard protein solutions used should be of the same composition as the test protein. The control of pH is important and although the reagent is heavily buffered, any samples that are very alkaline may alter the pH and so affect the result. Some detergents when present show significant interference resulting in increased absorbance values.

The method is capable of detecting as little as 5 μg protein and a calibration curve is necessary because of the variations between different proteins and the non-linearity of the absorbance–concentration relationship.

Procedure 11.4: Quantitation of protein using Coomassie brilliant blue dye binding

Reagent

Coomassie brilliant blue G250	0.1 g l^{-1}
Ethanol	47.0 ml l^{-1}
Orthophosphoric acid	85.0 g l^{-1}

Method
0.1 ml protein sample.
5.0 ml reagent.
Mix and measure absorbance at 595 nm.

Calculation
A calibration graph should be prepared using standard protein solutions ranging in concentration from 0.1 to 2.0 g l⁻¹.

Bromcresol green is frequently used for the quantitation of albumin, to which it binds selectively at pH 4.2 with a resulting increase in absorbance at 630 nm. The dye is initially a yellow colour but the resulting protein–dye complex is an intense blue colour (Figure 11.12). The method is relatively specific for albumin, bromcresol green being able to displace most substances that may be initially bound to the protein molecule.

The method is sensitive, showing a lower limit of detection of about 50 μg protein and a linear relationship between absorbance and protein concentration up to about 0.2 g.

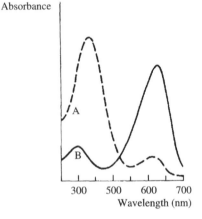

Figure 11.12
Absorption spectrum of the albumin–bromcresol green complex:
A, bromcresol green only;
B, albumin–dye complex.

Procedure 11.5: Quantitation of albumin using bromcresol green dye binding

Reagent
    Bromcresol green   0.15 mmol l⁻¹
    Succinate buffer, pH 4.2   0.075 mol l⁻¹

Method
5.0 ml dye reagent.
0.02 ml protein sample.
Mix and allow to stand for 10 min.
Measure absorbance at 630 nm.

Calculation
A calibration graph should be prepared using standard albumin solutions ranging in concentration from 0.1 to 2.0 g l⁻¹.

The use of bromcresol green for the measurement of albumin has been criticized on several counts. There is a tendency for the protein–dye complexes to precipitate at pH 4.2, which is very near the iso-ionic pH of albumin. It is claimed that the method is not absolutely specific for albumin and particularly with serum samples shows a positive bias in results. There is also some variability in the intensity of the colour produced with albumins from different sources, a fact which makes the choice of the standard material important.

Bromcresol purple has been suggested as showing improved specificity for albumin and less variation in colour intensity compared with bromcresol green. The dye–albumin complex shows an absorption maximum at 603 nm and the use of a reagent buffered at pH 5.2 appreciably reduces the tendency of the complex to precipitate.

Procedure 11.6: Quantitation of albumin using bromcresol purple dye binding

Reagent

Bromcresol purple	40 μmol l^{-1}
Acetate buffer, pH 5.2	0.1 mol l^{-1}

Method

5.0 dye reagent.
0.02 ml protein solution.
Mix and allow to stand for 10 min.
Measure absorbance at 603 nm.

Calculation

A calibration graph should be prepared using standard albumin solutions ranging in concentration from 0.1 to 2.0 g l^{-1}.

11.2.4 Immunological methods

> Immunoassay – see Section 7.3.

Antibodies provide a very convenient and specific analytical tool for protein determinations, a role considerably enhanced by the availability of a wide range of monoclonal antibodies. Many antibodies are capable of precipitating the target protein and the resulting turbidity can be measured by either turbidometric or nephelometric methods. The various types of alternative immunoassay offer increased sensitivity over the turbidometric methods.

11.3   Separation of proteins

Many of the methods described earlier do not differentiate between different proteins but it is often necessary to determine the amount of one particular protein in the presence of others. Although proteins are composed of amino acids, the problems involved in the separation of individual proteins are considerably increased compared with the separation of amino acids, and the large relative molecular mass means that some of the simpler separation techniques such as thin-láyer chromatography are inappropriate. The fact that the tertiary and quaternary structure of a protein can be seriously and often permanently altered by even fairly mild conditions presents an additional problem.

11.3.1 Precipitation

High concentrations of a variety of salts including sulphates, sulphites and phosphates can be used to precipitate proteins but each fraction produced still consists of a mixture of proteins and usually requires further purification. Excessive denaturation of the protein is avoided by the use of low temperatures.

Salt fractionation techniques prepare protein fractions by successively increased concentrations of the salt. Sufficient salt is added to the sample to give the lowest selected concentration and the resulting precipitate is removed, usually by centrifugation or filtration. More salt is then added to the sample to increase its concentration to the next selected level and the precipitate is again removed. The process can be repeated at increasing salt concentrations and a series of precipitates obtained which can be redissolved in a suitable buffer. The salt can subsequently be removed from the protein preparation by techniques such as dialysis or gel permeation chromatography.

Various alcohols may also be used and the classical Cohn fractions of serum proteins are separated using specific concentrations of ethanol under carefully controlled conditions of temperature and pH.

11.3.2 Electrophoresis

Electrophoresis, in all its forms, has a major application in the separation of proteins because charge and molecular size are important to both electrophoretic separation and to protein structure. The conventional techniques which use a flat supporting medium, e.g. cellulose acetate strip or open gel plate, are supplemented and often replaced by capillary techniques, which offer additional analytical features. In both techniques, the key factors in the separation of proteins are the choice of the operating pH and the electrophoretic conditions to be used.

▶ Capillary electrophoresis – see Section 3.3.5.

Capillary electrophoresis is an instrumental technique and is very dependent upon the availability of commercially produced equipment and reagents. The technical decisions are largely made by the manufacturer rather than the analyst. It is essential, however, that the analyst can identify the molecular feature of the analytical problem and then select an appropriate technique.

Technical aspects

Various factors must be considered when selecting a supporting medium. Filter paper has significant disadvantages, the most serious being the adsorption of proteins. Modified cellulose media such as cellulose acetate show significantly less adsorptive effects which, together with a very uniform pore structure, result in a greatly improved resolution, although the quality of cellulose acetate membrane does vary appreciably from one manufacturer to another.

The various other types of supporting media such as starch and polyacrylamide gels show improved resolution due to a molecular sieving effect.

Iso-electric focusing techniques probably give the best resolution and many of the resulting bands are due to specific proteins. They are used mainly as a qualitative or a semi-quantitative technique due primarily to the large number of bands that develop and are of particular value when successive samples from the same source need to be compared for the presence or absence of a particular protein or for investigation of physical properties, e.g. pI.

It is important to realize that the use of these different media for the same sample will result in separation patterns that cannot be easily compared with one another. The separation of serum proteins on cellulose acetate will result in 5–7 bands, while the use of polyacrylamide gel will give 17 bands.

The pH of the buffer used affects the charge carried by the protein and although in theory any pH may be used, in practice pH values greater than the iso-electric pH of the protein (resulting in the protein being negatively charged) give better separations than other pH values. In selecting the conditions for the separation of a particular protein mixture, a buffer pH that gives the greatest difference in the charge carried by each individual protein results in the greatest difference in velocities and hence in the final distances moved. In practice, buffers of pH 8.6 are most frequently used.

In addition to its pH, the concentration of a buffer also affects the mobility of proteins. At high concentrations the zeta potential of the protein is reduced resulting in a shorter distance of migration. However, because higher concentrations of buffer give improved resolution, a compromise concentration has to be found and buffers with ionic strength (μ) varying from 0.025 to 0.075 are frequently used.

The various types of capillary electrophoresis are performed either in free solution or in gels. The choice of method depends on the nature of the sample and the analytical objective but capillary gel electrophoresis, including iso-electric focusing and SDS electrophoresis, is particularly useful for protein applications.

A major problem with the use of fused silica capillaries is the adsorptive effect with proteins, which results in band broadening during the separation process. A number of techniques are available to minimize this effect, ranging from the use of extreme pHs to reduce the charge on the capillary, the use of capillaries coated with various hydrophilic groups to mask the ionized Si–OH groups in the silica and the addition of a range of substances to the buffer which are designed to compete effectively with the protein for the ionic sites.

Quantitative aspects

In conventional electrophoresis, which uses a solid supporting medium, the sample is applied as a streak. The zones or bands of protein that develop during electrophoresis can be precipitated in the pores of the supporting medium by trichloroacetic acid and stained using a suitable dye (Ponceau S, nigrosin, etc.). The amount of dye bound is often directly correlated to the amount of protein but this quantitative relationship is open to criticism for reasons discussed under dye-binding methods. However, it provides a convenient semi-quantitative method with the various fractions usually being expressed as a percentage of the total rather than in absolute amounts. If the total protein content of the original sample is determined using one of the general methods described earlier it is possible to calculate the amount of protein in each fraction.

The amount of dye bound by each fraction can be simply determined by cutting out the stained bands and eluting the dye into a fixed volume of a suitable solvent. The absorbance of each solution is measured and the sum of the absorbance values assumed to be proportional to the total amount of protein. Hence the amount of protein in each fraction can be calculated as a percentage of the total.

An alternative and more satisfactory method of quantitation is to scan the stained electrophoretic strip using a densitometer, which is a modified photometer in which the electrophoretogram replaces the usual glass cuvette. The strip is slowly moved across the light path and the signal from the photoelectric detector is plotted by a pen recorder, the chart speed of which is synchronized with the movement of the strip. The result is a trace that plots the absorbance value against the distance along the electrophoretogram and the area under the trace is proportional to the total protein content. From the area under each peak, the proportion of protein associated with that peak can be calculated (Figure 11.13).

A very narrow light path must be used in order to ensure that closely adjacent bands are resolved. If the instrument measures transmitted light it is necessary to make the supporting membrane translucent. This can be done by either impregnating the strip with an oil with a high refractive index or, for

> ▶ A densitometer is a photometer that is designed to measure the intensity of coloured stain on a thin-layer preparation.

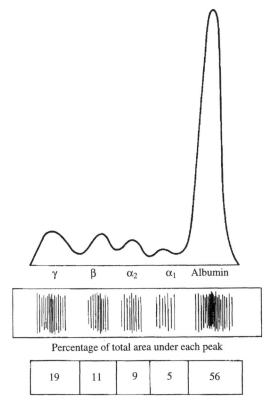

Percentage of total area under each peak

19	11	9	5	56

Figure 11.13 **Electrophoresis of human serum proteins.** The electrophoretogram, after staining with a suitable dye, can be scanned using a densitometer, which gives a trace of the absorbance pattern of the strip.

some types of cellulose acetate material, using an ethanol–acetic acid–ethylene-glycol reagent, which causes the collapse of the porous structure of the membrane rendering it transparent. Some densitometers are designed to measure reflected light rather than transmitted light and so enable the use of opaque strips. This simplifies the technique but does introduce a further factor in the relationship between absorbance and the amount of protein present in the sample. This relationship is still valid for reflectance measurements provided that the amount of protein involved is very small, and hence such instruments are usually designed for micro-analytical techniques.

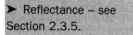 Reflectance – see Section 2.3.5.

Procedure 11.7: Location of proteins after electrophoresis on cellulose acetate medium

Reagent

Ponceau S	5.0 g l^{-1}
Trichloroacetic acid	75.0 g l^{-1}

Method

Immerse the electrophoretogram in the stain for 5 to 10 min.
Remove and clear the background of excess stain by soaking in dilute acetic acid (50 g l^{-1}).
Measure the absorbance at 520 nm using a densitometer.

▶ Polyacrylamide gel electrophoresis – see Section 3.3.2.

In the capillary electrophoresis of proteins it is essential that the buffer and the medium have good transparency in the ultraviolet to enable effective detection after the separation. Polyacrylamide gel does present a problem in this respect, interfering with protein detection at 214 nm. As a consequence of the commercial impact on capillary electrophoresis, a range of ready-to-use media is available from manufacturers.

SDS electrophoresis

Proteins can be dissociated into their constituent polypeptide chains by the detergent sodium dodecyl sulphate (SDS) after the reduction of any disulphide bonds. The SDS binds to the polypeptide chain producing a rod-shaped complex, the length of which is dependent upon the relative molecular mass of the protein. The large number of these strongly anionic detergent molecules bound by the protein (approximately equal to half the number of amino acid residues) effectively masks the native charge of the protein and at a neutral pH results in a relatively constant charge to mass ratio for all proteins. As a result, the electrophoretic mobility of all protein–SDS complexes is approximately equal but the molecular sieving effect of polyacrylamide gel results in a relative mobility which is inversely related to the size of the complex. Under certain conditions, this inverse relationship can be demonstrated by a linear plot of the relative mobility of the protein against the logarithm of its relative molecular mass (Figure 11.14). It is necessary to use a series of known proteins in order to produce a calibration curve and kits are available commercially for this purpose.

Prior to electrophoresis the sample is diluted in buffer containing SDS ($10\text{–}25 \text{ g l}^{-1}$) and β-mercaptoethanol ($10\text{–}50 \text{ ml l}^{-1}$), which reduces any disulphide bonds stabilizing the protein. It is then heated at 100 °C for 2–5 min in order to denature the protein and expose the total length of the polypeptide chain to the detergent. After cooling, electrophoresis is performed on poly-acrylamide gel and the bands subsequently visualized using an appropriate dye.

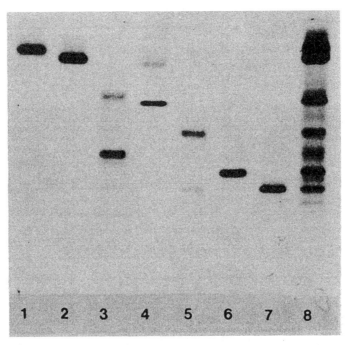

(a)

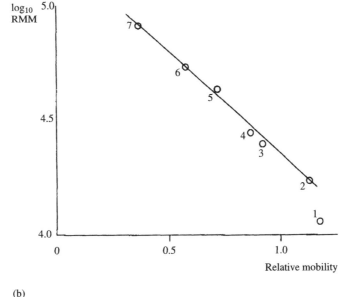

Figure 11.14
Determination of the relative molecular mass (RMM) of a protein by SDS electrophoresis.

(b)

Number	Protein	RMM	$\log_{10}$RMM
1	Cytochrome c (muscle)	11 700	4.068
2	Myoglobin (equine skeletal muscle)	17 200	4.236
3	γ-Globulin (L-chain)	23 500	4.371
4	Carbonic anhydrase (bovine)	29 000	4.462
5	Ovalbumin	43 000	4.634
6	Albumin (human)	68 000	4.832
7	Transferrin (human)	77 000	4.886

The photograph (a) shows these proteins separated on a 5% polyacrylamide gel after treatment with 0.1% SDS. A plot (b) of $\log_{10}$RMM against the relative mobility of each protein shows a linear relationship and provides the basis for the determination of the relative molecular mass of an unknown protein.

11.3.3   Immunological methods

➤ Immunoelectrophoresis – see Section 7.3.2.
➤ Immunoblotting techniques – see Section 7.3.4.

Separation of a mixture of proteins by electrophoretic techniques such as polyacrylamide gel, SDS polyacrylamide or iso-electric focusing usually results in a complex pattern of protein bands or zones. Interpretation of the results often involves a comparison of the patterns of test and reference mixtures and identification of an individual protein, even using immunoelectrophoresis (Figure 11.15), is very difficult. However, specific proteins can often be identified using an **immunoblotting** technique known as Western blotting. The prerequisite is the availability of an antibody, either polyclonal or monoclonal, against the test protein.

After the initial separation by a conventional electrophoretic technique in a gel, the proteins are transferred (or blotted) electrophoretically from the gel to a membrane, usually nitrocellulose. The gel and the membrane, which has been previously soaked in a suitable electrophoretic buffer, are sandwiched between two electrodes. A voltage is applied, e.g. 100 V, and the proteins migrate from

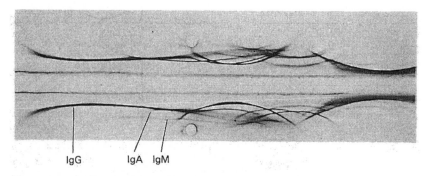

IgG   IgA IgM

Figure 11.15   **Immunoelectrophoresis of human serum proteins.** The proteins are separated electrophoretically from wells cut in a suitable gel. After electrophoresis, a trough is cut in the gel parallel to the direction of migration and filled with an antiserum. The components are allowed to diffuse for 24–48 hours for precipitation lines to develop. Human serum contains many proteins, among which the immunoglobulins can be identified.

the gel to the adjacent membrane. After about 1 h the membrane is removed and carefully washed with buffer and a dilute solution of bovine albumin to block any subsequent, non-specific adsorption of antibodies to the membrane.

The next step involves treating the membrane with a suitable dilution of the specific antibody and allowing the reaction to take place for at least 1 h. Excess antibodies are then washed from the membrane and the bound antibody which remains is detected using a second antibody against the first, e.g. anti-rabbit immunoglobulin. This second antibody has been previously labelled with, for instance, the enzyme horseradish peroxidase but other labels such as conjugated gold or an isotope such as ^{125}I may be used. The bands can then be visualized using an appropriate method. The resulting pattern of zones is then compared with the electrophoretogram prior to immunoblotting and the specific proteins pinpointed for any further investigation or separation.

The technique may also be done in a single stage if the initial antibody is labelled but this is less convenient and often more expensive than having a single labelled antibody preparation that can be used to detect all antibodies from a single species of donor.

➤ Immunoassays – see Section 7.4.

11.3.4   Immunoaffinity purification

Proteins are frequently powerful immunogens and the availability of specific antibodies, particularly monoclonal antibodies, makes the technique of affinity chromatography very useful in the separation and purification of individual proteins. The technique has been used to purify a wide range of proteins such as hormones, membrane receptors and complement proteins. However, it is not restricted to proteins and is potentially applicable to any immunogenic substance. The availability of suitable antibodies is essential and these may be raised by whole animal polyclonal techniques or by monoclonal cell culture. The former antibodies may need some prior purification before being immobilized.

➤ Affinity chromatography – see Section 3.5.1.

➤ Monoclonal antibodies – see Section 7.2.1.

The most frequently used immobilization technique involves the use of cyanogen bromide-activated sepharose media. The sepharose gel is activated by treatment with a cyanogen bromide solution for several minutes. It is then washed with ice-cold distilled water before being mixed with a dilution of the antibody to effect linking. After being allowed to react overnight, the gel is washed with Tris–HCl buffer to neutralize any remaining active groups.

The column is packed with the prepared gel and equilibrated with a suitable buffer, usually with a pH in the region of 8, before the sample is slowly passed through it. During this stage the test substance is bound by the immobilized antibody and held in the column. Large volumes of sample, typically about five times the column volume, can normally be accommodated.

The column is again washed with the buffer and finally the bound antigen is eluted. There is no general rule to enable the selection of a suitable eluting solution but buffers with relative extremes of pH, often pH 2.5, are frequently used. Alternatively, high concentrations of various solutes may be employed, e.g. urea, guanidinium salts and SDS. The antigen is displaced quickly from the column medium and collected in as small a volume as possible. After use, the medium can be washed and regenerated for further use.

11.3.5 Chromatographic methods

Various chromatographic techniques may be applied to the study of protein mixtures. Column techniques have the advantage that the resulting fractions are amenable to quantitation using the general method described earlier. Gel permeation chromatography is frequently used to separate protein mixtures but it is necessary to have some prior knowledge regarding the size of the proteins present in order to select the most suitable gel. Ion-exchange chromatography using the substituted cellulose ion-exchangers, diethylaminoethyl cellulose (DEAE) and carboxymethyl cellulose (CM), is frequently used but as with gel permeation chromatography the major applications are in the preparative aspects of protein analysis. Affinity chromatographic techniques, including those that employ antibodies as ligands, permit highly specific separation of proteins.

Reverse-phase HPLC can be used for the separation of peptides and proteins. Smaller peptides (less than 50 amino acid residues) may be satisfactorily separated on octadecylsilane (C-18) bonded phases whereas for adequate recovery of larger molecules, tetrylsilane (C-4) or octylsilane (C-8) is recommended. Porous column packing with gel permeation and reverse phase properties is usually required for proteins with relative molecular masses greater than 50 000.

Hydrophobic interaction chromatography (HIC) is a variation of reverse-phase chromatography which is particularly useful for protein separations. Instead of the strongly hydrophobic stationary phases normally used such as octadecylsilane (C-18), smaller and less hydrophobic phases, such as methyl, butyl and phenyl groups, are used. The mobile phases are aqueous and hydrophobic interactions between the proteins and the stationary phase are increased in high salt concentrations in the region of $1–2 \text{ mol l}^{-1}$. Typical separations involve equilibrating the column and applying the sample in a buffer with a high salt concentration and then sequentially eluting the proteins with a decreasing gradient of salt concentration. As with most separation techniques involving proteins, the pH and temperature of the eluting buffer are important.

> ➤ Hydrophobic interaction chromatography is a form of reverse-phase chromatography most appropriate for protein separations.

Self test questions

Sections 11.2/3

1. Proteins show an absorption maximum at about which wavelength?
   (a) 220 nm;
   (b) 260 nm;
   (c) 280 nm;
   (d) 340 nm.
2. All proteins contain approximately what percentage of nitrogen (w/w)?
   (a) 12%.
   (b) 16%.
   (c) 22%.
   (d) 27%.

Self test questions

3. Thin-layer chromatography is a suitable technique for separating proteins
   BECAUSE
   proteins are polar compounds and are adsorbed by silica gel.
4. Electrophoresis is a suitable technique for identification of a specific protein
   BECAUSE
   each protein has an iso-electric pH.

11.4   Further reading

Bollag, D.M. and Edelstein, S.J. (1991) *Protein methods,* John Wiley, UK.

Harris, E.L. and Angal, S. (eds) (1990) *Protein purification – a practical approach,* IRL Press, UK.

Hames, B.D. (ed.) (1990) *Gel electrophoresis of proteins – a practical approach,* 2nd edition, IRL Press, UK.

Kenny, A. and Fowell, S. (eds) (1992) *Practical protein chromatography,* Humana Press, USA.

Dunn, M.J. (1993) *Gel electrophoresis: proteins,* Bios Scientific, UK.

12 Lipids

➤ Lipids are substances that are soluble in organic solvents.

The word lipid is used loosely to describe the biological material that can be extracted from living tissues by organic solvents. This definition encompasses such chemically heterogeneous substances as fatty acids and their various derivatives, steroids, prostaglandins as well as the 'fat-soluble' vitamins A, D, E and K. However, a more restricted definition of lipids, which is generally more acceptable, is that of naturally occurring esters of long chain fatty acids that are soluble in organic solvents such as chloroform, diethyl ether or hydrocarbons but are insoluble in water. Under such a definition, the free forms of the fat-soluble vitamins are excluded, as are unesterified sterols including cholesterol, bile acids and steroid hormones. However, because in many cases they often occur naturally as esters of fatty acids and because there are similarities in the approach to their analysis, those of biological importance will be briefly considered.

Lipids have been classified in a variety of ways by different workers but generally two categories, simple and complex, are evident. In this text, substances that contain only a long chain fatty acid and an alcohol (which may be either glycerol, a long chain alcohol, a sterol or one of the fat-soluble vitamins) are considered under the general heading of simple lipids. Complex lipids are acyl esters of either glycerol or the long chain amino alcohol, sphingosine, which also contain a hydrophilic group such as a phosphate ester of an organic alcohol, a carbohydrate or a sulphate group.

➤ Simple lipids contain a long chain fatty acid and an alcohol.
➤ Complex lipids are acyl esters of an alcohol and contain a hydrophilic group.

Lipids rarely exist in an organism in the free state but are usually associated with proteins or combined with polysaccharides. They are important dietary constituents providing energy, vitamins and essential fatty acids and often give flavour and palatability to food. Lipids act as lubricants and insulators and the fat stores in the adipose tissue of the body are a rich source of energy. Combinations of lipid and protein are of particular cellular importance especially in membrane structures and also as a means of transporting lipids in the blood. The steroid hormones are derived from cholesterol and very small amounts of these exert potent physiological effects.

12.1   Fatty acids

> Fatty acids are long chain aliphatic monocarboxylic acids.

The major structural units of naturally occurring forms of both simple and complex lipids are the aliphatic, long chain monocarboxylic acids. These are of great importance because of their contribution to the physical and chemical characteristics of the compounds of which they are constituents. In the investigation of the structure of a lipid it may be necessary to determine not only the identity and amount of any fatty acids present but also their position within the molecule.

The most common fatty acids of plant and animal origin are linear chains with even numbers of carbon atoms (4 to 24 but especially 16 and 18) with one terminal carboxyl group. They may be fully saturated or contain one

Table 12.1   Examples of straight chain saturated fatty acids – $C_nH_{2n+1}.COOH$

Number of carbon atoms	Systematic name	Trivial name	Major occurrence
12	n-Dodecanoic	Lauric	Widely distributed. Major component of some seed oils
14	n-Tetradecanoic	Mystiric	Seed and nut oils
16	n-Hexadecanoic	Palmitic	Commonest saturated fatty acid in animals and plants
18	n-Octadecanoic	Stearic	Very common, particularly in complex lipids
20	n-Eicosanoic	Arachidic	Peanut oil
22	n-Docosanoic	Behenic	Seed triacylglycerols
24	n-Tetracosanoic	Lignoceric	Minor component of seed fats Found in cerebrosides
26	n-Hexacosanoic	Cerotic	Widespread as components of plant and insect waxes
28	n-Octacosanoic	Montanic	Plant waxes

Odd-numbered fatty acids do occur naturally with carbon numbers between 3 and 19.
Those with carbon numbers 15 to 19 are present in large amounts in certain species of fish and bacteria. Even-numbered fatty acids, 4 to 10, are mainly found in milk and butter fats.

or more (up to six) double bonds. Branched chain and more complex structures may also be found, particularly in fatty acids derived from plants or microorganisms.

12.1.1　Nomenclature

> ➤ The suffix -anoic indicates a saturated fatty acid component.
> ➤ The suffix -enoic indicates an unsaturated fatty acid component.

Fatty acids have been referred to by their trivial names for many years and these are still often used. The systematic nomenclature is based on naming the fatty acid after the saturated hydrocarbon with the same number of carbon atoms, the final -e being changed in saturated acids to -anoic (Table 12.1) and in the unsaturated acids to -enoic (Table 12.2), the carbon atoms being numbered from the carboxyl carbon. By an older system, carbon 2 is referred to as the α-carbon, carbon 3 as the β-carbon with the other carbon atoms lettered accordingly with the end methyl carbon being called ω-carbon (Figure 12.1). The convention for defining the position of any double bonds may use the symbol Δ with a superscript numeral. Other shorthand versions are also widely used, which may give rise to confusion because the same compound can be written in a variety of ways (Table 12.3).

Table 12.2　Some naturally occurring mono-unsaturated fatty acids

Number of carbon atoms	Systematic name	Trivial name	Major occurrence
12	cis-5-Dodecenoic	Denticetic	Bacteria
	cis-9-Dodecenoic	Lauroleic	
14	cis-5-Tetradecenoic	Physeteric	Bacteria
	cis-9-Tetradecenoic	Myristoleic	Bacteria and plants
16	cis-5-Hexadecenoic	—	Plants
	cis-7-Hexadecenoic	—	Plants, algae, bacteria
	cis-9-Hexadecenoic	Palmitoleic	Widespread
	cis-11-Hexadecenoic	Palmitvaccenic	Seed oils
	trans-3-Hexadecenoic	—	Plants
18	cis-6-Octadecenoic	Petroselenic	Plants, seed oils
	cis-9-Octadecenoic	Oleic	Very widespread
	cis-11-Octadecenoic	cis-Vaccenic	Bacteria
	trans-9-Octadecenoic	Elaidic	Ruminant animal fats
	trans-11-Octadecenoic	trans-Vaccenic	Ruminant animal fats
20	cis-9-Eicosenoic	Gadoleic	Seed and fish oils
	cis-11-Eicosenoic	—	Seed and fish oils
22	cis-13-Docosenoic	Erucic	Seed and fish oils
	trans-13-Docosenoic	Brassidic	Seed and fish oils
24	cis-15-Tetracosenoic	Nervonic	Brain cerebrosides

Trans-isomers are much rarer than *cis*-isomers. Many different positional isomers of monoenoic acids may be present in a single, natural lipid and this is not a comprehensive list. Palmitoleic and oleic acids are quantitatively the commonest unsaturated fatty acids in most organisms. Odd-chain monoenoic acids are minor components of animal lipids but are more significant in some fish and bacterial lipids.

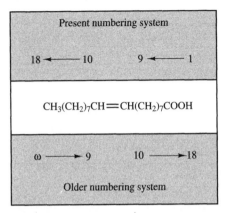

Figure 12.1 Fatty acid carbon numbering system. The carbon atoms are numbered in sequence with the carboxyl carbon atom as number 1. By this system the terminal methyl carbon of oleic (octadecenoic) acid is number 18. An older system names this methyl carbon as the ω-carbon and the adjacent carbon atom as number 2. Numbering then continues towards the carboxyl carbon and values are given for n to denote the number of carbon atoms from the ω-carbon to any double bonds. See Table 12.3.

Table 12.3 Fatty acid nomenclature

Trivial name	Oleic acid	Linoleic acid
Number of carbon atoms	18	18
Structure	$CH_3(CH_2)_7CH{=}CH(CH_2)_7COOH$	$CH_3(CH_2)_4CH{=}CH.CH_2.$ $CH{=}CH(CH_2)_7COOH$
Systematic names	cis-9-Octadecenoic acid	cis,cis,-9,12-Octadecadienoic acid
	(Δ-)9-Octadecenoic acid	(Δ-)9,12-Octadecadienoic acid
	Δ^9-Octadecenoic acid	$\Delta^{9,12}$-Octadecadienoic acid
	*ω-9-Octadecenoic acid	ω-6,9-Octadecadienoic acid
	*(n-9)-Octadecenoic acid	(n-6,n-9)-Octadecadienoic acid
Symbols	18:1	18:2
	18:1; 9	18:2; 9,12
	$C_{18:1}$	$C_{18:2}$
	$\Delta^9 C_{18:1}$	$\Delta^{9,12} C_{18:2}$
	$\Delta 9{-}18{:}1$	$\Delta 9,12{-}18{:}2$
	$18{:}1^{\Delta 9}$	$18{:}2^{\Delta 9,12}$
	*ω-9-18:1	ω-6,9-18:2
	*$18{:}1\omega 9$	$18{:}2\omega 6,9$
	*9-18:1	6,9-18:2
	*18:1(n-9)	18:2(n-6)

* Based on older terminology where the numbering is from the ω-carbon (the terminal methyl group) and n denotes the number of carbon atoms from the last double bond to the ω-carbon. More acceptable nomenclature designates the carboxyl carbon as carbon 1.

12.2	**Simple lipids**

12.2.1 Waxes

➤ Waxes are esters of fatty acids and long chain primary alcohols.

Waxes are esters of fatty acids with long chain primary alcohols (Figure 12.2). The fatty acid is usually straight chain which may be saturated or mono-unsaturated although occasionally branched chain or hydroxy acids are found. They are extremely non-polar compounds and are relatively inert chemically but they can be hydrolysed using a strong alkali, such as potassium hydroxide, a process called saponification.

Waxes are found in animal and insect secretions and the cell walls of some bacteria. The stored fat of marine animals has a high wax component which forms an energy reserve. Waxes are extracted and used commercially in the preparation of creams, cosmetics, polishes, lubricants and protective coatings for surfaces.

$$CH_3(CH_2)_n - \overset{\overset{\displaystyle O}{\|}}{C} - O - CH_2(CH_2)_mCH_3$$

$$CH_3(CH_2)_{14} - \overset{\overset{\displaystyle O}{\|}}{C} - O - CH_2(CH_2)_{14}CH_3$$

Figure 12.2 **Structure of waxes.** The general structure of a wax is shown in which *n* and *m* are usually between 8 and 18. The structure of hexadecyl hexadecanoate (cetyl palmitate), a wax found in sperm whale oil, is shown. This is an ester of hexadecanoic (palmitic) acid and hexadecanol (cetyl alcohol).

12.2.2 Acylglycerols

➤ Acylglycerols are fatty acid esters of glycerol.
➤ Triacylglycerols are often called triglycerides.

Acylglycerols are fatty acid esters of the trihydric alcohol, glycerol, and are often called glycerides. A prefix (mono-, di-, tri-) indicates the number of fatty acids esterified to glycerol which, in triacylglycerols, are nearly always different. Triacylglycerols are probably the most important single class of lipid and are major components of fats and oils of both animals and plants, whereas monoacylglycerols and diacylglycerols appear in only trace amounts in these tissues.

If positions 1 and 3 on the glycerol molecule contain different fatty acids a centre of symmetry is created and the triacylglycerol will exist in different enantiomorphic forms. Several systems of nomenclature have been used in attempts to describe the positioning of the fatty acids on the glycerol molecule but the stereospecific numbering system (*sn*) is now favoured. In this system, which gives an unambiguous description of glycerol derivatives, the secondary hydroxyl group on carbon 2 of glycerol is shown to the left in the Fischer projection and the carbon atom shown above this is then called carbon 1 and the one below, carbon 3 (Figure 12.3). In older nomenclature the symbols α and α' were used to indicate the position of the primary hydroxyl groups and β was used to specify the secondary hydroxyl group.

In some naturally occurring acylglycerols one or two of the fatty acids are esterified to an alkyl ether of glycerol (Figure 12.4). The alkyl group of these 1-alkyl-2,3-diacyl-*sn*-glycerols usually contains 16 or 18 carbon atoms. They are found mainly in the liver oil of marine animals and only in very small amounts in animal and plant tissues. They sometimes occur together with a related group of compounds in which the carbon 1 of the glycerol is linked by a vinyl ether bond to an alkyl group (Figure 12.4). These substances, the neutral plasmalogens, are 1-alkenyl-2,3-diacyl-*sn*-glycerols.

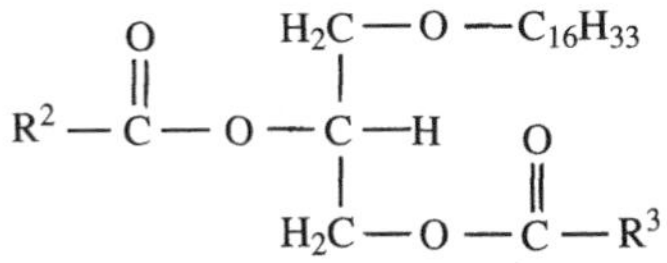

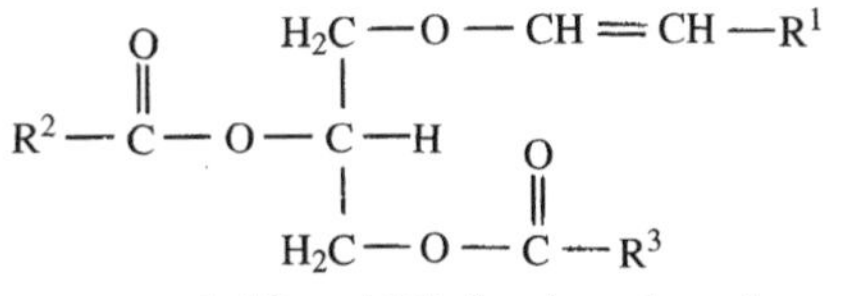

The fatty acids with an even number of carbon atoms most frequently involved in acylglycerols are palmitic, stearic, oleic and linoleic, while odd-numbered and branched chain fatty acids only occur infrequently. Animal acylglycerols contain mainly saturated fatty acids although many fish oils are

1,2,3-Triacyl-*sn*-glycerol

Glycerol

Figure 12.3
***sn* Nomenclature of triacylglycerols.** R^1, R^2 and R^3 are acyl chains.

1-Hexadecyl-*sn*-glycerol
(chimyl alcohol)

1-Alkyl-2,3-diacyl-*sn*-glycerol

1-Alkenyl-2,3-diacyl-*sn*-glycerol
(a neutral plasmalogen)

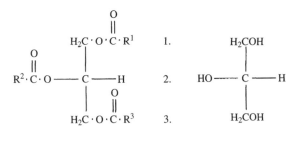

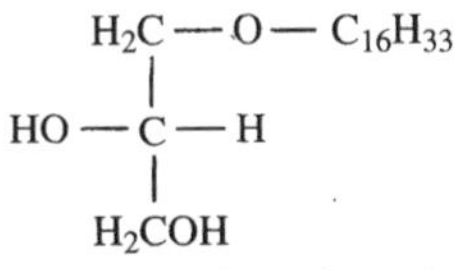

Figure 12.4
Glyceryl ethers and their esters. R^1, R^2 and R^3 are acyl chains.

very rich in polyunsaturated acids. Plant acylglycerols contain slightly increased proportions of unsaturated fatty acids, many of which show higher degrees of unsaturation. Milk fats and some seed and nut oils have a higher proportion of short chain fatty acids (C-4–C-10).

12.2.3   Sterol esters

> ➤ Sterols are alcohols with a steroid nucleus.

Sterols are compounds that have a steroid structure (Figure 12.5) and contain a hydroxyl group, which is capable of forming an ester. They are widespread in nature and commonly exist as mixtures of the free sterol and esters with fatty acids. Cholesterol (Figure 12.5) is by far the commonest sterol in animals, a high concentration being found in brain, nervous tissue and membranes. Plants

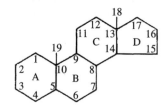

Cyclopentanoperhydrophenanthrene nucleus

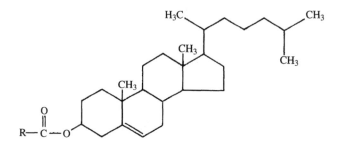

Cholesterol ester

Figure 12.5   **Cholesterol ester.** Cholesterol is a derivative of the cyclopentanoperhydrophenanthrene condensed ring system and is therefore a member of the group of compounds called steroids. It often occurs esterified at position 3 with a long chain fatty acid, which may be saturated or unsaturated and commonly contains 16 or 18 carbon atoms. Steroids that form esters in this way are more correctly called sterols.

contain the related steroids, β-sitosterol, ergosterol and stigmasterol (Figure 12.6) but bacteria cannot synthesize sterols. Many biologically important substances have similar structures and the steroid sex hormones and adrenocortical hormones, bile acids and cardiac glycosides fall into this category. The different attachments to the steroid nucleus result in compounds with a wide range of biological actions.

The D vitamins (Figure 12.7) are a group of compounds that are derived from a steroid nucleus. They are involved in calcium and phosphate metabolism.

The active vitamins are produced by conversion of provitamins by ultraviolet light. Ergosterol, a yeast sterol, is converted to its active form, ergocalciferol (vitamin D$_2$), and 7-dehydrocholesterol, which is found in many natural foods and is also synthesized in man, is converted to cholecalciferol (vitamin D$_3$). Fish liver oils are virtually the only source of vitamin D$_3$ in nature. The most active form of vitamin D$_3$ is 1,25-dihydroxycholecalciferol and this is produced by the hydroxylation of cholecalciferol at position 25 in the liver and then at position 1 in the kidney.

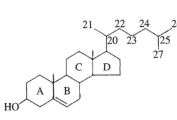

Cholesterol

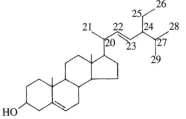

Stigmasterol

Figure 12.6   **Sterols.**

β-Sitosterol

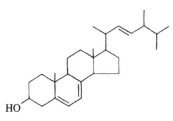

Ergosterol

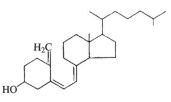

D$_2$ Ergocalciferol

D$_3$ 7-Cholecalciferol

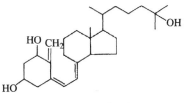

Figure 12.7
**Active forms of the
D vitamins.**

1,25-Dihydroxycholecalciferol

12.2.4   Vitamins A, E and K

The α-, β- and γ-carotenes, which are found in most plants, are vitamin A provitamins and are converted to vitamin A alcohol (all-*trans*-retinol), which is usually called vitamin A_1 (Figure 12.8) by oxidative mid-point cleavage. Retinol and its fatty acid esters are the main forms in which vitamin A is stored in animals and humans, and its oxidation product, 11-*cis*-retinal (vitamin A_1 aldehyde), is required for the visual process.

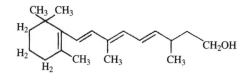

All-*trans*-retinol

Figure 12.8 **Vitamin A_1.** All-*trans*-retinol (Vitamin A alcohol) is usually referred to as Vitamin A_1.

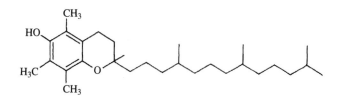

Figure 12.9 **α-Tocopherol.** This is the form of Vitamin E which is the most active biologically, the β, γ and δ tocopherols being less potent.

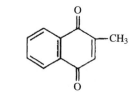

2-Methyl-1,4-naphthoquinone

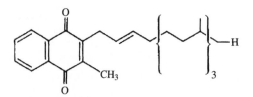

2-Methyl-3-phytyl-1,4-naphthoquinone

Figure 12.10 **Active forms of vitamin K.** Although 2-methyl-1,4-naphthoquinone has vitamin K activity, the more important compounds have an isoprene chain at position 3.

Several different tocopherols are known to have vitamin E activity, but α-tocopherol, a trimethyltocol (Figure 12.9) is the most biologically active. Other less potent forms are the β-, γ- and δ-tocopherols, which contain fewer methyl groups. They all have antioxidant properties and a deficiency results in a lack of protection of the unsaturated fatty acids in the membrane phospholipids against oxidation by molecular oxygen.

Compounds showing vitamin K activity are substituted naphthoquinones. The parent compound, 2-methyl-1,4-naphthoquinone, does show some biological activity as do other similar but synthetic compounds. The production of the complete naturally active forms is thought to depend upon the addition of an isoprene chain at position 3 on the aromatic ring. Differences in this side chain produce the various K vitamins (Figure 12.10). A most important physiological role of vitamin K is in the synthesis of the blood clotting factors, II (prothrombin), VII, IX and X.

12.3   Complex lipids

> Complex lipids are also called conjugated or polar lipids.

Complex lipids are considered here to be acyl esters of glycerol or sphingosine which also contain a hydrophilic group. They are sometimes called conjugated, compound or polar lipids. The latter name stems from the fact that they all contain at least one hydrophilic group, which gives the molecule a polar region (head). The remainder of the molecule, however, is non-polar owing to the presence of the hydrocarbon chains.

Because of the overlap in terminology, it is difficult to divide these lipids into distinct groups but it is convenient to classify them on the basis of the

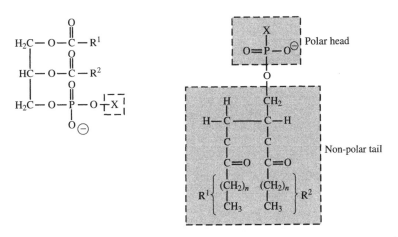

Figure 12.11 **Phosphoglyceride structure.** The members of this group are derivatives of the parent compound, 1,2-diacyl-*sn*-glycerol-3-phosphate (phosphatidic acid) in which X is a hydrogen atom. This is replaced by either an amino alcohol or a polyhydroxy residue. In phosphoglycerides derived from animal tissues R^1 is usually a saturated acyl chain of between 16 and 20 carbon atoms and R^2 is usually unsaturated. Polyunsaturated acyl chains containing 16 or 18 carbon atoms predominate in leaf phosphoglycerides and those of bacterial origin are often more complex.

hydrophilic group (phosphate ester, carbohydrate or sulphate group) and the presence of glycerol or sphingosine. In such a system, any lipid that contains a phosphate group, regardless of its other polar components, is called a phospholipid. Similarly, a sulpholipid is any lipid with a sulphate group, while a glycolipid contains a carbohydrate but has no phosphate group.

12.3.1   Phospholipids

Phospholipids in which the phosphate group and fatty acids are esterified to glycerol are named phosphoglycerides (glycerophospholipids or phosphatides) while in phosphosphingolipids (sphingophospholipids) the alcohol is the long chain alcohol sphingosine.

Phosphoglycerides

These are the most common class of complex lipid (Figure 12.11) and contain a phosphoric acid residue (phosphate group) and two fatty acids esterified to glycerol. Attached to the phosphate group is an amino alcohol, sometimes referred to as the nitrogenous base, which may be either serine, choline or ethanolamine or sometimes the monomethyl or dimethyl derivatives of ethanolamine (Table 12.4). Alternatively, a polyhydroxy compound which is either glycerol, *myo*-inositol or one of their derivatives is attached instead

Table 12.4   Nitrogen-containing phosphoglycerides

Name	Nitrogenous base*	Major occurrence	Comments
Phosphatidyl cholines	$-CH_2CH_2\overset{+}{N}.(CH_3)_3$	Very abundant in animal tissues. Also found in plants and microorganisms. Important in lung alveoli	Commonly termed lecithins
Phosphatidyl ethanolamines	$-CH_2CH_2\overset{+}{N}H_3$	Large amounts in animals, plants and microorganisms	Old trivial name is cephalin. Methyl and dimethyl ethanolamine derivatives also occur
Phosphatidyl serines (or threonine)	$\overset{+}{N}H_3$ $\mid$ $-CH_2CH.COOH$	Widely distributed but in small amounts. Is a major component of brain and red cell lipids	Usually isolated as salts with K^+, Ca^{2+}, Na^+ or Mg^{2+}

*Represented by X in Figure 12.11.

Table 12.5   Structural variations of polyhydroxy phosphoglycerides

The basic structure of phosphatidic acid shown in Figure 12.11 is modified by the substitution of various polyhydric compounds in place of X

Glycerol
Glycerol × 2
Glycerol–amino acid
Glycerol-3-phosphate
Myo-inositol
Myo-inositol-4-phosphate
Myo-inositol-4,5-diphosphate

(Table 12.5). Phosphoglycerides are derivatives of glycerophosphoric acid (1,2-diacyl-*sn*-3-phosphate) which is also called phosphatidic acid.

The different phosphoglycerides are often named by placing the constituent attached to the phosphate group after 'phosphatidyl', e.g. phosphatidyl choline (3-*sn*-phosphatidylcholine or 1,2-diacyl-*sn*-glycero-3-phosphorylcholine). There are many phosphoglycerides because of the possible variation in the fatty acid chains, and when the full chemical structure is known, it should be used (e.g. 1-palmitoyl-2-oleoyl-phosphatidylcholine). Nomenclature that entails the use of the DL system should be avoided.

The phosphoglycerides are good emulsifying agents because of the polar phosphate group and the presence of a group attached to it which may also be charged, making one end of the molecule highly hydrophilic while the rest of the molecule is composed of large hydrophobic groups. They are significant components of cell membranes and plasma lipoproteins and also have very important physiological roles including the prevention of the collapse of lung alveoli, which is made possible by the surfactant properties of mainly phosphatidyl choline.

Phospholipids are also found which have similar structures (Table 12.6). These include lysophospholipids, which have only one of the two possible positions of glycerol esterified, almost invariably at carbon 1, and the plasmalogens, in which there is a long chain vinyl ether at carbon 1 instead of a fatty acid ester. These compounds also contain an amino alcohol, which may be either serine, ethanolamine or choline. Other rarer phospholipids are the monoacyl monoether, the diether and the phosphono forms.

Phosphosphingolipids

Sphingolipids contain the long chain alcohol sphingosine, or one of its derivatives. The sphingolipids that contain a phosphate group, i.e. phosphosphingolipids, are normally referred to as sphingomyelins. They are esters of a ceramide and phosphoryl choline (Figure 12.12) and named ceramide-1-phosphorylcholines. The ceramide is usually an amide of a fatty acid and the long chain amino alcohol, sphingosine (4-sphingenine), although derivatives of sphingosine, e.g. dihydrosphingosine (sphinganine), are sometimes found. The fatty acid may be saturated or monoenoic and is often a hydroxy acid.

Sphingomyelins were first isolated from brain and nervous tissue, where they are abundant, but they also occur in many other animal tissues.

Table 12.6   Variations in phosphoglyceride structure

Structural form	Trivial name	Structure	Common nature of X
Monoacyl	Lysophospholipids	CH_2—O—CO—R^1	Serine
		CH—OH	Ethanolamine
		CH_2—O—P—O—X	Choline
Monoacyl, l-alkenyl	Plasmalogens	CH_2—O—CH=CH—R^1	Serine
		CH—O—CO—R^2	Ethanolamine
		CH_2—O—P—O—X	Choline
Monoacyl, monoether	—	CH_2—O—R^1	Ethanolamine
		CH—O—CO—R^2	Serine
		CH_2—O—P—O—X	
Diether	—	CH_2—O—R^1	Glycerol
		CH—O—R^2	Glycerol-3-phosphate
		CH_2—O—P—O—X	
Carbon–phosphorus bond	Phosphonolipids	CH_2—O—CO—R^1	Ethanolamine
		CH—O—CO—R^2	
		CH_2—O—P—CH_2—CH_2—$NH_3^{\oplus}$	

R^1 and R^2 represent saturated or unsaturated acyl chains.

12.3.2 Glycolipids

Glycolipids may also be classified according to whether they contain glycerol or sphingosine. However, the latter group (glycosphingolipids) have varied and complex structures requiring subdivision into smaller groups.

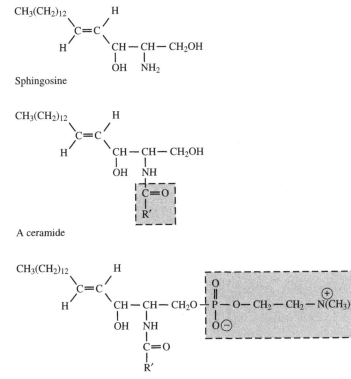

Figure 12.12 **Sphingomyelins.** Sphingomyelins are esters of a ceramide and phosphoryl choline. However, similar compounds are ceramide-1-phosphoryl ethanolamines and phosphono forms of sphingolipids. Ceramides (*N*-acyl-sphingosines) are amides of a long chain di- or trihydroxy base containing 12 to 22 carbon atoms, of which sphingosine (4-sphingenine) is the commonest, and a long chain fatty acid whose acyl chain is shown by R^1. This may contain up to 26 carbon atoms.

Glycosyl diacylglycerols

These are major components of plants and micro-organisms although small amounts can be detected in some animal tissues. They contain glycerol with two fatty acids esterified at positions 1 and 2, and one or more carbohydrate residues linked by a glycosidic bond to position 3 (Figure 12.13). They are named 1,2-diacyl,3-glycosyl-*sn*-glycerols. Those found in plants often contain a highly unsaturated fatty acid and one or two molecules of galactose. Bacterial glycosyl diacylglycerols often contain a branched chain fatty acid and up to four carbohydrates, which may include glucose, mannose, galactose, rhamnose and glucuronic acid.

Glycosphingolipids

Glycosphingolipids in which a ceramide unit (*N*-acyl sphingosine) is linked through position 1 on the long chain base to either glucose or galactose are called monoglycosyl ceramides (cerebrosides) (Figure 12.14). Compounds containing more than one carbohydrate residue should be named appropriately as di-, tri- or tetra-oligoglycosylceramides and not called

➤ Cerebrosides contain a ceramide unit and either a glucose or galactose residue.

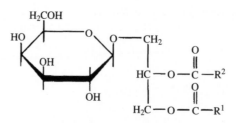

Figure 12.13   Glycosyl diacylglycerol structure. The carbohydrate, which may be either a monosaccharide or a disaccharide residue, is attached by a glycosidic linkage to the *sn*-3 position of glycerol.

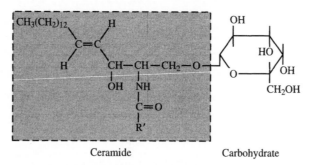

Ceramide   Carbohydrate

Figure 12.14   Monoglycosyl ceramides (cerebrosides). R' is an acyl chain. The carbohydrate residue of galactose is shown although this may be replaced by another hexose. Glycosyl ceramides containing between two and six carbohydrate residues are also found but only the mono derivatives are referred to as cerebrosides.

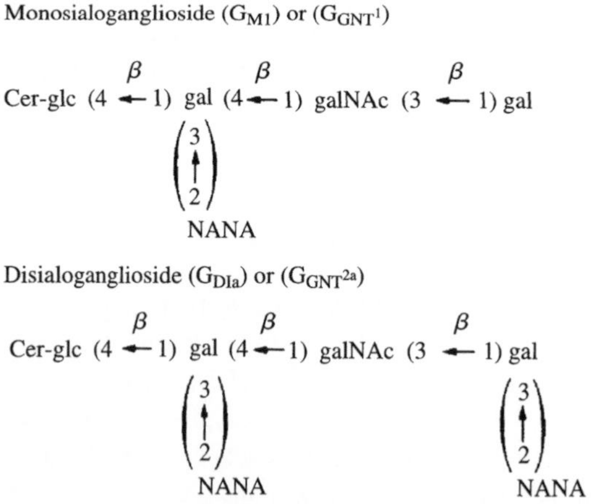

Monosialoganglioside (G_{M1}) or (G_{GNT^1})

$$\beta \qquad \beta \qquad \beta$$

Cer-glc (4 ← 1) gal (4 ← 1) galNAc (3 ← 1) gal

$$\left(\begin{array}{c} 3 \\ \uparrow \\ 2 \end{array}\right)$$

NANA

Disialoganglioside (G_{DIa}) or ($G_{GNT^{2a}}$)

$$\beta \qquad \beta \qquad \beta$$

Cer-glc (4 ← 1) gal (4 ← 1) galNAc (3 ← 1) gal

$$\left(\begin{array}{c} 3 \\ \uparrow \\ 2 \end{array}\right) \qquad\qquad \left(\begin{array}{c} 3 \\ \uparrow \\ 2 \end{array}\right)$$

NANA   NANA

Figure 12.15   Gangliosides. The prefix mono-, di, tri- or tetra- denotes the number of sialic acid (*N*-acetyl neuraminic acid, NANA) residues present in the molecule. The many different gangliosides have complex structures and for convenience shorthand notations are used. The two commonest were introduced by Svennerholm, who named the 'parent' compound G_{M1}, and Wiegnandt, who named it G_{GNT^1}. The latter system gives enough information for the structures to be worked out from the shorthand form once the symbols have been learnt. GalNAc represents *N*-acetyl galactosamine in the above structure.

cerebrosides. Glycosyl ceramides were first found in the brain (hence the name cerebroside) and these mainly contain galactose, whereas ceramide glucosides are more common in other animal tissues and in plants. Cerebroside sulphates (sulphatides) are found mostly in the brain and in these a sulphate group is attached to position 3 of the galactose.

Glycosphingolipids which contain one or more *N*-acetylneuraminic acid (sialic acid) molecules, each linked to one or more of the carbohydrate residues of an oligoglycosylceramide, are called gangliosides (Figure 12.15). They contain glucose, galactose and *N*-acetyl galactosamine in varying proportions in different tissues. They are found on the outer surface of cell membranes, especially the ganglion cells of the brain and nervous system.

> ➤ Gangliosides are complex amino glycolipids.

12.4   Lipid–protein structures

Lipids are rarely present in cells in a 'free' form except as storage reserves but are usually found in combination with proteins or the carbohydrate components of macromolecules. The two major lipid–protein structures are cell membranes and plasma lipoproteins.

12.4.1   Cell membranes

All biological membranes contain at least 25% lipid in association with protein in a complex, organized structure. Several different proteins may be present in a particular membrane and while some have structural roles others possess enzymic properties. The distribution of lipids throughout the membrane may be variable but generalizations can be made about the proportions and type of lipids present. The lipoprotein from myelin sheath of nervous tissue contains about 80% lipid, whereas the protein content is significantly higher in cells that are involved in active metabolism. The membranes of animal cells contain cholesterol, phosphoglycerides, sphingolipids and glycolipids. Bacterial cells do not contain cholesterol and neither animal nor bacterial cells contain sulpholipids, which are found in high concentrations in plant chloroplasts together with glycolipids. There is a predominance of phospholipids in animal cell membranes but the proportions of different lipids vary considerably from one membrane to another.

The general structure is that of a bilayer in which the hydrophobic sections of a variety of lipids are orientated towards the centre of the bilayer, with the hydrophilic areas towards the aqueous medium on the inside and outside of the cell. In the accepted fluid mosaic model, the lipids have lateral movement within the structure.

12.4.2   Plasma lipoproteins

The lipoproteins that circulate in blood plasma are complexes of varying proportions of lipids and proteins (Table 12.7). It is the presence of phospholipids and proteins that makes these aggregates water-soluble and it is in this form that otherwise insoluble lipids are carried in an aqueous environment between

Table 12.7 Plasma lipoprotein fractions

Fraction	Percentage lipid	Percentage composition by dry weight				Transport role
		Triacylglycerols	Cholesterol and cholesterol esters	Phospholipids	Proteins	
Chylomicron	99	80–95	2–6	3–9	~1	Transport of fat absorbed in the intestine to the liver and adipose tissues (exogenous triacylglycerol)
Very low density lipoprotein (VLDL)	90	40–80	10–40	15–20	5–10	Transport of fat synthesized in the liver (endogenous triacylglycerol)
Low density lipoprotein (LDL)	78	10	45–50	~20	20–25	Transport of cholesterol, mostly as the acyl ester, synthesized in the liver
High density lipoprotein (HDL)	50	1–5	~20	~30	45–55	Transport of cholesterol from peripheral tissues to the liver for catabolism

► Ultracentrifugation –
see Section 3.4.3.

tissues. They are micellar and consist of hydrophobic centres of acylglycerols and cholesterol esters surrounded by hydrophilic coats of protein and phospholipids. The lipoproteins are usually grouped into classes based on their density (Table 12.8). Each class includes lipoproteins whose physical and chemical characteristics and lipid composition are similar and whose densities, which are related to the lipid : protein ratio, fall within a specified range. The variations in density permit the different classes to be separated, if not completely, by ultracentrifugation. Alternative methods include precipitation by various salts and, more frequently, electrophoresis. The electrophoretic mobilities are comparable to some plasma proteins (α-, β- and pre-β-globulins) and this has been used for classification purposes. The investigation of patients with suspected lipoprotein abnormalities may include the analysis of the lipid content of a particular lipoprotein class, e.g. HDL cholesterol levels.

Table 12.8 Plasma lipoproteins

Fraction	Density range (gl^{-1})	Svedberg flotation rate (S_f)*	Approximate diameter (nm)**
Chylomicrons	<0.95	>400	100–1000
VLDL (pre-β)	0.95–1.006	20–400	30–80
LDL (β)	1.006–1.063	0–20	15–25
HDL (α)	1.063–1.210	—	7.5–10.0
***Albumin:NEFA	>1.281	—	Smallest

* Svedberg flotation rate (analogous to a negative sedimentation coefficient) is the rate at which each lipoprotein floats up through a solution of sodium chloride of specific gravity 1.063.

** Diameters vary extensively in each group and if the extreme values of each group were given they would demonstrate the overlap between the groups.

*** The albumin and non-esterified fatty acid complexes (99:1) constitute only about 5% of all plasma lipoprotein.

Self test questions

Sections 12.1/2/3/4
1. What is the systematic name for a mono-unsaturated fatty acid containing 18 carbon atoms and a double bond at carbon 6?
   (a) 5-octadecenoic acid.
   (b) 6-octadecenoic acid.
   (c) 5-octadecanoic acid.
   (d) 6-octadecadienoic acid.
2. Which of the following are complex lipids?
   (a) Triacylglycerols.
   (b) Phosphoglycerides.
   (c) Glycerosphingolipids.
   (d) Waxes.

3. Lipids are transported in blood plasma bound to proteins, in complexes called lipoproteins
   BECAUSE
   lipoproteins are classified according to their density.
4. Triacylglycerols are the main component of animal fats
   BECAUSE
   triacylglycerols are known as saturated fats.

12.5   Sample preparation and handling

Quantitation of lipids may require an initial extraction step. This should neither degrade the lipids nor extract any non-lipid components, such as carbohydrates, amino acids, etc. Individual requirements will dictate how rigorous any extraction and purification procedure must be but several fundamental precautions must always be taken in order to minimize the possibility of errors.

Ideally samples from living organisms should be extracted without any delay to prevent autoxidative or enzymic deterioration of their lipid constituents. If this is not feasable the sample should be frozen immediately and stored at $-20°C$ in a glass container under nitrogen. Often lipids will be extracted into an organic solvent and during this and subsequent steps in the analytical procedure minimal exposure of the lipids to air, light and heat is very important to prevent oxidation or destruction of the lipids.

The general rules that should therefore be observed include the use of a blanket of nitrogen whenever possible and evaporation of solvents at the lowest feasible temperatures, which must not exceed $50°C$. The addition of an antioxidant such as butylated hydroxytoluene (2,6-di-t-butyl-4-methylphenol) to the extraction solvents (0.1 g l^{-1}) might be necessary to prevent deterioration of unsaturated lipids but it is essential for storage of lipid extracts at about 0.1% of the weight of lipid. Inactivation of lipolytic enzymes may usually be achieved by addition of an alcohol such as methanol or, in some cases, isopropanol. The latter is recommended for some more stable enzymes sometimes found in plant tissues. Alternatively the plant may be briefly immersed in boiling water.

Contamination is a major problem in lipid analysis and the use of plastic containers and stoppers, rubber bungs or tubing, and any grease on stopcocks, etc., must be avoided. All solvents used should be of the purest grade and be peroxide-free and all glassware should be scrupulously clean.

12.5.1   Extraction procedures

Lipids from liquid samples or cell suspensions may be extracted directly into organic solvents but solid samples will require prior treatment, such as homogenization or ultrasonication. The solvents are chosen with reference to the nature of the lipids present and to some extent, the type of sample, e.g. animal or vegetable tissue. Ethanol–diethyl ether (3 : 1) with subsequent extraction into a petroleum hydrocarbon is useful for non-polar lipids but variations of the Folch methanol–chloroform–water mixture are more widely used. The methanol disrupts the lipid–protein membrane complexes and inactivates the lipases, and the chloroform dissolves the lipids. A salt solution (KCl, NaCl or $CaCl_2$) is often added and the non-lipid substances are taken up into this aqueous phase by vigorous shaking. They can be removed in the upper aqueous layer following centrifugation to separate the phases. Repeated extraction and washing will improve the isolation of uncontaminated lipids in the lower chloroform layer, which can then be analysed by the method of choice.

12.6   Quantitative methods

12.6.1   Total lipids

The total lipid content of a sample can be assessed gravimetrically if the lipid is first extracted and then completely dried until a constant weight is recorded. For accurate results, the whole procedure must be designed to minimize the loss of lipids at every stage. This includes using highly efficient extraction and purification techniques under appropriate analytical conditions.

Other methods, while being technically less demanding, also provide questionable results because they may only measure certain lipid components. Thus the pink colour produced by unsaturated lipids with a reagent containing sulphuric acid, phosphoric acid and vanillin may be used as a screening procedure to assess the total lipid content without preliminary extraction but the result will only be a reflection of the unsaturated lipid content of the sample. The sample is heated with concentrated sulphuric acid and after cooling a phosphovanillin reagent is added, which contains 1.2 g vanillin in 1 litre of 80% phosphoric acid. The absorbance is read at 530 nm after 30 min and compared with known concentrations of lipid treated in the same way. Exact quantitation is not feasible, although a factor, derived with reference to the particular lipid used for preparation of the standard curve, may be useful in some circumstances.

12.6.2   Cholesterol

➤ Total cholesterol measurements include the free and esterified fractions.

In many tissues cholesterol and other sterols exist as a mixture of the free alchohol and its long chain fatty acid ester (esterified at position 3 of the steroid nucleus). The determination of the cholesterol content of a sample may involve the measurement of either of these two fractions individually or the total cholesterol. It is possible to precipitate free cholesterol by adding an equal volume of digitonin (1 g l^{-1} in 95% ethanol), a naturally occurring glucoside, to form a complex that is insoluble in most solvents, including water.

The chemical methods for the quantitation of cholesterol measure total cholesterol, i.e. free and esterified, and so a digitonin precipitate must be prepared if free cholesterol is to be measured. Enzymic methods do not measure the esters and a hydrolysis stage, either chemical or enzymic (using cholesterol ester hydrolase, EC 3.1.1.13), is necessary for the measurement of total cholesterol.

Colorimetric method

The colour reaction of cholesterol and cholesterol esters with acetic anhydride and concentrated sulphuric acid provides the basis of the method attributed to Liebermann and Burchard. This reaction in not entirely specific for cholesterol or its esters because other sterols will also react. In its original form the reagent consists of acetic anhydride, concentrated sulphuric acid and glacial acetic acid and the intensity of the green colour is affected by the proportions of the reagents and the amount of water present. It is possible to achieve an increase in sensitivity if the reagent contains ferric ions. Various modifications of reagent composition have been used and some methods are fluorimetric.

Procedure 12.1:   Chemical method for the quantitation of total cholesterol

Reagents
1. Ferric chloride 2.5 g l^{-1} in 85% phosphoric acid
2. Glacial acetic acid
3. Concentrated sulphuric acid.

Method
To 0.2 ml of sample or standard
Rapidly add 5.0 ml isopropanol.
Mix thoroughly and centrifuge.
To 1.0 ml of the supernatant
Add 3.0 ml ferric chloride reagent.
Mix well and carefully add 3.0 ml concentrated sulphuric acid.
Mix well and allow to stand for 10 min.
Measure the absorbance at 560 nm.

Standards
A series of standard solutions of cholesterol in isopropanol ($0-5$ mmol l^{-1}) should be treated in exactly the same manner as the sample and a calibration graph drawn using the results.

Calculation
The concentration of cholesterol in the sample is read off the calibration graph using the absorbance value obtained for the sample.

Enzymic method

The enzymic methods show improved accuracy and reliability compared with the colorimetric procedures. The assay using cholesterol oxidase is analogous to the measurement of glucose using glucose oxidase but is not completely specific for cholesterol, and all sterols with a 3-β-hydroxyl group and a double bond in the 4–5 or 5–6 position will react. The method is based on the oxidation of cholesterol to cholest-4-ene-3-one with the simultaneous production of hydrogen peroxide. Total cholesterol can be determined if preliminary hydrolysis of the esters is undertaken. The concentration of the reaction products will be proportional to the amount of cholesterol originally present and either may be measured. Cholest-4-ene-3-one has an absorption maximum at 240 nm and its concentration can be calculated from the molar absorption coefficient, thus

> ➤ Hydrogen peroxide assay – see Section 9.3.1.

obviating the need for cholesterol standards. The alternative approach is to measure the hydrogen peroxide produced and although this may be achieved in a variety of ways, the usual procedure employs a peroxidase and a chromogenic oxygen acceptor:

$$\text{cholesterol} + O_2 \xrightarrow[\text{pH 7-8}]{\substack{\text{cholesterol}\\\text{oxidase}}} \text{cholest-4-ene-3-one} + H_2O_2$$

$$H_2O_2 + \text{4-aminoantipyrene} + \text{phenol} \xrightarrow{\text{peroxidase}} \text{quinoneimine} + H_2O$$

Procedure 12.2: Enzymic method for the quantitation of free cholesterol

Reagents

1. Enzyme reagent prepared in buffer pH 6.8

Cholesterol oxidase (EC 1.1.3.6)	1000 units l^{-1} (17 μkatal l^{-1})
Peroxidase (EC 1.11.1.70)	5000 unit l^{-1} (85 μkatal l^{-1})
4-aminoantipyrene	1 mmol l^{-1}
Phenol	32 mmol l^{-1}

Method

To 3.0 ml reagent

Add 0.1 ml sample or standard solution.

Incubate at 37°C for 10 min.

Measure the absorbance at 500 nm.

Standards

A series of standard solutions of cholesterol in isopropanol (0–5 mmol l^{-1}) should be treated in exactly the same manner as the sample and a calibration graph drawn using the results.

Calculation

The concentration of cholesterol in the sample is read off the calibration graph using the absorbance value obtained for the sample.

12.6.3 Acylglycerols

> ➤ Hydrolysis of acylglycerols splits the fatty acids from the glycerol.
>
> ➤ Saponification is the term given to the chemical hydrolysis of acylglycerols.

Mono-, di- and triacylglycerols may all be measured by determination of the amount of glycerol released by hydrolysis. The lipid is first extracted into chloroform–methanol (2 : 1) and saponification is performed under conditions that will not affect any phosphate ester bonds, otherwise glycerol originating from phosphoglycerides would also be measured. Heating at 70°C for 30 min with alcoholic potassium hydroxide (0.5 mol l^{-1}) has been shown to be satisfactory. However, the phospholipids may be removed prior to saponification either by extraction or by adsorption on activated silicic acid.

To compensate for any free glycerol present in the sample, the assay is performed both before and after the chosen hydrolysis step, the difference in results giving the glycerol derived from the acylglycerols.

Colorimetric method

Glycerol can be oxidized to formaldehyde by periodic acid and the formaldehyde measured spectrophotometrically at 570 nm after reaction with chromotropic acid, or fluorimetrically after the addition of diacetylacetone and ammonia. The chromotropic acid reagent consists of 8-dihydroxynaphthalene-3,6-disulphonic acid dissolved in 50% sulphuric acid.

Enzymic method

> ➤ Lipases are enzymes that hydrolyse acylglycerols to release free fatty acids and glycerol.
> ➤ NADH methods – see Section 8.3.2.

Enzymic assays have been developed for the measurement of glycerol and commercial kits are available, some of which include a microbial lipase to effect enzymic hydrolysis of acylglycerols. In a coupled assay method glycerol kinase (EC 2.7.1.30) is used to convert glycerol to glycerol-3-phosphate, which is coupled to the oxidation of NADH using pyruvate kinase (EC 2.7.1.40) and lactate dehydrogenase (EC 1.1.1.27). The decrease in absorbance at 340 nm is related to the concentration of glycerol and therefore to the amount of acylglycerol originally present:

$$\text{glycerol} + \text{ATP} \xrightarrow{\text{glycerol kinase}} \text{glycerol-3-phosphate} + \text{ADP}$$

$$\text{ADP} + \text{phosphoenolpyruvate} \xrightarrow{\text{pyruvate kinase}} \text{pyruvate} + \text{ATP}$$

$$\text{pyruvate} + \text{NADH} \xrightarrow{\text{lactate dehydrogenase}} \text{lactate} + \text{NAD}^+$$

An alternative method uses glycerol dehydrogenase (EC 1.1.1.6) to produce NADH. This is measured either by its absorbance at 340 nm or by using a diaphorase (EC 1.6.4.3), which catalyses the reduction of a dye, *p*-iodonitrotetrazolium violet (INT) to produce a coloured complex with an absorption maximum at 540 nm:

$$\text{glycerol} + \text{NAD}^+ \xrightarrow{\text{glycerol dehydrogenase}} \text{NADH} + \text{dihydroxyacetone}$$

$$\text{NADH} + \text{INT} \xrightarrow{\text{diaphorase}} \text{NAD}^+ + \text{INT(H)}$$

Procedure 12.3: Enzymic method for the quantitation of acylglycerols

Reagents

1. Enzyme reagent prepared in buffer pH 7.6

Lactate dehydrogenase (EC 1.1.1.27)	5000 units l^{-1} (85 μkatal l^{-1})
Pyruvate kinase (EC 2.7.1.40)	1200 units l^{-1} (20 μkatal l^{-1})
NADH	0.2 mmol l^{-1}
ATP	1.0 mmol l^{-1}
Phosphoenolpyruvate	0.3 mmol l^{-1}
Magnesium sulphate	4.0 mmol l^{-1}

2. Glycerol kinase reagent prepared in buffer pH 7.6

Glycerol kinase (EC 2.7.1.30)	500 units l^{-1} (8.5 μkatal l^{-1})

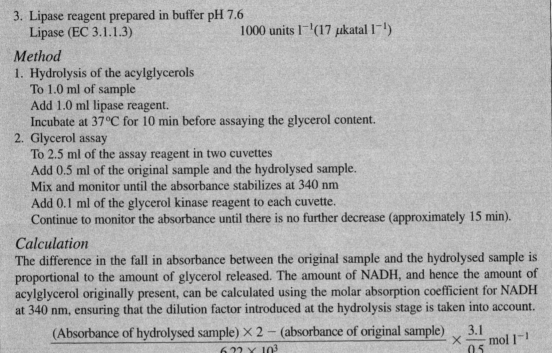

3. Lipase reagent prepared in buffer pH 7.6
   Lipase (EC 3.1.1.3)   1000 units l^{-1} (17 μkatal l^{-1})

Method

1. Hydrolysis of the acylglycerols
   To 1.0 ml of sample
   Add 1.0 ml lipase reagent.
   Incubate at 37 °C for 10 min before assaying the glycerol content.
2. Glycerol assay
   To 2.5 ml of the assay reagent in two cuvettes
   Add 0.5 ml of the original sample and the hydrolysed sample.
   Mix and monitor until the absorbance stabilizes at 340 nm
   Add 0.1 ml of the glycerol kinase reagent to each cuvette.
   Continue to monitor the absorbance until there is no further decrease (approximately 15 min).

Calculation

The difference in the fall in absorbance between the original sample and the hydrolysed sample is proportional to the amount of glycerol released. The amount of NADH, and hence the amount of acylglycerol originally present, can be calculated using the molar absorption coefficient for NADH at 340 nm, ensuring that the dilution factor introduced at the hydrolysis stage is taken into account.

$$\frac{(\text{Absorbance of hydrolysed sample}) \times 2 - (\text{absorbance of original sample})}{6.22 \times 10^3} \times \frac{3.1}{0.5} \text{ mol } l^{-1}$$

12.7   Separation of lipid mixtures

12.7.1 Solvent extraction

Lipids can be extracted from biological samples using a variety of organic solvents. A chloroform–methanol solvent is suitable for all lipids but it is possible to extract different classes of lipid selectively on the basis of their solubility in different organic solvents. This may be achieved by the addition of a solvent that will effect either the precipitation or the extraction of the lipids of interest. An example of the former is the precipitation of high concentrations of phospholipids with cold, dry acetone, and of the latter, the extraction of fatty acids into ether or heptane at an acid pH. However, like all solvent extraction procedures these are not entirely specific.

The partition of different lipids between two immiscible solvents (countercurrent distribution) is useful for crude fractionation of lipid classes with greatly differing polarities. Repeated extractions in a carefully chosen solvent pair increase the effectiveness of the separation but in practice mixtures of lipids are still found in each fraction. A petroleum ether–ethanol–water system can be used to remove polar contaminants (into the alcoholic phase) when interest lies in the subsequent analysis of neutral glycerides, which may be recovered from the ether phase. Carbon tetrachloride–methanol–water is useful in the analysis of plant lipids for separating non-polar carotenes and chlorophylls from more polar phospholipids and glycolipids. The lipid content of each phase can be easily monitored using thin-layer chromatography and the extraction process repeated until the desired separation is achieved.

Table 12.9 Lipid classes in order of increasing polarity

Least polar

Saturated hydrocarbons	
Unsaturated hydrocarbons	
Wax esters	
Steryl esters	
Long chain aldehydes	
Triacylglycerols	Neutral lipids
Long chain alcohols	
Free fatty acids	
Quinones	
Sterols	
Diacylglycerols	
Monoacylglycerols	
Cerebrosides	
Glycosyl diacylglycerols	
Sulpholipids	
Acid glycerophosphatides	
Phosphatidyl ethanolamine	
Lecithin	
Sphingomyelin	
Lysolecithin	

Most polar

12.7.2 Column techniques

> ➤ Adsorption chromatography – See Section 3.2.1.

Adsorption chromatography on a column of silicic acid (silica gel) may be used to separate mixtures of lipids on a preparative scale. Alternative adsorbents are Florisil (magnesium silicate), acid-washed Florisil or alumina and these offer advantages in some cases, e.g. Florisil is useful for the isolation of glycolipids. The extent of the separation generally achieved will depend on many factors including the complexity of the lipid mixture, the dimensions of the column, the type of adsorbent used and the choice of eluting solvents. Thin-layer techniques can be used to purify further the collected fractions as well as being useful to monitor the composition of the column effluent.

The lipids are bound to the fine particles of the adsorbent by polar, ionic and van der Waals forces, the most polar being held most tightly, and separation takes place according to the relative polarities of the individual lipids (Table 12.9). It is usual to elute the lipids from the column with solvents of increasing polarity, when the least polar lipids will emerge in the early fractions. The solvents and the sequence in which they are used will be dictated by the components of the sample. Initial separation of a diverse mixture into the three broad groups of neutral lipids, glycolipids and phosphatides can be effected using a chloroform–acetone–methanol solvent sequence. The collected fractions may be then separated into their individual components by column or preparative thin-layer techniques.

Silicic acid columns can be used for the separation of mixtures of neutral lipids by elution with hexane containing increasing proportions of diethyl ether (0–100%). The lipids are eluted in the following order:

Hydrocarbons
Cholesterol esters
Triacylglycerols
Free fatty acids
Cholesterol
Diacylglycerols
Monoacylglycerols.

Samples that are rich in glycolipids and sulpholipids (e.g. plant extracts) can be separated by eluting the glycolipids with chloroform–acetone (1 : 1) followed by the sulpholipids using acetone. Fractionation of extracts containing high concentrations of several phospholipids (e.g. animal tissues) can be achieved using chloroform to which increasing proportions of methanol have been added (95 : 5 increasing to 50 : 50) resulting in the following typical sequence:

Phosphatidic acid
Phosphatidyl ethanolamine
Phosphatidyl serine
Phosphatidyl choline
Phosphatidyl inositol
Sphingomyelin.

➤ Ion-exchange chromatography – see Section 3.3.1.

Ion-exchange cellulose chromatography with DEAE (diethylaminoethyl) or TEAE (triethylaminoethyl) is a useful technique for the separation of complex lipids on a large scale, the former having the widest range of application. It can sometimes be used after silicic acid chromatography of total lipids to resolve the lipids that are eluted together, e.g. phosphatides and glycolipids or glycolipids and sulpholipids. Separation takes place primarily on the basis of ion-exchange of those lipids with ionic groups, although non-ionic polar groups, e.g. hydroxyl, also exert an influence owing to adsorption by the cellulose. The DEAE cellulose is used in the acetate form, and charged, non-ionic lipids (e.g. phosphatidyl ethanolamine or choline, sphingomyelin, cerebrosides) are eluted with chloroform containing varying proportions of methanol. Weakly acidic lipids (e.g. phosphatidyl serine) can be eluted with glacial acetic acid and strongly acidic or highly polar lipids (e.g. phosphatidyl glycerol, phosphatidyl inositol, sulphatides and sulpholipids) with chloroform–methanol containing ammonium acetate and dilute ammonia.

12.7.3   Thin-layer chromatography

➤ Thin-layer chromatography – see Section 3.2.1.

Silica gel thin-layer plates may be used to separate lipids on either a preparative or an analytical scale. They are sometimes used to fractionate the lipids into classes prior to removal from the plate and further analysis by GLC. In this case the appropriate area of silica gel is scraped off and the lipid extracted into chloroform or diethyl ether containing 1–2% methanol for simple lipids or into chloroform–methanol–water (5 : 5 : 1) for polar lipids.

The polarity of the lipid determines the degree of adsorption to the silica, the most polar being most strongly held. The mobility of the lipids during chromatography will be affected by the polarity of the solvent and increasingly polar solvents will break the adsorptive bonds with greater efficiency, resulting in greater R_F values. When the sample contains compounds of widely varying polarities, a particular solvent may only be able to separate the lipids into general classes, e.g. fatty acids, acylglycerols, cholesterol, phospholipid. It may be possible to resolve the components of such groups by careful selection of solvents.

Double development in the same direction using the same or two different solvents sometimes aids the separation of non-polar lipids. However, two-dimensional techniques are often necessary to resolve mixtures of complex polar lipids (e.g. phospholipids, glycolipids, sphingolipids). This involves carrying out the chromatography in one solvent and after drying the plate and turning it through 90°, running it in another solvent.

Silica gel G contains a binder, calcium sulphate, to help it to adhere to the plate, but silica gel H does not and is preferable for some lipid separations, particularly polar mixtures. Some commercially prepared plates contain an alternative organic binder that does not interfere in the same way as calcium sulphate but can present some difficulty with location methods, particularly charring. The degree of hydration of the adsorbent and the particle size will effect the separation and because these cannot be guaranteed, authentic standards should always be run at the same time as the samples.

➤ Silver nitrate chromatography is also known as argentation or coordinate bond chromatography.

Silver nitrate may be incorporated in the adsorbent slurry (25 g l^{-1}) giving a final concentration of about 5% in the dry plate. The silver ions bind reversibly with the double bonds in the unsaturated compounds, resulting in selective retardation, and the lipids are separated according to the number and configuration (*cis* or *trans*) of their double bonds. This technique is extremely useful in fatty acids, mono-, di- and particularly triacylglycerol analyses when even positional isomers may be resolved. Borate ions may also be incorporated in the silica gel and these plates are used to separate compounds with adjacent free hydroxyl groups.

Non-polar lipids

Various combinations of hexane or light petroleum (40–60°C, bp) and diethyl ether, usually with a small amount of acetic acid (e.g. 90 : 10 : 1) or diisopropyl ether and acetic acid (98.5 : 1.5) are commonly used. The greater mobility is demonstrated by cholesterol esters followed by triacylglycerols, free fatty acids, cholesterol, diacylglycerols and monoacylglycerols, with complex polar lipids remaining unmoved. Double development in two solvents, e.g. diisopropyl ether–acetic acid (94 : 4) followed by hexane–diethyl ether–acetic acid (90 : 10 : 1) or diethyl ether–benzene–ethanol–glacial acetic acid (40 : 50 : 2 : 0.2) followed by diethyl ether–hexane (6 : 94) may improve the quality of separation.

Fatty acids

Only a very small proportion of the fatty acids are present in the free, unesterified form and the vast majority are components of other lipids. Nevertheless it is important to be able to measure and identify the free fatty acids present in either form and for this they must be first extracted into an organic solvent and then usually converted to their methyl ester. The simplest method of methylation, which is applicable to both esterified and non-esterified fatty acids, is to heat the lipid sample for 2 h under a current of nitrogen at 80–90°C with 4% sulphuric acid in methanol. After cooling and the addition of water, the resulting methyl esters are extracted several times into hexane and the combined extracts are dried over sodium carbonate and anhydrous sodium sulphate. The solvent fraction is then reduced in volume by a stream of nitrogen.

> Fatty acids are frequently converted to their methyl esters prior to separation.

Methylation of the fatty acids of acylglycerols may be achieved by dissolving the sample in dichloromethane and heating at 50 °C with sodium methoxide in methanol in a stoppered vial for 15 min. This process effects both the release of the fatty acids from the acylglycerol and their methylation. It is known as transesterification. After acidifying with glacial acetic acid and diluting with water, the fatty acid methyl esters are extracted into diethyl ether. The extract is washed with sodium bicarbonate solution (20 g l^{-1}), dried and reduced in volume with nitrogen. Free fatty acids are not methylated by this procedure.

An alternative procedure involves the release of the fatty acids by alkaline hydrolysis (saponification) by refluxing the extracted sample with dilute alcoholic potassium hydroxide for 1 h. After cooling, adding water and acidifying, the fatty acids are extracted into diethyl ether. The methyl esters can then be prepared by treatment with diazomethane, which may also be used directly on free fatty acids. Saponification is less satisfactory, because it is a lengthy procedure and often results in the loss of lipid components.

Argentation thin-layer chromatography is an extemely useful procedure for the separation of methyl esters of fatty acids. Saturated fatty acids have the highest R_F values, which decrease with the increasing degree of unsaturation, and for a particular acid, the *trans* isomer usually travels ahead of its corresponding *cis* isomer. The solvents most commonly used contain hexane and diethyl ether (9 : 1) although a mixture of 4 : 6 is used to separate compounds with more than two double bonds. In order to separate positional isomers of the same acid, conditions must be carefully controlled and multiple development in toluene at low temperatures is often necessary.

Acylglycerols

> Acyl migration refers to the movement of a fatty acid residue from one position on glycerol to another.

A wide range of mono-, di- and triacylglycerols may be separated on silver nitrate silica plates using a chloroform–acetic acid solvent (99 : 0.5) (Figure 12.16). For mixtures of monoacylglycerols and diacylglycerols, borate-impregnated plates offer better resolution and a chloroform–acetone solvent (96 : 4) can be used to separate monoacylglycerols from the diacylglycerols, 1,2 isomers from the 1,3 isomers and the 1 isomer from the 3 isomer in

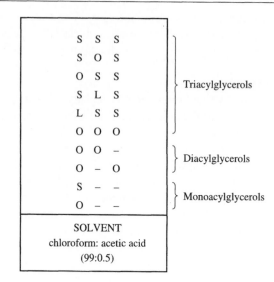

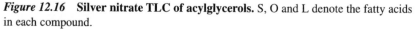

Figure 12.16 **Silver nitrate TLC of acylglycerols.** S, O and L denote the fatty acids in each compound.

S, stearyl, 18 : 0

O, oleoyl, 18 : $1^{\Delta 9}$

L, linoleyl, 18 : $2^{\Delta 9,12}$

Improved resolution and mobility of mixtures of monocylglycerols and diacylglycerols can be achieved using a chloroform–acetone solvent and borate-impregnated plates.

monoacylglycerols. Confusing results may be obtained owing to acyl migration occurring during the extraction even though care is taken to avoid the use of heat and alcoholic solvents. The production of the acetate derivatives prevents acyl migration and still permits separation in the usual way.

Triacylglycerols can be separated in a similar manner on silver nitrate-impregnated plates but different solvents may be required to resolve mixtures containing compounds with widely differing numbers of double bonds. Solvents such as hexane–diethyl ether (80 : 20) or chloroform–methanol (99 : 1) are applicable when up to four double bonds are present, whereas more polar solvents, e.g. diethyl ether or chloroform–methanol (96 : 4) are required for higher degrees of unsaturation and even chloroform–methanol–water (65 : 25 : 4) for the highly unsaturated, i.e. 6 to 12 double bonds (Figure 12.17).

Complex lipids

It is difficult to separate the wide range of complex lipids in a single solvent system but the task is simplified if the sample has already been partially purified (e.g. by a column technique) and only one class of lipid is present. Even so, it is often necessary to perform two-dimensional chromatography and silica gel without binder is often preferred.

Mixtures of phosphoglycerides can be separated using a chloroform–methanol–water mixture, the proportions of which may be varied to suit the sample constituents (e.g. 65 : 25 : 4). Acetone is sometimes included in the solvent and silver nitrate-impregnated plates can be used. Acetic acid (1–4%) is also a useful additive to effect the separation of neutral phosphoglycerides

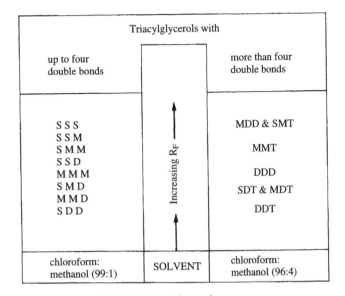

Figure 12.17 **Silver nitrate TLC of triacylglycerols.**
S, saturated fatty acid
M, monoenoic fatty acid
D, dienoic fatty acid
T, trienoic fatty acid
Triacylglycerols with similar total numbers of double bonds can be separated using a chloroform–methanol solvent. The proportions of the solvent are adjusted to achieve resolution of compounds with either a total of up to four double bonds or more than four.

Table 12.10 Destructive staining procedures

Reagent	Method	Comments
30–50% sulphuric acid in water (v/v)	Spray the plate and heat at 180°C for 5 min on a hot plate or 30 min in an oven	Detects all lipids by carbonization of organic material producing black charred areas. At lower temperatures (~80°C) sterols give a purple–red
Chromic acid 6 g l^{-1} potassium dichromate in 55% sulphuric acid	Spray the plate and heat at 180°C for 5 min on a hot plate or 30 min in an oven	All lipids give black charred spots
50% orthophosphoric acid in water	After spraying, heat on a hot plate at 300°C for 5 min	All lipids give black charred spots

from acidic phosphoglycerides (phosphatidyl serine and phosphatidyl inositol). Other modifications which may improve the quality of the separation include the addition of sodium carbonate or ammonium sulphate in the preparation of the silica gel or the preliminary removal of the simple lipids to the top of the plate by running it in acetone–hexane solvent (1 : 3). When the extract is rich in acidic phosphoglycerides it is preferable to use a solvent containing ammonia.

Table 12.11   Examples of non-destructive stains

Reagent	Method	Comments
Iodine vapour	Place the dried plate in a sealed tank containing a few iodine crystals	Dark yellow–brown spots appear within a few minutes where lipids have absorbed the iodine. Unsaturated lipids are more intensely stained. Glycolipids do not stain significantly
Rhodamine 6G (0.1 g l^{-1} in water)	Spray or dip plate and view under UV light	A versatile stain for all lipid classes. Pink–purple–red spots. Most useful with alkaline solvents. The dye may be incorporated in the plate
2′,7′-dichloro-fluorescein (1–2 g l^{-1} in 95% ethanol)	Spray or dip plate and view under UV light	Does not detect phospholipids or glycolipids. Other lipids give yellow spots on purple background. Useful for acid solvents. Dye may be incorporated in the plate
Water	Spray the plate to saturate	Opaque spots appear on a translucent background. Not very sensitive and only applicable to preparative work

Two-dimensional techniques are usually employed if both phosphoglycerides and glycolipids are present, but it is possible to resolve members of both classes using a diisobutylketone–acetic acid–water mixture (40 : 25 : 5). A solvent composed of acetone, acetic acid and water (100 : 2 : 1) will separate the mono- and di-galactosyldiglycerides, which are particularly abundant in plant extracts, from phosphoglycerides, which remain at the origin.

Many solvent combinations have been described for two-dimensional separation and, in general, if an alkaline or neutral solvent is chosen for the first dimension then the second solvent should be acidic. Also it may be useful for one solvent to contain acetone, which will enhance the movement of glycolipids relative to the phosphoglycerides. Acetic acid should not be used in the first solvent, because it is difficult to remove completely and affects the quality of the separation in the second dimension. Difficulties are also encountered in the removal of butanol, which interferes with the charring process often used for the location of the spots.

Detection and quantitation

Lipids may be visualized on thin-layer plates using a general stain but this does not usually indicate the nature of the lipid present. Many of these are destructive (Table 12.10) and the lipid cannot be analysed further after elution from the plate, but other stains are non-destructive (Table 12.11). There

Table 12.12   Specific locating reagents

Reagent	Lipids detected	Comments
Orcinol–sulphuric acid 2 g orcinol in a litre of 75% sulphuric acid	Glycolipids	Heat the plate at 120°C. Blue–violet spots on a white background. Approximately 10 min
α-Naphthol 5 g l^{-1} α-Naphthol in methanol : water (1:1). Spray the plate, air dry and respray with 95% sulphuric acid	Glycolipids	Heat the plate at 120°C. Blue–purple spots. Other polar lipids give yellow spots and cholesterol gives a grey–red spot
Periodate–Schiff Spray with 10 g l^{-1} sodium periodate. Leave for 10 min for oxidation. Remove excess with sulphur dioxide. When plate is colourless, spray with 10 g l^{-1} p-rosaniline HCl in water previously decolorized by saturation with sulphur dioxide	Glycolipids Phosphatides with a vicinal hydroxyl group	Phosphatidyl glycerol gives a purple spot almost immediately and phosphatidyl inositol a yellow spot after about 40 min. Glycolipids give blue spots, the intensity increasing up to 24 hours
Dragendorff stain A. Potassium iodide 400 g l^{-1} in water. B. Bismuth subnitrate 17 g l^{-1} in 20% acetic acid. Mix A and B and water (1:4:15) immediately before use	Choline containing	Orange–red spots within a few minutes with gentle warming
Ninhydrin 2 g l^{-1} in acetone	Containing free amino groups	Purple spots appear with phosphatidyl ethanolamine and phosphatidyl serine, for instance, after heating at 100°C
Molybdenum blue Sodium molybdate 100 g l^{-1} in 4 mol l^{-1} HCl. Hydrazine HCl 10 g l^{-1} in water. Mix 100 ml of each and heat in boiling water for 5 min. Cool and dilute to 1 litre	Containing phosphorus	Blue spots appear immediately
Resorcinol A. 20 g l^{-1} resorcinol in water B. Concentrated HCl C. 0.1 mol l^{-1} copper sulphate D. Water Mix A,B,C,D (10:80:0.5:10)	Gangliosides	Compounds containing sialic acid residues react to give blue–violet spots. Other glycolipids give yellow spots
Acidic ferric chloride reagent 0.5 g ferric chloride, 5 ml glacial acetic acid and 5 ml concentrated sulphuric acid in 1 litre of water	Cholesterol and its esters	Heat to 100°C. Cholesterol appears as a red–violet spot in a few minutes followed by the esters. Continued heating causes charring of all lipids

are specific reagents that will react with one of the chemical groups in the lipid (Table 12.12), e.g. ninhydrin for lipids containing a free amino group.

In some instances quantitation, or at least comparison of the proportions of the separated lipids, may be possible by scanning the plate directly by reflectance densitometry, e.g. after charring. Alternatively, the spot may be scraped off the plate and the colour or fluorescence of the eluted solution measured. The eluted compounds can be submitted to further analysis, by GLC for instance, to enable quantitation or identification. For reliable and meaningful results, known standards should always be run and analysed under identical conditions because of variable colour yields with different lipids. Even so, results obtained by different methods will often vary.

12.7.4   Gas–liquid chromatography

> ➤ Gas–liquid chromatography – see Section 3.2.3.

Gas–liquid chromatography is a very useful technique in lipid analysis, particularly for the separation of very similar compounds within classes. Because of the wide variations in structure and properties between classes it is not usually possible to resolve members of different classes on the same column. GLC is useful for both quantitative and qualitative analysis and also in the investigation of lipid structure.

However, full structural analysis of a lipid will often necessitate further analysis of the collected column effluent for a single GLC peak. Infrared and NMR spectroscopy and mass spectrometry are all useful techniques which will give information for identification purposes, including the position and configuration of any double bonds.

Acylglycerols

Acylglycerols may be separated as intact molecules by GLC after the preparation of acetate or trimethylsilyl (TMS) derivatives to reduce the polarity and prevent acyl migration. Acetylation can be carried out at room temperature by the addition of a solution of acetic anhydride in pyridine (5 : 1). After allowing the reaction to take place overnight, the excess reagent is removed with a stream of nitrogen and gentle warming. TMS derivatives may be prepared by treatment with excess N,O-bis(trimethylsilyl) acetamide (10% v/v in hexane) or with a mixture of hexamethyldisilazane and trimethylchlorosilane in hexane. Silylation is complete in a few minutes at room temperature.

TMS ether derivatives of monoacylglycerols are sufficiently volatile to be analysed on a polyester column (e.g. EGA, PEGA), which permits separation on the basis of chain length and degree of saturation of the fatty acid component. Non-polar silicone stationary phase (e.g. SE-30, OV-1), which are thermally stable up to 350 °C are, however, necessary for diacylglycerols and triacylglycerols. The compounds are separated solely on the total number of fatty acid carbon atoms present, known as the carbon number, and those differing in their carbon number by two can be resolved. The temperature required for separation of diacylglycerols is approximately 270–310 °C with temperature programming between 2 and 4 °C per minute (Figure 12.18). A higher temperature may be required for triacylglycerols whereas monoacylglycerols, which may also be analysed on these columns, require much lower temperatures.

Qualitative analysis may involve the identification of an intact acylglycerol molecule or it may be a study of its composition and structure. GLC is particularly useful in the investigation of the fatty acids present in a triacylglycerol and their position in the molecule. Chemical hydrolysis can be used to release all the fatty acids whereas the enzyme pancreatic lipase (EC 3.1.1.3) will, in theory, only hydrolyse the fatty acids esterified at positions 1 and 3 leaving the 2-monoacylglycerol. In practice, hydrolysis at position 2 does eventually occur and the procedure must be carefully controlled or misleading results will be obtained. In a similar manner, specific phospholipases are also available which will cleave a particular bond in a phospholipid permitting an individual component to be analysed.

Wax esters

Wax esters have similar relative molecular masses to diacylglycerols and are eluted under comparable gas chromatographic conditions. Alternatively, the alcohol and fatty acid moieties can be released by saponification and their methyl esters subjected to gas chromatography.

Cholesterol esters

These can be analysed using the columns applicable to triacylglycerols. It may be possible with temperature programming to separate compounds that differ

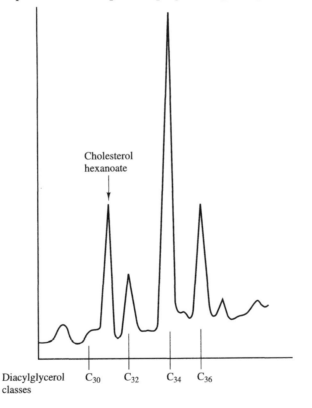

Figure 12.18 **GLC of diacylglycerols.** Schematic representation of the separation of the trimethylsilyl derivatives of diacylglycerols on an SE-30 stationary phase with temperature programming. Cholesterol hexanoate is the internal standard.

by only one carbon atom whereas isothermal analysis will normally resolve compounds with the same carbon number but which contain fatty acids with different degrees of unsaturation. These columns are also applicable to glycosyldiglyceride analysis.

Fatty acids

It is usual to convert fatty acids to their methyl ester derivatives before separation by GLC, although it may be possible to analyse those with short chain lengths (two to eight carbon atoms) as the free fatty acids. Polar or non-polar stationary phases can be used and capillary (open-tubular) or SCOT columns will separate positional and geometric isomers. The *cis* isomers have shorter retention times than the corresponding *trans* isomers on a non-polar phase and visa versa on a polar phase.

The non-polar columns have either saturated paraffin hydrocarbon (Apiezon grease) or silicone greases as the liquid phase. Some suitable silicone greases include SE-30, OV-101, JXR 9 (all methyl silicones) and OV-1 (dimethyl silicone). Separation takes place largely on the basis of relative molecular mass (chain length) and these greases are useful for resolving mixtures of saturated fatty acids, usually isothermally. They can, however, differentiate between saturated and unsaturated components with the same chain length, unsaturated acids having shorter retention times than the corresponding saturated acids, as is also the case with branched chain structures compared with straight chains.

Table 12.13   Polar stationary phases for GLC of fatty acid (methyl esters)

Polarity	Stationary phase	
Highly polar	EGS	Ethyleneglycol succinate
	DEGS	Diethyleneglycol succinate
	EGSS-X	Copolymer of EGS with methyl silicone
Medium polarity	EGA	Ethyleneglycol adipate
	BDS	Butanediol succinate
	EGSS-Y	Copolymer of EGS with a higher proportion of methyl silicone than EGSS-X
Low polarity	NPGS	Neopentylglycol succinate
	EGSP-Z	Copolymer of EGS and phenyl silicone

The liquid phases of polar columns are usually the heat-stable polymers of ethyleneglycol and the dibasic acids, succinic or adipic (Table 12.13). Fatty acids are separated on the basis of both chain length and the degree of unsaturation and some columns are capable of resolving fatty acids with the same chain length but different numbers of double bonds (0–6). The saturated fatty acids show the shortest retention times followed by the monoenoic, dienoic, etc. (Figure 12.19).

In some instances it may be possible to identify an unknown compound with some degree of certainty by showing that it has a retention time identical to that of a known fatty acid. The two substances must undergo

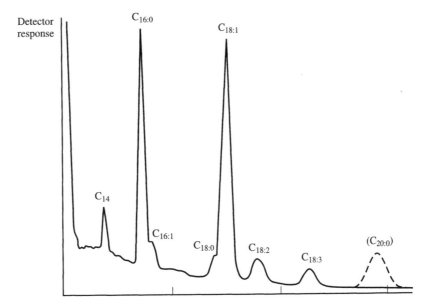

Figure 12.19   **GLC of fatty acid methyl esters.** Schematic representation of isothermal separation on a PEGA stationary phase.

co-chromatography on at least two columns which have stationary phases of different polarity, e.g. EGSS-X and EGSS-Y. When this is not feasible, the chain length of the test substance may be determined using its relative retention time compared with the relative retention times of known fatty acids. The relative retention time of a substance is its own retention time compared with that of a chosen standard of known chain length, e.g. methyl palmitate (16 : 0).

$$\text{Relative retention time of X} = \frac{\text{retention time of methyl ester X}}{\text{retention time of methyl palmitate}}$$

To determine the chain length of saturated acids, the relative retention times of several fatty acids of known chain length are found. A plot of the logarithm of the relative retention time against chain length results in a straight line graph which can be used to find the chain length of the test substance.

The identification of unsaturated compounds is more involved but a useful approach is to determine the relative retention times of several different long chain fatty acids with known numbers of double bonds on both a polar and a non-polar column. A plot of the relative retention times of each reference compound on the polar phase against those demonstrated on the non-polar phase results in a series of parallel straight lines, each line corresponding to a homologous series of saturates, monoenes, dienes, trienes, etc. This graph can then be used to give information about the structure, chain length and number of double bonds present in the unknown acid.

Conversion of an unsaturated acid to the saturated compound permits the chain length to be more easily determined by GLC. This can be achieved by hydrogenation and if double bonds were originally present the resulting substance will alter its retention time to that of the saturated acid.

Unsaturated acids may be split chemically at their double bonds. Permanganate–periodate oxidation has been used to produce the corresponding carboxylic acids, while an alternative technique of ozonolysis results in the formation of aldehydes and aldehyde esters. All these reaction products may be identified by GLC and the information used to determine the position of the double bond in the original fatty acid.

Self test questions

Sections 12.5/6/7

1. Saponification
   (a) is a process of reacting an esterified lipid with an alkali;
   (b) releases carbohydrates from a glycosphingolipid;
   (c) will not affect phosphoglycerides;
   (d) is a transamination reaction?
2. Analytical problems associated with the extraction of lipid from tissue with organic solvents are:
   (a) evaporation;
   (b) oxidation;
   (c) inactivation;
   (d) contamination?
3. Enzymic methods for the quantitation of triacylglycerols use an enzyme to measure glycerol
   BECAUSE
   glycerol is released from a triacylglycerol by the enzyme lipase.
4. The most polar lipids will elute last from a column of silicic acid if solvents of increasing polarity are applied
   BECAUSE
   non-polar lipids are eluted from a silicic acid column by polar organic solvents.

12.8   Further reading

Gurr, M.I. and Harwood, J.L. (1991) *Lipid biochemistry, an introduction*, 4th edition, Chapman & Hall, UK.

Christie, W.W. (1989) *Gas chromatography and lipids: a practical guide*, Oily Press Ltd, UK.

Christie, W.W. (1987) *HPLC and lipids*, Pergamon Press, UK.

Kates, M. (1986) *Techniques of lipidology*, 2nd edition, Elsevier Science, Netherlands.

13 Nucleic acids

Key topics

- Nucleic acid composition and structure
- Isolation and purification of nucleic acids
- Methods of nucleic acid analysis
- Vectors
- DNA sequencing

All cells in living organisms contain the large nucleic acid molecules of ribonucleic acid (RNA) and deoxyribonucleic acid (DNA). Both these molecules are polymers of nucleotides. DNA is found in chromosomes, and genes are unique sequences of DNA nucleotides. The genes contain the inheritable information which together with RNA directs the synthesis of all the cell's proteins.

The realization that DNA, and thus the inheritable information, can be transferred from one organism to another has led to the important field of recombinant DNA technology. The ability to open up the DNA molecule and insert another gene is fundamental to these manipulations and requires the use of two special classes of enzymes. Restriction endonucleases cut the DNA molecule between specific nucleotides while ligases rejoin pieces of cut DNA. Plasmids, the non-chromosomal DNA found in microorganisms, are the usual acceptors for these transferred pieces of DNA and the new molecule that is formed is known as a recombinant DNA molecule. In addition, another group of enzymes, the DNA polymerases, which synthesize a new strand of DNA using an existing DNA template, are vitally important in the analysis of gene structure and expression.

Recombinant DNA technology is a major growth industry and is being exploited commercially to produce a range of products, including vaccines, drugs and enzymes.

13.1   Nucleic acid composition and structure

The nucleotides of RNA and DNA consist of three components: a carbohydrate, a phosphate group and an organic nitrogenous base. There are two types of carbohydrate molecule in nucleic acids, both of which are D-pentoses, i.e. contain five carbon atoms. The carbohydrate in RNA is ribose, while DNA contains deoxyribose, which has a hydrogen atom instead of a hydroxyl group attached to the carbon in the 2 position (Figure 13.1).

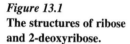

▶ Pentoses – see Section 9.1.

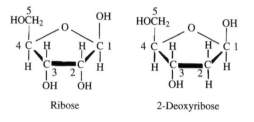

Ribose   2-Deoxyribose

Figure 13.1
The structures of ribose and 2-deoxyribose.

There are five common bases found in nucleic acids. Adenine (A), guanine (G) and cytosine (C) are found in both DNA and RNA. Uracil (U) is found only in RNA and thymine (T) only in DNA. The structures of these bases are shown in Figure 13.2. Adenine and guanine are purine bases while uracil, thymine and cytosine are the pyrimidine bases.

Alkaline hydrolysis splits the nucleotide into its phosphate and sugar–base residues. The sugar–base is known as a nucleoside. The nucleosides are named according to the type of base present. If a purine base is present it will end -osine, e.g. adenosine, while if a pyrimidine is present the name will end -idine, e.g. uridine.

Pyrimidines

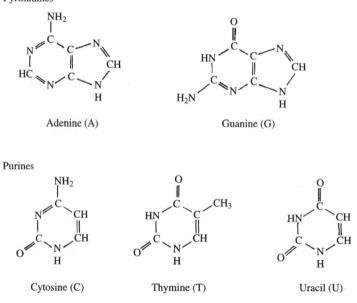

Adenine (A)   Guanine (G)

Purines

Figure 13.2
The structures of the pyrimidine and purine bases of nucleic acids.

Cytosine (C)   Thymine (T)   Uracil (U)

The nucleotides are named according to the nucleoside and the position on the sugar molecule to which the phosphate is attached. To differentiate between the sugar ring system and the base ring system the sugar positions are designated with a superscript ′. The phosphate can be attached to the sugar residue at either the 3′ or the 5′ position (Figure 13.3). A nucleotide containing the base adenine, with phosphate attached to the 5′ position of ribose, would therefore be called adenosine-5′-phosphate. The names of the nucleotides found in DNA and RNA are shown in Table 13.1.

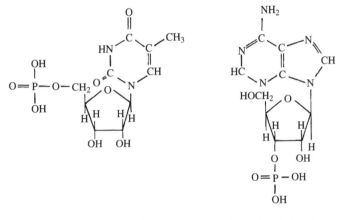

Thymidine-5′-phosphate   Adenosine-3′-phosphate

Figure 13.3 **The structure and naming of nucleotides.** The 5′ indicates that the phosphate group is attached to the 5 position of the sugar ring.

Table 13.1   The names of nucleotides in DNA and RNA

Base	Ribonucleotide	Deoxyribonucleotide
Adenine	Adenosine-5′-phosphate (AMP)	Deoxyadenosine-5′-phosphate (dAMP)
Guanine	Guanosine-5′-phosphate (GMP)	Deoxyguanosine-5′-phosphate (dGMP)
Cytosine	Cytidine-5′-phosphate (CMP)	Deoxycytidine-5′-phosphate (dCMP)
Thymine		Deoxythymidine-5′-phosphate (dTMP)
Uracil	Uridine-5′-phosphate (UMP)	

Individual nucleotides are joined together to form oligonucleotides or nucleic acids. The linkage is from the sugar of one nucleotide to the sugar of the next nucleotide through the phosphate group. This linkage is called a phosphodiester link. The phosphate between the sugars is linked to the 5′ position of one sugar molecule and the 3′ position of the second molecule. This repeating sugar–phosphate sequence forms the backbone of the nucleic acid. The five different bases, attached to the sugars at the 1′ position, stick out away from the sugar–phosphate backbone (Figure 13.4). By convention, the end of the nucleic acid containing the free 5′ hydroxyl or phosphate group is written to the left and is called the head, while the end with the free 3′ phosphate is the tail and is written to the right.

It is the sequence of different bases in the oligonucleotide that is of functional importance and therefore the nucleic acid chain is usually expressed simply as its base sequence, e.g.

AGCTAAGGTCCATG...

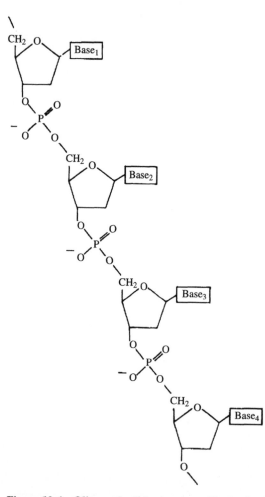

Figure 13.4   Oligonucleotide structure. The nucleotides are joined via the phosphodiester bridges between sugar residues. The bases stick out away from the sugar–phosphate backbone.

13.1.1   The structure of DNA

Watson and Crick showed that the normal structure of DNA consists of a double helix made from two single oligonucleotide strands. The two sugar–phosphate strands of the helix run in opposite (anti-parallel) directions and the bases point inwards and pair very precisely with bases on the opposite strand. Base pairing occurs between a purine and pyrimidine base; guanine always pairs with cytosine and adenine with thymine (Figure 13.5).

This helical structure is held together by hydrogen bonding between the base pairs. The hydrogen bond is formed between a pair of electrons on the keto group or ring nitrogen of one base and a hydrogen atom on a ring nitrogen or amino group of another base (Figure 13.6). Three such bonds are formed between a cytosine/guanine base pair and two bonds between an adenine/thymine pair. Individual hydrogen bonds are very weak but the presence of a large number of bonds in the double helix produces an extremely stable configuration. The structure results in the base pairs being stacked parallel to each other, thus allowing hydrophobic interactions between the stacked base pairs, which also helps to stabilize the structure.

> A DNA molecule is said to be supercoiled when the ends of a DNA double helix join to form a closed circle which then winds back upon itself.

In most DNA molecules the two ends of the DNA double helix are joined to form a closed circle. When this occurs the DNA then forms a supercoil by winding back on itself. Supercoiled DNA depends on the integrity of the closed circle. Any breaks in the circle will cause unwinding of the supercoil.

The two anti-parallel DNA strands are complementary to each other. Thus where there is an adenine residue in one strand there is a thymine in the

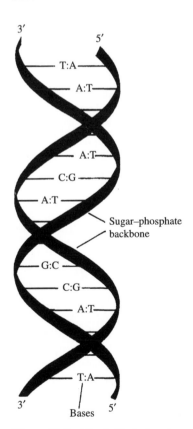

Figure 13.5 **The helical structure of DNA.** Bases on the opposite strands pair. The pairing is always between thymine and adenine and cytosine and guanine.

> ➤ Hybridization occurs when two complementary strands of DNA or RNA come together as a result of the hydrogen bonding between base pairs.
> ➤ Plasmids are extra-chromosomal DNA molecules present in the cytoplasm of many bacteria, which are able to replicate independently of the chromosome.

other, while guanine and cytosine complement each other. When two complementary strands of nucleic acids (DNA or RNA) are mixed in the test-tube the hydrogen bonding between the base pairs of opposite strands will bring the strands together and produce a double-stranded stable structure. This process is known as hybridization and is used extensively in the analysis of DNA. The strength of the hybridization will depend on the number of correctly complemented base pairs, i.e. the homology between the two strands.

As well as chromosomal DNA many bacteria contain closed circular double-stranded DNA structures in their cytoplasm called plasmids. These plasmids are able to replicate independently of the chromosome and usually carry one or more genes that are responsible for a useful characteristic displayed by the host bacteria, e.g. antibiotic resistance. The number of plasmid molecules in each bacterium is known as the copy number and varies between different species of bacteria and types of plasmid.

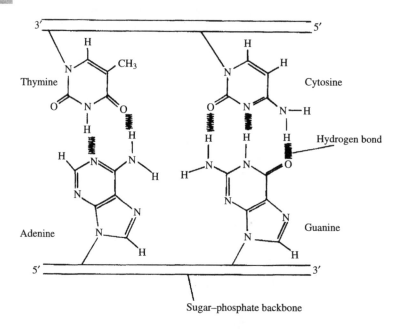

Figure 13.6
Hydrogen bonding between the base pairs.

13.1.2 The structure of RNA

There are three types of RNA: messenger RNA (mRNA), ribosomal RNA (rRNA) and transfer RNA (tRNA). Of these only mRNA is used in recombinant DNA technology. mRNA is a single-stranded oligonucleotide which is copied (transcribed) from one of the strands of the DNA helix and therefore has a base sequence complementary to the original DNA sequence. It differs from DNA and other species of RNA because of a sequence of adenine bases attached to its 3′ end (the poly-A tail). The single-stranded mRNA does not form a precise secondary structure like that of DNA. However, some base pairing between nucleotides on the same strand probably occurs to produce short helical areas which are separated by extensive non-helical regions.

> ➤ mRNA is a single-stranded nucleic acid molecule with a base sequence that is complementary to its corresponding DNA sequence.

13.2   Isolation and purification of nucleic acids

Of all the various nucleic acid molecules within the cell the three that are most often isolated are chromosomal DNA, mRNA from tissue or cells and plasmid DNA from bacteria. In all cases the principles for isolation are similar and can be divided into three stages.

1. The cells or tissue are broken open to release their contents.
2. The cell extract is treated to remove all components except the DNA or RNA.
3. The resulting DNA or RNA solution is concentrated.

13.2.1   Isolation of total cellular DNA

The tissue or cell sample is firstly homogenized in a buffer containing a detergent such as Triton X-100 and sodium deodecyl sulphate (SDS), which disrupts the cell and dissociates DNA–protein complexes. Protein and RNA are then removed by sequential incubations with a proteolytic enzyme (usually proteinase K) and ribonuclease. Finally the DNA is extracted into ethanol. Ethanol only precipitates long chain nucleic acids and so leaves the single nucleotides from RNA digestion in the aqueous layer.

Procedure 13.1: Isolation of total cellular DNA

Reagents

Lysis buffer	Tris buffer (10 mmol l^{-1}) pH 7.6 containing 1% v/v Triton X-100 and 1% w/v SDS
Tris buffer	10 mmol l^{-1} pH 7.6
Proteinase K	10 mg ml^{-1}
Phenol–chloroform	1 : 1 v/v
Ribonuclease	10 mg l^{-1} in Tris buffer pH 7.6
Absolute ethanol	

Method

Homogenize tissue sample in 6 vols of lysis buffer.
Pellet cell debris by centrifugation at $2500\,g$ for 20 min at 4 °C.
Incubate supernatant with proteinase K at 37 °C for 2–4 h.
Remove the protein by extraction into phenol/chloroform.
Precipitate nucleic acids (both DNA and RNA) from the aqueous phase by addition of 2 vols of absolute ethanol. Centrifuge.
Resuspend nucleic acids in 1 ml Tris buffer and add 10 μl ribonuclease.
Incubate for 1 hour at 37 °C.
Precipitate DNA with 2 vols of absolute ethanol. Centrifuge.
Resuspend in buffer.
Measure absorbance at 260 nm.

Calculation

A solution containing 50 μg ml^{-1} DNA has an absorbance of 1.0.
Concentration of DNA in the sample is therefore;

$$\text{absorbance} \times 50 \ \mu g \ ml^{-1}$$

Caesium chloride density gradient centrifugation

➤ Density gradient centrifugation – see Section 3.4.3.

An alternative method for DNA isolation is caesium chloride (CsCl) density gradient centrifugation, which separates protein, RNA and DNA according to their buoyant densities. A density gradient is produced by centrifuging a solution of CsCl at high speed. Macromolecules present in the CsCl solution during centrifugation will form bands at distinct points in the gradient depending on their buoyant density. DNA has a buoyant density of 1.7 g ml^{-1} and will therefore migrate to this point in the gradient. Protein has a much lower buoyant density and will float at the top of the tube, while RNA will pellet at the bottom (Figure 13.7).

Density gradient centrifugation in the presence of ethidium bromide (EtBr) is more often used in the separation of plasmid DNA from chromosomal DNA. The binding of EtBr to DNA causes partial unwinding of the double helix. This unwinding results in a decrease in buoyant density of about 0.125 g ml^{-1} for linear DNA. However, supercoiled DNA binds only a small amount of EtBr and the decrease in buoyant density is only about 0.085 g ml^{-1}. During extraction of DNA from bacteria the chromosomal DNA is broken into small fragments. This process causes the DNA to relax into a linear state. The much smaller plasmid DNA is not broken and remains in a supercoiled form. Therefore the supercoiled plasmid molecules will band in an EtBr–CsCl gradient at a different position to linear chromosomal DNA (Figure 13.7). The position of the DNA bands can be seen using ultraviolet light, which causes the bound EtBr to fluoresce.

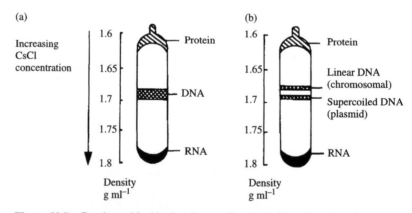

Figure 13.7   **Caesium chloride density gradient centrifugation** for (a) the separation of DNA from RNA and protein and (b) the separation of linear DNA and supercoiled DNA.

13.2.2   Isolation of RNA

The isolation of RNA is less straightforward than DNA. The sample is easily contaminated by ribonuclease causing breakdown of the RNA. Endogenous ribonuclease activity is prevented by addition of inhibitors in the early stages

➤ Polysomes are organelles consisting of ribosomes and mRNA created when several ribosomes are attached, at about 100 nucleotide intervals, along a single mRNA molecule.

while exogenous activity is minimized by using pure chemicals and sterile glassware. A second problem is the tight association of RNA with protein as in polysomes. Harsh treatments are required to release the RNA from these complexes. Many methods have been described for the isolation of RNA including those shown below.

Proteinase K method

In this method the cells are lysed by incubation in a hypotonic solution, which leaves the nuclei intact. The cell debris and nuclei are pelleted by centrifugation leaving the cytoplasmic RNA free from DNA in the supernatant. The RNA is released from the polysomes by incubation with proteinase K and the protein extracted into phenol/chloroform. The RNA is then precipitated from the aqueous phase using ethanol.

Guanidine thiocyanate method

The cells are lysed in a buffer containing strong chaotropic reagents such as guanidine thiocyanate and 2-mercaptoethanol, which completely denatures any ribonuclease present. The supernatant is then placed on a cushion of CsCl (5.7 mol l^{-1}) and centrifuged at $100\,000\,g$ for 18 h. The RNA passes through the CsCl and is pelleted, while the DNA and protein remain in the aqueous solution. The RNA pellet is dissolved in buffer and concentrated by precipitation in cold ethanol.

➤ Affinity chromatography – see Section 3.5.1.

The mRNA is then isolated from this total cellular extract by affinity chromatography using oligo-dT-cellulose or poly(U)-sepharose.

The isolation of DNA and RNA from tissue has been simplified in recent years by the use of kits that enable the extraction to take place in a single step and therefore speed up the process. These reagents promote the formation of complexes of RNA with guanidine and water molecules and abolish hydrophilic interactions of DNA and protein. In effect, DNA and proteins are effectively removed from the aqueous phase while RNA remains in this phase. After centrifugation to separate the two phases, the RNA is precipitated from the aqueous phase by ethanol or isopropanol. After the removal of the aqueous phase, DNA and protein in the organic phase can be recovered by sequential precipitation with ethanol and isopropanol. All types of RNA are isolated by this method. If necessary, mRNA can be isolated by affinity chromatography using oligo-dT-cellulose or poly(U)-sepharose.

13.2.3 Electrophoretic methods

➤ Agarose gel electrophoresis – see Section 3.3.2.

The standard method used to separate, identify and purify DNA (and RNA) fragments is electrophoresis through agarose gels. The technique is simple, rapid to perform and capable of resolving mixtures of DNA fragments to a high degree.

At neutral or alkaline pH the phosphate groups of DNA give rise to a uniform negative charge per unit length of DNA molecule, which therefore moves towards the anode in an electric field. Molecules of different sizes can be separated by carrying out the electrophoresis through a supporting matrix. The size of DNA molecules depends on the chain length and so DNA molecular sizes are quoted as numbers of base pairs. The matrix offers less resistance to small molecules, which therefore move faster. The rate of movement of a molecule of a particular length in the matrix depends on the pore size of the gel, which is determined by the concentration of agarose. Concentrations between 0.5 and 2% are commonly used covering a DNA size range from 200 to 50 000 base pairs.

The separation can be done in vertical or horizontal apparatus in rods or slabs. Because of the ease of use, horizontal slab electrophoresis systems are most commonly used in which the whole gel is submerged in buffer (Figure 13.8).

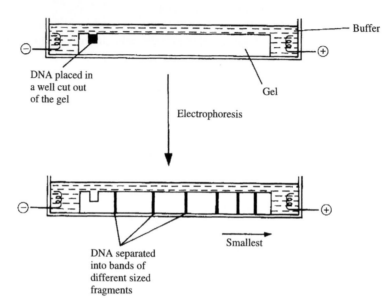

Figure 13.8
The separation of DNA by agarose gel electrophoresis.

The DNA within the gel can be directly located by including a low concentration of the dye ethidium bromide in the gel, which binds to the DNA, causing it to fluoresce. Adequate safety precautions must always be taken when using ethidium bromide owing to its hazardous nature. As little as 1 ng of DNA can be detected by direct examination of the gel in ultraviolet light. As ultraviolet light may cause eye damage, goggles must be worn when observing gels. To avoid prolonged exposure to ultraviolet light, which can cause damage to the DNA as well as to the person, it is usual to photograph the gels for subsequent analysis. A transilluminator shines ultraviolet light of wavelength 300 nm on the gel from below and the fluorescent emission at 590 nm is photographed by a Polaroid camera. A red filter can be used to enhance the contrast between the fluorescent band and background (Figure 13.9).

For the determination of DNA size suitable markers are included in one of the lanes. A straight line calibration graph can be obtained by plotting distance migrated against $\log_{10}$ base pairs of the markers.

Alternatives to agarose gel electrophoresis

Agarose gel electrophoresis can only be used for DNA molecules that are greater than 200 base pairs in size. For molecules smaller than this polyacrylamide gel electrophoresis is required. This type of electrophoresis works on the same principle as agarose electrophoresis and separates molecules according to their size. Gels containing 12% polyacrylamide are used to resolve fragments between 20 and 100 base pairs and 3.5% polyacrylamide for fragments between 80 and 200 base pairs.

Capillary gel electrophoresis is increasingly being used for the separation of DNA fragments. Polyacrylamide gel-filled capillaries are usually employed because agarose gels are unable to withstand the heat generated by the high voltages used. This technique is used particularly to separate oligonucleotides and DNA sequence products, but double-stranded DNA, such as the products of restriction endonuclease digestion and PCR, can be separated with physical gels. These are relatively low viscosity polymer solutions, e.g. alkylcellulose.

> **Capillary electrophoresis – see Section 3.3.5.**

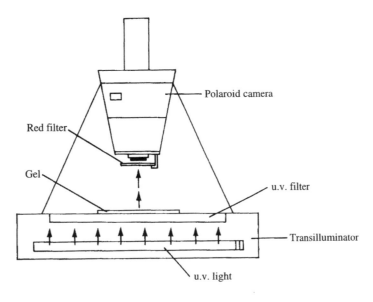

Figure 13.9
Transilluminator and camera for the photography of DNA in gels.

Extraction of DNA from gels

In a number of cases, e.g. for cloning and DNA sequencing, the DNA must be recovered from the gel in as high a yield as possible and without damage to the molecules. A number of methods have been tried, none entirely satisfactory, but electroelution is most often used. The two major problems with this method are, firstly, that the agarose is often contaminated by sulphated polysaccharides, which are extracted from the gel together with the DNA, and are potent inhibitors of many enzymes and, secondly, that the yield of DNA, particularly for large molecules, is relatively low.

The DNA band is cut from the gel and placed in a dialysis bag containing a small volume of buffer. The bag is placed in an electrophoretic tank and, when the current is switched on, the DNA passes out of the gel. The polarity of the current is reversed for a few moments to cause any DNA actually on the dialysis membrane to move back into the buffer within the bag. The DNA in the buffer is then precipitated with ethanol.

13.2.4   Chromatographic methods

High performance liquid chromatography

➤ HPLC – see Section 3.2.2.

An alternative to nucleic acid isolation by electrophoresis followed by electrophoretic elution is HPLC. There are several advantages in the use of HPLC. It is simple and easy to use and has a high degree of resolution, it is faster than electrophoresis and contamination of DNA fragments by impurities from the agarose gel is avoided.

The two types of HPLC most commonly used are reverse phase and ion exchange. Reverse-phase HPLC relies on the differences in hydrophobicity between nucleic acids of different lengths and base composition. Its major application is in the separation of small pieces of DNA (less than 50 base pairs) such as the products of oligonucleotide synthesis. Gradient elution with increasing amounts of acetonitrile is commonly used with the smaller, more polar nucleotides eluting first. The resolution is excellent and separation of oligonucleotides differing in length by only one base pair is easily achieved. Furthermore, the difference in hydrophobicity of the different bases means that oligonucleotides of the same length but with different base composition can also be separated. The order of polarity and therefore elution is $G \gg C > A > T$ (Figure 13.10).

➤ Oligonucleotide synthesis is used to produce small specific sequences of DNA. These are particularly important as primers in the polymerase chain reaction.

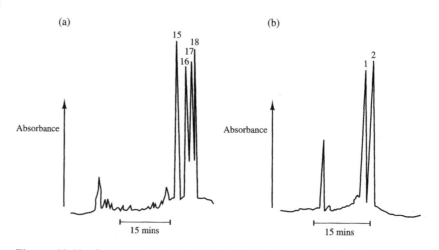

Figure 13.10 **Separation of oligonucleotides by reverse phase HPLC.** (a) Resolution of a mixture of oligonucleotides of 15, 16, 17 and 18 nucleotides long. (b) Resolution of two oligonucleotides each of 18 nucleotides but differing in base composition. Peak 1 is TCACAGTCTGATCTCACC. Peak 2 is TCACAGTCTGATCTC-GAT.

In ion-exchange chromatography nucleic acids are separated on the basis of the difference in charge. Each nucleic acid has a different net charge based on the number of phosphate groups in the molecule (base length) and the respective charges on the bases (base composition). The latter effect is relatively minor and so separation depends almost exclusively on size. Separation is accomplished by slowly increasing the ionic strength of the mobile phase using NaCl, KCl or KH_2PO_4. The larger, more charged oligonucleotides elute later than the shorter ones.

Ion-exchange HPLC can also be useful in the separation of larger nucleic acid molecules. One such application is as an alternative to CsCl density gradient centrifugation in the preparation of plasmids. Plasmid molecules typically consist of between 1000 and 10 000 base pairs. The plasmid is first isolated from the bacterial cell by alkaline lysis and pure plasmid obtained from this crude extract by a one-step chromatographic separation.

Affinity chromatography for the isolation of mRNA

> ➤ Poly (A) tails are lengths of nucleotides containing adenine bases present on the 3' end of the eukaryotic mRNA molecules.

Most eukaryotic mRNA molecules have up to 250 adenine bases at their 3′ end. These 'poly (A) tails' can be used in the affinity chromatographic purification of mRNA from a total cellular RNA extract. Under high salt conditions, poly (A) will hybridize to oligo-dT-cellulose or poly(U)-sepharose. These materials are polymers of 10 to 20 deoxythymidine or uridine nucleotides covalently bound to a carbohydrate support. They bind mRNA containing poly (A) tails as short as 20 residues. rRNA and tRNA do not possess poly (A) sequences and will not bind. After washing the mRNA can be eluted with a low salt buffer.

Self test questions

Sections 13.1/2

1. Which of the following are present in molecules of DNA?
   (a) Poly (A) tails.
   (b) Deoxyadenosine-5′-phosphate.
   (c) Cytidine-5′-phosphate.
   (d) Phosphodiester linkages.
2. Which of the following techniques used in the separation of RNA and DNA molecules rely on the fact that RNA and DNA are negatively charged molecules?
   (a) Electrophoresis.
   (b) Reverse-phase HPLC.
   (c) Ion-exchange HPLC.
   (d) Affinity chromatography.
3. RNA and DNA are isolated from tissues using caesium chloride density gradient sedimentation
   BECAUSE
   the binding of DNA to ethidium bromide causes unwinding of its helical structure.
4. Two molecules of DNA with lengths of 2000 and 3000 base pairs can be separated by agarose gel electrophoresis
   BECAUSE
   in agarose gel electrophoresis, larger molecules move faster through the gel than smaller molecules.

13.3   Methods of nucleic acid analysis

The method used to determine DNA concentration will depend on the nature of the sample. In recombinant DNA technology relatively pure samples of DNA or RNA are used and so simple spectrophotometric methods are adequate. However, when there are large amounts of interfering substances present, such as in a tissue or cell homogenate, a chemical method will be necessary.

13.3.1   Spectrophotometric methods

➤ Absorption of radiation – see Section 2.1.1.

The presence of conjugated double-bond systems in the purine and pyrimidine bases means that DNA and RNA absorb light in the ultraviolet region at 260 nm. For approximate determinations it can be assumed that a 50 μg ml^{-1} solution of double-stranded DNA (dsDNA) has an absorbance of 1 at 260 nm. More exact quantitation can be obtained by comparing the ratio of the absorbance of the sample at 260 and 280 nm. The term optical density (OD) is often used in place of absorbance. Pure DNA preparations should have an OD 260/OD 280 of 1.8. Ratios less than this may indicate protein contamination while higher ratios may indicate the presence of RNA.

➤ T_m is the temperature at which dsDNA separates into two single strands.

Bases that are not hydrogen bonded absorb more strongly than when base paired. Therefore treatments that disrupt the hydrogen bonding between the base pairs, such as heat or alkali, will increase the absorbance of DNA. When a solution of double-stranded DNA (dsDNA) is heated slowly, initially there is little change in the absorption until the 'melting temperature' (T_m) is reached. Here the hydrogen bonds are broken, producing a rapid increase in

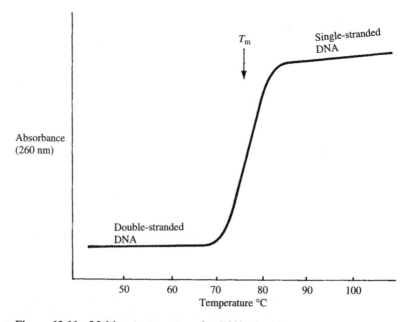

Figure 13.11 **Melting temperature for DNA.** At this temperature the absorbance increases because the two strands of DNA come apart.

absorbance to a higher value, which is not significantly changed on further heating (Figure 13.11).

DNA is often present in amounts too small to be detected by direct spectroscopy. In this case, the fluorescent dye EtBr can be used to amplify the absorption. EtBr binds to the DNA molecule by intercalating between adjacent base pairs. It absorbs ultraviolet light at 300 nm and emits light at 590 nm in the red/orange region of the visible spectrum. The method can be used to determine the amount of DNA in a test-tube by comparing the EtBr-mediated fluorescence of the sample with that of standards of known amounts of DNA.

13.3.2   Chemical methods

Burton method

This is a spectrophotometric assay based on the reaction of diphenylamine with the deoxyribose moiety of DNA to produce a complex that absorbs at 600 nm. The reaction is specific for deoxyribose and RNA does not interfere. It can be used on relatively crude extracts where direct spectrophotometric determinations of DNA concentration are not possible.

Procedure 13.2:   The Burton method for the quantitation of DNA

Reagents
Tissue sample
Standard DNA samples ranging from 1 to 50 μg ml^{-1}
Perchloric acid 0.5 mol l^{-1}
Burton reagent
    diphenylamine   15 g l^{-1}
    sulphuric acid   0.25 mol l^{-1}
    acetaldehyde   0.05 ml l^{-1}

Method
1. Add enough perchloric acid to cover the tissue sample. Add the same volume of perchloric acid to the standard samples of DNA.
2. Heat both samples to 70–80°C for between 20 and 90 min.
3. Centrifuge at 300 g to remove the cellular debris.
4. Add 1 ml of sample or standard supernatant to 2 ml of reagent.
5. Mix and leave for 18 hours at 30°C.
6. Measure the absorbance at 600 nm.

Calculation
1. Prepare a calibration graph from the absorbance values of the standard DNA samples.
2. Read off the DNA content of the tissue samples.

DABA fluorescence assay

Diaminobenzoic acid (DABA) reacts with aldehydes of the form RCH CHO to produce a strongly fluorescent compound. Acid-catalysed removal of the purine base from the nucleic acid exposes the 1′ and 2′ carbons of deoxyribose and produces such an aldehyde group. Deoxyribose is the predominant alde-hyde present in mammalian cells and essentially the only one present in acid precipitates of aqueous extracts. Hence, no purification is required and RNA does not interfere. The method can be used on very small samples (e.g. 100 μl).

Procedure 13.3: The DABA fluorescence method for the quantitation of DNA

Reagents
   Tissue sample
   Standard DNA samples ranging from 1 to 50 μg ml^{-1}
   Hydrochloric acid   1.0 mol l^{-1}
   Diaminobenzoic acid   320 g l^{-1}

Method
1. Spot 100 μl of homogenized tissue sample or DNA standard onto the bottom of a polypropylene tube and dry at 50°C.
2. Add 100 μl diaminobenzoic acid reagent and incubate at 55°C for 45 min.
3. Add 1.0 ml hydrochloric acid.
4. Measure fluorescence (excitation wavelength 410 nm: emission wavelength 510 nm).

Calculation
1. Prepare a calibration graph from the absorbance values of the standard DNA samples.
2. Read off the DNA content of the test samples.

13.3.3   Enzymic methods

Most DNA manipulations require the use of purified enzymes. Within cells these enzymes are used for DNA replication and transcription, breakdown of foreign DNA, repair of mutated DNA and recombination between different DNA molecules. Most of the purified enzymes are available commercially. They are usually supplied in Tris buffer containing 50% glycerol and are stored at –20°C. The glycerol is included in the buffer to prevent freezing but must be removed prior to use as it alters the activity of some enzymes.

For each enzyme, an optimized buffer solution is provided. Most of the buffers are based on Tris–HCl and have a pH range of 7.4–8.0. All contain magnesium ions as activators for the enzymes. The ionic strength of the buffers is important and different enzymes require different ionic strengths for optimal activity. In general, three strengths of buffer are used: low, medium and high, which contain 10, 50 and 100 mol l^{-1} sodium or potassium chloride respectively.

Restriction endonucleases

Restriction endonucleases are enzymes isolated from bacteria. *In vivo* these enzymes are involved in recognising and cutting up foreign DNA, thus pro-tecting the bacteria against phage and virus infection. They recognise specific base sequences in double-stranded DNA and break the phosphodiester bonds between two nucleotides within the sequence. These recognition sequences may be four, five or six nucleotides long. Over 350 different enzymes have been isolated so far and each recognises a different base sequence. Some of the more commonly used enzymes with their recognition sequences are shown in Table 13.2.

Some enzymes, e.g. *Eco* RI, cut the DNA at a different position in the two strands, producing a single stranded overhang or a 'sticky end'. Other enzymes, e.g. *Pvu*II, produce blunt-ended fragments with no sticky ends. The formation of sticky ends is useful because, by base pair matching of the sticky ends, any DNA fragment produced by the action of the same restriction enzyme can be joined together in a specific manner.

> ➤ A sticky end is an extension of nucleotides on one strand only of a dsDNA molecule.

Table 13.2 Recognition sequences of a selection of restriction endonucleases. The double-stranded DNA is cut at the positions arrowed

Enzyme	Organism	Recognition sequence
Eco RI	*Escherichia coli*	G̓AATTC CTTAA̗G
Bam HI	*Bacillus amyloliquefaciens*	G̓GATCC CCTAG̗G
Pvu I	*Proteus vulgaris*	CGAT̓CG GC̗TAGC
Pvu II	*Proteus vulgaris*	CAG̓CTG GTC̗GAC
Hin dIII	*Haemophilus influenzae* Rd	A̓AGCTT TTCGA̗A
Alu I	*Arthrobacter luteus*	AG̓CT TC̗GA
Hae III	*Haemophilus aegyptius*	GG̓CC CC̗GG
Hin fI	*Haemophilus influenzae*	G̓ANTC CTNA̗G

DNA ligases

Ligases are found in all cells but the two that are used in DNA analysis are either DNA ligase from *Escherichia coli*, which requires the nucleotide NAD as a cofactor, or T4 DNA ligase from the T4 phage, which requires adenosine triphosphate (ATP) as a cofactor. Both enzymes catalyse the synthesis of a phosphodiester bond between the 3′ hydroxyl group of one nucleotide and the 5′ phosphoryl group of the next nucleotide in a dsDNA molecule. Ligases are most often used to produce a covalent bond between nucleotides in DNA molecules joined by base pair matching of sticky ends (Figure 13.12).

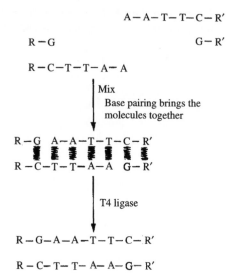

$$A-A-T-T-C-R'$$

$$R-G \qquad\qquad\qquad G-R'$$

$$R-C-T-T-A-A$$

Mix
Base pairing brings the
molecules together

$$R-G\ \ A-A-T-T-C-R'$$

$$R-C-T-T-A-A\ \ G-R'$$

T4 ligase

$$R-G-A-A-T-T-C-R'$$

$$R-C-T-T-A-A-G-R'$$

Figure 13.12   **The joining of two molecules of DNA by the enzyme T4 ligase.** The enzyme produces covalent bonds between nucleotides in a strand of DNA. The DNA molecules are brought together by base pairing of their sticky ends.

> ➤ Primers are short sequences of nucleotides used to initiate replication of DNA molecules by DNA polymerase enzymes.

DNA polymerases

DNA polymerases synthesize a new strand of DNA complementary to an existing DNA or RNA strand (Figure 13.13). Short oligonucleotides complementary to the ends of the original DNA or RNA strand, known as primers, are added to the incubation mixture.

1. **DNA polymerase 1**. For most applications DNA polymerase 1 from *E. coli* is used. The enzyme attaches to a short single-stranded region in a dsDNA molecule and then synthesizes a new strand of DNA, degrading the existing strand as it proceeds. When used *in vitro* this incubation is carried out at 12–15 °C to prevent more than one round of replication occurring. It is used for *in vitro* labelling of DNA by the nick translation method (described below).

2. **The Klenow fragment**. The DNA polymerase 1 molecule contains its polymerase and nuclease activities on different parts of the enzyme molecule. These two parts can be separated by treatment with the enzyme **subtilisin**. The part which retains the polymerase function is known as the Klenow fragment. This enzyme sythesizes a new DNA strand complementary to the single strand of DNA (the template) only. It is used to create blunt ends in dsDNA and in the dideoxy method of DNA sequencing.

3. **Reverse transcriptase**. This enzyme is involved in the replication of retroviruses *in vivo*. It synthesizes a complementary DNA (cDNA) strand using RNA instead of DNA as its template. It is widely used to create a strand of cDNA from mRNA extracted from cells or tissue for cloning or for PCR analysis.

4. **Taq polymerase** is a thermostable DNA polymerase which was originally isolated from the bacterium *Thermus aquaticus,* which lives in hot springs.

This enzyme has a maximum activity at 70–80°C and remains active at temperatures up to 90°C. It is used extensively in the analysis of DNA using PCR.

DNA-modifying enzymes

Many other enzymes can be used to remove or add groups to the ends of the DNA molecule. The three major ones are alkaline phosphatase, which removes a phosphate group from the 5′ terminus; polynucleotide kinase, which adds a phosphate group to a free 5′ terminus; and terminal deoxynucleotidyl transferase, which adds one or more deoxynucleotides to the 3′ terminus.

(a) DNA Polymerase 1

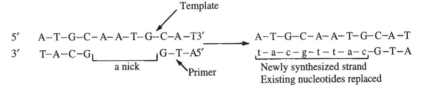

(b) The Klenow Fragment

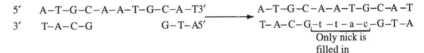

(c) Reverse Transcriptase

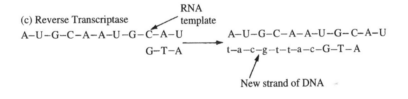

Figure 13.13
The actions of the DNA polymerases used in recombinant DNA technology.

13.3.4   Detection of specific sequences of DNA

If a single-stranded DNA molecule is placed with a complementary single DNA sequence the two molecules will hybridize. This hybridization forms the basis of a number of very powerful techniques for detecting and quantifying specific nucleic acid sequences. The hybridization may be carried out either in solution or more commonly with the DNA immobilized on nitrocellulose filters. The complementary DNA sequence is known as a cDNA probe. Probes for a large number of important nucleic acid sequences are now available.

Production of labelled DNA probes

Before using the probe for hybridization, it must be labelled to enable subsequent detection. One method of labelling is by the incorporation of ^{32}P-labelled nucleotides, a process that may be carried out in two ways.

Nick translation uses the enzyme DNA polymerase 1. Most pieces of DNA contain some nicked areas. DNA polymerase 1 can attach here and catalyse a strand replacement reaction. When the reaction is carried out in the presence of radioactively labelled deoxynucleotides the replaced DNA strand becomes labelled. Usually only one labelled deoxynucleotide is included.

End filling is a more gentle method but can only be used on DNA molecules that have sticky ends. It uses the Klenow fragment, which fills in the sticky ends by synthesizing the complementary strand. Again if the reaction is carried out in the presence of labelled deoxynucleotides the result is a strand of DNA which is radioactively labelled.

Non-radioactive labelling systems are also used. These incorporate biotinylated or fluorescein-conjugated nucleotides into the cDNA probe in place of the ^{32}P-nucleotide. The radio-labelled probe can be detected directly by autoradiography, whereas the biotinylated or fluorescein-conjugated probes are detected using alkaline phosphatase or peroxidase-conjugated antibodies directed against the biotin or fluorescein. The binding of antibody to the nucleotide can then be visualized by incubation with the enzyme substrate and a colour reagent. Although non-radioactive labelling systems are advantageous with regards to safety and disposal, the visualization procedure requires several more incubation steps, which may be time consuming.

➤ Autoradiography – see Section 5.2.3.

➤ Immunoassay – see Section 7.4.

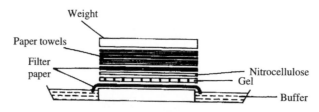

Figure 13.14 **The transfer of DNA from a gel to a nitrocellulose sheet.** Buffer soaks through the gel to the nitrocellulose and takes the DNA with it.

Southern blotting

The method most often used in the detection of specific DNA sequences is Southern blotting. The DNA is fragmented by digestion with a restriction endonuclease and the fragments separated by agarose gel electrophoresis. The DNA bands are then transferred to a nitrocellulose membrane so that individual DNA molecules are retained during the hybridization process. The apparatus used for this transfer is shown in Figure 13.14. The nitrocellulose is placed on the gel and buffer is allowed to soak through, carrying DNA from the gel to the membrane where the DNA is bound. The result is that the nitrocellulose membrane carries a replica of DNA bands from the agarose gel. The DNA on the membrane is then incubated with labelled DNA probe, which will hybridize to any bands which contain sequences complementary to it. After washing, this hybridization can be visualized by autoradiography or by using an antibody detection system (Figure 13.15).

The specificity of the hybridization between probe and DNA depends on the temperature and ionic strength of the buffer. At high temperatures and low salt concentration (high stringency), hybridization is very specific. At lower stringencies (low temperatures and high salt concentration) the hybridization is less specific and the cDNA probe will bind to many sequences. In Southern blotting, incubation with the probe is carried out at low stringencies, which allows hybridization to a large number of fragments. The membrane is then washed at high stringency so that the probe remains bound only to highly homologous sequences.

Another similar technique to Southern blotting is Northern blotting. Here, instead of DNA fragments, mRNA fragments are probed with a labelled cDNA probe after separation by electrophoresis and transfer to nitrocellulose membranes. Northern blotting is used to detect and quantify mRNA from tissue extracts.

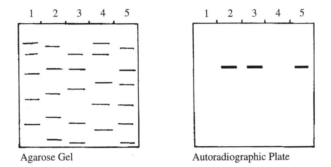

Agarose Gel   Autoradiographic Plate

Figure 13.15 **Detection of a specific sequence of DNA by hybridization to a ^{32}P-labelled cDNA probe.** DNA is transferred to nitrocellulose and incubated with the probe. After washing, specific binding is visualized by autoradiography. The DNA sequence detected by the probe is present in lanes 2, 3 and 5 but not 1 and 4.

The polymerase chain reaction

Although with the use of cDNA probes containing many molecules of [32]P (high specific activity) as little as 20 pg of DNA or mRNA can be detected by Southern or Northern blotting, many DNA or RNA sequences within tissues are below this level of detection. The polymerase chain reaction allows massive replication of a target sequence of DNA and allows detection of DNA or RNA sequences that are present as only one or two copies per cell. The target sequence of DNA is extracted from the cell and denatured into its single strand by heating to 90°C. Two oligonucleotide primers complementary to the opposing ends of the single strands of the target DNA are synthesized and annealed. Primers for PCR consist of a 20–30 nucleotide sequence specific for the target DNA and result in the amplification of that sequence of DNA only. The primed single strand is then extended using DNA polymerase 1. The newly synthesized double-stranded sequence is then denatured by heating to 90°C and new oligonucleotide primers added. This process is then repeated through 25 to 30 cycles and results in a large increase in DNA (Figure 13.16).

The reaction depends upon the use of Taq polymerase, which is stable at high temperatures and can survive the repeated heating to 90°C without loss of activity. The amplification process is carried out in a thermocycler, which allows the automatic control of the rate of heating and cooling, the incubation times at each temperature and the number of cycles. This information can be programmed into the thermocycler at the start of the reaction process, which is carried out in the presence of labelled nucleotides. The PCR product is detected using agarose gel electrophoresis and autoradiography.

An extension of the PCR method is RT-PCR. Here mRNA is extracted from the cells or tissue, converted to cDNA using the enzyme reverse transcriptase and PCR is then carried out. This method is used to detect the expression of specific mRNA sequences in cells or tissues.

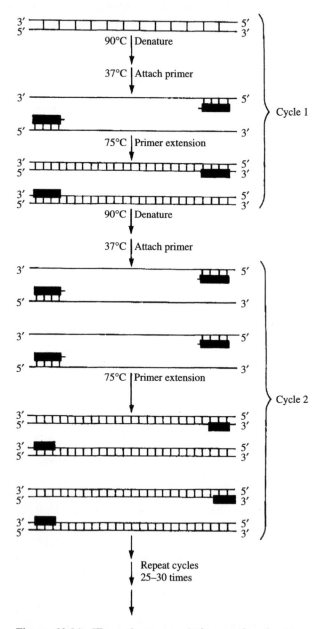

Figure 13.16 **The polymerase chain reaction for the amplification of DNA sequences.** DNA is heated to separate the two strands. A primer is attached to the 5′ end of each strand and extended using DNA polymerase 1. The two new strands are separated as before and the cycle repeated up to 30 times.

Self test questions

Section 13.3

1. Which of the following enzymes are used in RT-PCR?
   (a) Restriction endonucleases.
   (b) Taq polymerase.
   (c) Reverse transcriptase.
   (d) T4 DNA ligase.
2. Which of the following occur in the analysis of DNA by Southern blotting?
   (a) Separation of DNA molecules by electrophoresis.
   (b) Amplification of DNA sequences.
   (c) Hybridization between complementary strands of DNA.
   (d) Incubation with specific primers.
3. A 15 μg ml^{-1} solution of purified double-stranded DNA will have an absorbance of approximately 0.3 at 260 nm
   BECAUSE
   an increase in temperature disrupts the hydrogen bonding between base pairs.
4. The enzyme DNA polymerase 1 is used in the production of labelled cDNA probes
   BECAUSE
   the enzyme DNA polymerase 1 adds nucleotides to the end of the cDNA molecule

13.4   Vectors

Pieces of DNA, for example genes or cDNA probes, are multiplied using a method based on *in vivo* replication in bacterial cells. The sequence of DNA is incorporated into a vehicle or vector which transports it into a cell host, usually *E. coli*. As the bacteria culture grows, the vector also replicates so producing more of the required sequence of DNA. Two of the naturally occurring types of DNA molecule which can be used as vectors are plasmids and viral chromosomes.

13.4.1 Plasmids

Plasmids are the circular double-stranded DNA molecules found in the cytoplasm of bacterial cells. To be a useful vector the plasmid must be small. The smaller the plasmid, the easier it is taken up into the bacteria. Most vector plasmids are about 3–4 kilobases (kb) in length. This allows incorporation of sequences of DNA up to 10–12 kb without impairing plasmid uptake. It is also helpful if the vector has a high copy number. The larger the copy number, the more molecules of DNA will be produced.

➤ Transformation is the uptake of DNA into bacterial cells to produce a recombinant.

The uptake of the plasmid into the bacterial cell is called transformation and in the laboratory can be induced in two ways. In one method the bacteria and DNA are placed in ice-cold CaCl$_2$ (50 mmol l^{-1}). This induces the DNA to stick to the outside of the bacteria. The temperature is then increased to 42°C and the DNA enters the cell. The second method is by electroporation, which involves the application of a brief high voltage pulse to a suspension of cells and DNA. This results in a transient membrane permeability and the subsequent uptake of DNA into the bacteria.

13.4.2   Viral chromosomes

The chromosomes of bacteriophages, which are viruses that specifically infect bacteria, can also be used as vectors. These are linear double-stranded DNA molecules which contain the genes required for viral replication within the host cell. During infection the DNA molecules are inserted into the bacterial host. The most widely used bacteriophage is lambda (λ) which infects *E. coli*. Within the *E. coli* lambda has two alternative modes of replication: the lytic and the lysogenic cycles. In the lytic cycle the DNA is incorporated into the *E. coli* chromosome. Replication takes place, resulting in the production of a large number of mature phage particles, which are released from the cell by cell lysis. In the lysogenic mode the lambda DNA is incorporated into the chromosomal DNA in such a way that replication of the lambda DNA occurs but it is not packaged into mature particles and there is no cell lysis. Instead the lambda genome is inherited by each daughter bacterial cell at cell division. Lambda is maintained in this prophage state through the action of a repressor protein, cI. If the repressor protein is inactivated the lambda DNA is excised from the host chromosome and a lytic cycle is activated.

> ➤ In lysogenic cycles, replication of the bacteriophage lambda occurs without lysis of the infected cell.

In one mutated form of lambda the repressor protein is inactive at temperatures above 37°C but active at lower temperatures. When this lambda is used as a vector the infected *E. coli* are grown first at 32°C to allow replication of the DNA in the lysogenic cycle. The temperature is then increased to 37°C to inactivate the repressor. This results in the excision of the lambda genome and release of lambda particles by lysis.

Another viral chromosome that can be used as a vector is that of the filamentous phage M13. The M13 chromosome is a single-stranded DNA molecule which when inserted into the bacterial host replicates outside the bacterial chromosome in the cytoplasm. The virus is then reassembled and released from the bacterial cell without cell lysis.

13.4.3   Selection of transformed bacteria

The procedure of inserting DNA into a vector and its subsequent introduction into the bacterial host is extremely inefficient and will result in the majority of bacteria in the culture either containing none of the required vector or containing a vector in which no DNA is inserted. However, vectors have been produced that, by placing the bacteria in specific media, allow only the growth of those containing the vector and the inserted DNA.

One such plasmid vector is pBR322 (Figure 13.17). It contains two genes coding for resistance to the antibiotics ampicillin and tetracycline. Bacteria that take up the vector (transformed bacteria) will be resistant to the antibiotic and will grow in its presence, while non-transformed bacteria will not. The bacteria are therefore grown on media containing either ampicillin or tetracycline. The resultant colonies must contain bacteria that have taken up the vector.

Within the genes coding for ampicillin and tetracycline are unique

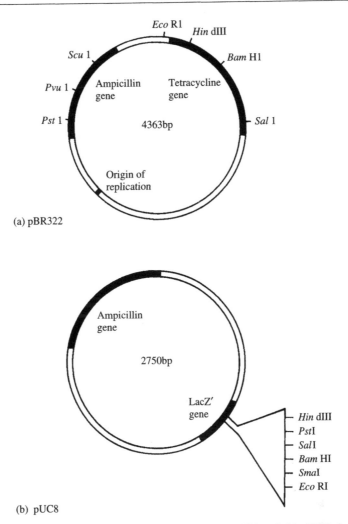

Figure 13.17   Maps of the plasmids (a) pBR322 and (b) pUC8 showing restriction endonuclease sites in genes used to select recombinants.

recognition sites for many different restriction enzymes. DNA fragments are inserted into these sites using the appropriate enzyme. Insertion of the DNA fragment results in disruption of the gene and the resistance to the antibiotic is lost. Therefore any bacteria that have taken up the vector with the fragment of DNA inserted (recombinants) will not grow in the presence of that antibiotic. This process is called insertional inactivation.

To select for a bacterium that contains a fragment of DNA inserted into the tetracycline gene of pBR322, for example after using the *Bam* Hl restriction enzyme, the bacteria are grown on a medium containing ampicillin. Colonies that grow contain the plasmid but only a few contain the recombinant pBR322 molecules. The colonies are then replica plated onto media containing tetracycline. The colonies that grow are resistant to tetracycline and so have an intact tetracycline gene. Those that contain the recombinant molecule will not grow but can be taken from the original plate and grown up in bulk (Figure 13.18).

➤ Insertional inactivation is the inactivation of a gene by the insertion of another DNA sequence.

Another widely used form of insertional inactivation is the *lacZ'* gene which was first used in the pUC8 vector (Figure 13.17). The *lacZ* gene codes for the enzyme β-galactosidase, which breaks down lactose and is present in the *E. coli* chromosome. Some strains of *E. coli* have a modified *lacZ* gene (*lacZ'*) which has a large segment missing. These mutants can only synthesize the whole enzyme when the missing *lacZ* segment is supplied by a plasmid such as pUC8. This *lacZ* segment can be inactivated by insertion of a DNA fragment using an appropriate restriction enzyme. pUC8 also contains an ampicillin resistance gene and so selection involves firstly growing the bacteria on ampicillin to select for transformants followed by transfer to media containing X-gal. X-gal is a synthetic substrate for the enzyme β-galactosidase and gives a blue product. Bacteria containing a recombinant pUC8 molecule will not supply the extra piece of enzyme, so X-gal is not broken down and the colonies are white. Bacteria that contain the intact pUC8 molecule will supply the β-galactosidase fragment, so X-gal is broken down and blue colonies are formed. The white colonies are the ones required and can be taken directly from this plate.

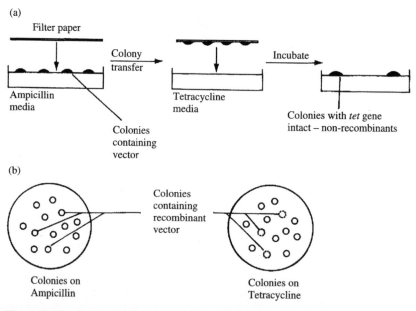

Figure 13.18   Replica plating for the selection of bacteria containing a recombinant vector. (a) The method used to transfer colonies of bacteria between plates. (b) Comparison of the distribution of the colonies between the plates enables identification of the colonies containing the recombinant vector. The dotted circles represent colonies that do not grow on tetracycline.

13.5   DNA sequencing

It is the sequence of the nucleotide bases in the DNA molecule that is fundamental to nucleic acid function. Hence the techniques that determine this sequence are extremely important in recombinant DNA technology. Two similar methods have been developed. Both depend on the production of a mixture of oligonucleotides labelled either radioactively or with fluorescein, with one common end and differing in length by a single nucleotide at the other end. This mixture of oligonucleotides is separated by high resolution electrophoresis on polyacrylamide gels and the position of the bands determined.

Known DNA sequences associated with specific genes are stored in computer libraries which are freely accessible. This allows the comparison of newly sequenced genes with those coding for known proteins.

13.5.1   The Maxam and Gilbert method

In this method the single-stranded DNA fragment to be sequenced is end-labelled by treatment with alkaline phosphatase to remove the 5' phosphate, followed by reaction with ^{32}P-labelled ATP in the presence of polynucleotide kinase, which attaches ^{32}P to the 5' terminal. The labelled DNA fragment is then divided into four aliquots, each of which is treated with a reagent which modifies a specific base as follows.

Aliquot 1 is treated with dimethyl sulphate, which methylates guanine residues.

Aliquot 2 is treated with formic acid, which modifies adenine and guanine residues.

Aliquot 3 is treated with hydrazine, which modifies thymine and cytosine residues.

Aliquot 4 is treated with hydrazine in the presence of NaCl (5 mol l^{-1}), which makes the reaction specific for cytosine.

After these reactions the four aliquots are incubated with piperidine, which cleaves the sugar–phosphate backbone of DNA next to the residue that has been modified. The modification reaction is carried out under conditions that allow only a limited number of bases in each DNA molecule to react. In perfect conditions this will be one base per DNA molecule. This means that cleavage with piperidine will result in the production of a mixture of oligonucleotides each having the same 5' end but extending to the nucleotide next to the one that was modified (Figure 13.19). Any base in the original DNA molecule is as likely to react as any other with the reagent which is specific for it. Therefore these reactions evenly distribute the radioactivity among the cleavage products.

► SDS electrophoresis – see Section 11.3.2.

The oligonucleotides in each aliquot are placed in four different but adjacent wells of a polyacrylamide gel containing SDS and separated by electrophoresis. The separated fragments are detected by autoradiography. The resolution of these gels is such that oligonucleotides differing by only one nucleotide can be distinguished. The smallest fragments travel the furthest. Hence the band that has travelled the furthest contains only one nucleotide, while the next band up the gel contains two nucleotides, etc. The lane in which the band is found indicates the next base in the sequence. The DNA sequence is therefore obtained by reading the gel from bottom to top (Figure 13.19).

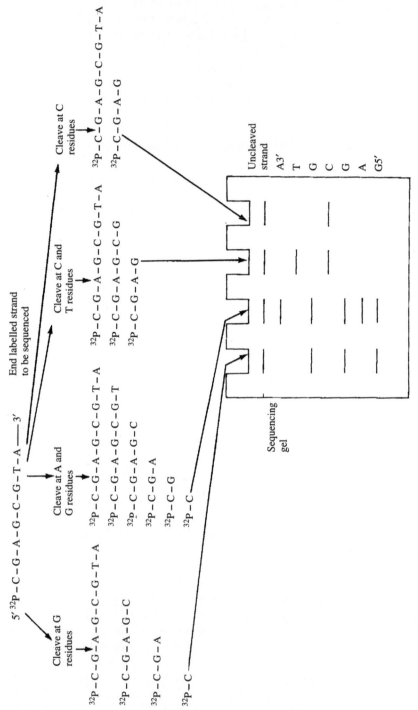

Figure 13.19 **The Maxam and Gilbert method of DNA sequencing.** The labelled DNA strand is divided into four aliquots. Each is treated to cleave the strand next to a different base resulting in a mixture of different length nucleotides. These nucleotides are separated by polyacrylamide gel electrophoresis and the DNA sequence read from the gel.

13.5.2 The dideoxy method

This method was developed by Sanger and his colleagues and is also referred to as the chain termination method. The procedure requires a single-stranded DNA template and a short primer complementary to the 3′ end of the region of DNA to be sequenced. A complementary copy of the DNA strand is produced

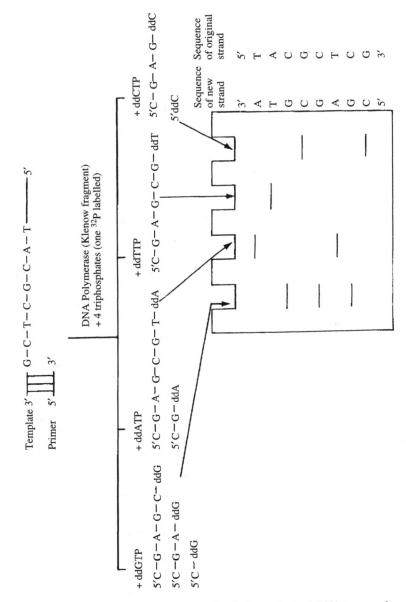

Figure 13.20 **The dideoxy (chain termination) method of DNA sequencing.** A complementary copy of the DNA strand is synthesized in the presence of dideoxy analogues of each base. Chain termination occurs where the analogue is inserted in place of the true base. The newly synthesized strands of DNA are separated by polyacrylamide gel electrophoresis and the sequence deduced from the pattern of bands.

by the Klenow fragment of DNA polymerase 1. Synthesis is carried out in the presence of the four deoxyribonucleotide triphosphates, one of which is labelled with ^{32}P. In the presence of competing dideoxyribonucleotide triphosphates (ddNTPs), specific termination of DNA synthesis occurs where the dideoxy derivative is incorporated instead of the deoxyribonucleotide. Four incubations are carried out, each in the presence of a different dideoxy derivative. Each incubation generates a heterogeneous population of labelled oligonucleotides terminating with the same nucleotide (Figure 13.20).

Urea is added to each incubation to separate the two strands of DNA and the single strands are separated electrophoretically in adjacent lanes of an SDS polyacrylamide gel. Urea is also present in the gel to ensure that the strands of DNA stay separated. The gel is then autoradiographed. The DNA sequence can be deduced from the ascending order of the bands in the four adjacent lanes.

The single-stranded template DNA required for this method of sequencing is usually provided by using the bacteriophage M13 as a vector. The most widely used method of DNA sequencing using this system is the 'shotgun approach'. A random series of fragments is produced from the DNA to be sequenced. This may be done by sonication or digestion with restriction enzymes. Each fragment is incorporated into the M13 vector and its sequence obtained. Comparison of the sequences of each fragment will show overlapping sections and allow deduction of the sequence of the original piece of DNA.

Fluorescein-labelled deoxyribonucleotide triphosphates may be used in place of those labelled with ^{32}P. Once the DNA sequences are separated by electrophoresis, the resulting DNA bands fluoresce and are analysed by a fluorogram imager, which produces a picture of the fluorescent bands similar to the autoradiography produced when using ^{32}P-labelled nucleotides.

Kits containing all the reagents required for DNA sequencing including the DNA polymerase enzyme, M13, vector, primers, buffers and labelled and unlabelled nucleotides are available from a number of manufacturers. All the components of these kits have been optimized for use with each other.

Self test questions

Sections 13.4/5

1. Which of the following statements are true of the dideoxy method for sequencing DNA?
   (a) New chains of DNA are synthesized that are terminated at selective bases.
   (b) Chain termination is brought about by incubation with dideoxyribonucleotide triphosphate.
   (c) The DNA strand is broken next to selected bases by incubation with piperidine.
   (d) The DNA strands produced are separated by gel electrophoresis.
2. Which of the following statements are true?
   (a) Insertion of a piece of DNA into a particular gene causes activation of that gene.
   (b) Bacteria transformed with pBR322 plasmid with DNA inserted into the ampicillin resistance gene will grow in a medium containing tetracycline.

(c) Bacteria transformed with the pUC8 plasmid with DNA inserted into the *LacZ'* gene will not grow in a medium containing ampicillin.

(d) Plasmid molecules are cleaved at particular sites using restriction endonucleases.

3. Bacteria containing the pUC8 plasmid with a gene inserted into the *LacZ'* gene grown on medium containing X-gal are blue in colour

   BECAUSE

   the *LacZ'* gene codes for a component of the galactosidase enzyme which converts X-gal to a blue-coloured product.

4. The bacteriophage lambda (λ) contains the genes coding for resistance to ampicillin and tetracycline

   BECAUSE

   the bacteriophage lambda (λ) contains genes that enable it to undergo both lytic and lysogenic modes of replication.

13.6   Further reading

Brown, T.A. (1995) *Gene cloning*, 3rd edition, Chapman & Hall, UK.

Old, R.W. and Primrose, S.B. (1994) *Principles of gene manipulation*, 4th edition, Blackwell Law, UK.

Emery, A.E. and Malcolm, S. (1995) *An introduction to recombinant DNA in medicine*, 2nd edition, Wiley-Liss Inc., USA.

Walker, J. (ed.) (1989) *Nucleic acid techniques in methods in molecular biology*, Humana Press Inc., USA.

Brown, T.A. (1991) *Essential molecular biology – a practical approach*, IRL Press, UK.

Appendix: Numbering and classification of enzymes

1. Oxidoreductases

1.1 Acting on the CH–OH group of donors

1.1.1 With NAD^+ or $NADP^+$ as acceptor
1.1.2 With a cytochrome as acceptor
1.1.3 With oxygen as acceptor
1.1.99 With other acceptors

1.2 Acting on the aldehyde or oxo group of donors

1.2.1 With NAD^+ or $NADP^+$ as acceptor
1.2.2 With a cytochrome as acceptor
1.2.3 With oxygen as acceptor
1.2.4 With a disulphide compound as acceptor
1.2.7 With an iron–sulphur protein as acceptor
1.2.99 With other acceptors

1.3 Acting on the CH–CH group of donors

1.3.1 With NAD^+ or $NADP^+$ as acceptor
1.3.2 With a cytochrome as acceptor
1.3.3 With oxygen as acceptor
1.3.7 With an iron–sulphur protein as acceptor
1.3.99 With other acceptors

1.4 Acting on the CH–NH$_2$ group of donors

1.4.1 With NAD^+ or $NADP^+$ as acceptor
1.4.2 With a cytochrome as acceptor
1.4.3 With oxygen as acceptor
1.4.4 With a disulphide compound as acceptor
1.4.7 With an iron–sulphur protein as acceptor
1.4.99 With other acceptors

1.5 Acting on the CH–NH group of donors

1.5.1 With NAD^+ or $NADP^+$ as acceptor
1.5.3 With oxygen as acceptor
1.5.99 With other acceptors

1.6 Acting on NADH or NADPH

1.6.1 With NAD^+ or $NADP^+$ as acceptor
1.6.2 With a cytochrome as acceptor
1.6.4 With a disulphide compound as acceptor
1.6.5 With a quinone or related compound as acceptor
1.6.6 With a nitrogenous group as acceptor
1.6.7 With an iron–sulphur protein as acceptor
1.6.99 With other acceptors

1.7 Acting on other nitrogenous compounds as donors

1.7.2 With a cytochrome as acceptor
1.7.3 With oxygen as acceptor
1.7.7 With an iron–sulphur protein as acceptor
1.7.99 With other acceptors

1.8 Acting on a sulphur group of donors

1.8.1 With NAD^+ or $NADP^+$ as acceptor
1.8.2 With a cytochrome as acceptor
1.8.3 With oxygen as acceptor

1.8.4   With a disulphide compound as acceptor

1.8.5   With a quinone or related compound as acceptor

1.8.7   With an iron–sulphur protein as acceptor

1.8.99   With other acceptors

1.9   Acting on a haem group of donors

1.9.3   With oxygen as acceptor

1.9.6   With a nitrogenous group as acceptor

1.9.99   With other acceptors

1.10   Acting on diphenols and related substances as donors

1.10.1   With NAD^+ or $NADP^+$ as acceptor

1.10.2   With a cytochrome as acceptor

1.10.3   With oxygen as acceptor

1.11   Acting on hydrogen peroxide as acceptor

1.12   Acting on hydrogen as donor

1.12.1   With NAD^+ or $NADP^+$ as acceptor

1.12.2   With a cytochrome as acceptor

1.12.7   With an iron–sulphur protein as acceptor

1.13   Acting on single donors with incorporation of molecular oxygen (oxygenases)

1.13.11   With incorporation of two atoms of oxygen

1.13.12   With incorporation of one atom of oxygen (internal monooxygenases or internal mixed function oxidases)

1.13.99   Miscellaneous (requires further characterization)

1.14   Acting on paired donors with incorporation of molecular oxygen

1.14.11   With 2-oxoglutarate as one donor, and incorporation of one atom each of oxygen into both donors

1.14.12   With NADH or NADPH as one donor, and incorporation of two atoms of oxygen into one donor

1.14.13   With NADH or NADPH as one donor, and incorporation of one atom of oxygen

1.14.14   With reduced flavin or flavoprotein as one donor, and incorporation of one atom of oxygen

1.14.15   With a reduced iron–sulphur protein as one donor, and incorporation of one atom of oxygen

1.14.16   With reduced pteridine as one donor, and incorporation of one atom of oxygen

1.14.17   With ascorbate as one donor, and incorporation of one atom of oxygen

1.14.18   With another compound as one donor, and incorporation of one atom of oxygen

1.14.99   Miscellaneous (requires further characterization)

1.15   Acting on superoxide radicals as acceptor

1.16   Oxidizing metal ions

1.16.3   With oxygen as acceptor

1.17   Acting on —CH_2 groups

1.17.1   With NAD^+ or $NADP^+$ as acceptor

1.17.4   With a disulphide compound as acceptor

1.18   Acting on reduced ferredoxin as donor

1.18.1   With NAD^+ or $NADP^+$ as acceptor

1.18.2   With dinitrogen as acceptor

1.18.3   With H^+ as acceptor

1.19   Acting on reduced flavodoxin as donor

1.19.2   With dinitrogen as acceptor

1.97   Other oxidoreductases

2.   Transferases

2.1   Transfering one-carbon groups

2.1.1   Methyltransferases

2.1.2   Hydroxymethyl-, formyl- and related transferases

2.1.3   Carboxyl- and carbamoyltransferases

2.1.4   Amidinotransferases

2.2   Transferring aldehyde or ketonic residues

3.3   Acting on ether bonds

3.3.1   Thioether hydrolases

3.3.2   Ether hydrolases

3.4   Acting on peptide bonds (peptide hydrolases)

3.4.11 α-Aminoacylpeptide hydrolases

3.4.13   Dipeptide hydrolases

3.4.14   Dipeptidylpeptide hydrolases

3.4.15   Peptidyldipeptide hydrolases

3.4.16   Serine carboxypeptidases

3.4.17   Metallo-carboxypeptidases

3.4.21   Serine proteinases

3.4.22   Thiol proteinases

3.4.23   Carboxyl (acid) proteinases

3.4.24   Metalloproteinases

3.4.99   Proteinases of unknown catalytic mechanism

3.5   Acting on carbon–nitrogen bonds, other than peptide bonds

3.5.1   In linear amides

3.5.2   In cyclic amides

3.5.3   In linear amidines

3.5.4   In cyclic amidines

3.5.5   In nitriles

3.5.99   In other compounds

3.6   Acting on acid anhydrides

3.6.1   In phosphorus-containing anhydrides

3.6.2   In sulphonyl-containing anhydrides

3.7   Acting on carbon–carbon bonds

3.7.1   In ketonic substances

3.8   Acting on halide bonds

3.8.1   In C–halide compounds

3.8.2   In P–halide compounds

3.9   Acting on phosphorus–nitrogen bonds

3.10   Acting on sulphur–nitrogen bonds

3.11   Acting on carbon–phosphorus bonds

4.   Lyases

4.1   Carbon–carbon lyases

4.1.1   Carboxyl lyases

4.1.2   Aldehyde lyases

4.1.3   Oxo-acid lyases

4.1.99   Other carbon–carbon lyases

4.2   Carbon–oxygen lyases

4.2.1   Hydro-lyases

4.2.2   Acting on polysaccharides

4.2.99   Other carbon–oxygen lyases

4.3   Carbon–nitrogen lyases

4.3.1   Ammonia lyases

4.3.2   Amidine lyases

4.4   Carbon–sulphur lyases

4.5   Carbon–halide lyases

4.6   Phosphorus–oxygen lyases

4.99   Other lyases

5.   Isomerases

5.1   Racemases and epimerases

5.1.1   Acting on amino acids and derivatives

5.1.2   Acting on hydroxy acids and derivatives

5.1.3   Acting on carbohydrates and derivatives

5.1.99   Acting on other compounds

5.2   *Cis–trans* isomerases

5.3   Intramolecular oxidoreductases

5.3.1   Interconverting aldoses and ketoses

5.3.2   Interconverting keto- and enol- groups

5.3.3   Transposing C=C bonds

5.3.4   Transposing S−S bonds

5.3.99   Other intramolecular oxidoreductases

5.4   Intramolecular transferases

5.4.1   Transferring acyl groups

5.4.2   Transferring phosphoryl groups

Self Test Questions: Instructions & answers

Self Test Questions
Instructions

Each set of self test questions consists of four questions. The first two are multiple choice questions offering four alternative answers to the question and you should select the correct answers. Any number of them or none of them might be correct. The last two questions each consist of two statements joined by the word BECAUSE. Consider each statement separately and decide whether it is true or false. If both questions are true, decide whether the whole sentence is true, i.e. does the second statement provide a correct explanation for the first statement.

Self Test Questions
Answers

For each multiple choice question the correct answers are indicated.
For each relationship analysis question, the two statements are identified as either TRUE or FALSE. When both are true, the relationship between the two statements is identified as RELATED or NOT RELATED.

Answers to Self Test Questions.
For an explanation, see the instructions on page xiii.

Section	Q1	Q2	Q3		Related	Q4		Related
1.1	a	c	False	True		True	True	Yes
1.2	b	a	True	True	Yes	True	True	Yes
1.3	b c	c						
2.1	a	b	True	True	Yes	False	True	—
2.2	d	a	True	True	Yes	True	True	No
2.3	a b	c	True	True	No	False	True	—
2.4/5/6	c d	c	True	False	—	True	False	—
3.1	a b d	a b	True	False	—	True	True	Yes
3.2	c	a d	True	True	Yes	False	True	—
3.3	a c d	c d	True	False	—	False	True	—
3.4/5	b d	a c	False	True	—	True	True	No
4.1	c	a	True	False	—	True	True	Yes
4.2/3/4/5	d	b	False	True	—	True	True	Yes
5.1	c	b	True	True	Yes	True	False	—
5.2/3	b c	a	False	False	—	False	False	—
6.1	c	b	True	True	No	False	True	—
6.2	c d	c d	True	True	Yes	True	False	—
7.1/2	b d	b d	True	True	No	True	True	Yes
7.3	a b	c d	False	True	—	True	False	—
7.4	a b c d	a d	True	False	—	True	False	—
8.1	a c	b	True	True	Yes	True	False	—
8.2	c	a	True	True	No	False	True	—
8.3	a	a	True	True	Yes	True	True	No
8.4/5/6/7	a b d	c	False	True	—	False	False	—
9.1	DPDM	a c	True	False	—	True	False	—
9.2/3/4	b c	c	True	False	—	False	False	—
10.1	c	b-B c-A	True	True	Yes	True	True	Yes
10.2/3/4	c d	b c	False	True	—	False	False	—
10.5/6	b	a b d	True	True	Yes	True	False	—
11.1	b	a b	True	True	No	False	True	—
11.2/3	a c	b	False	True	—	False	True	—
12.1/2/3/4	b	c	True	True	No	True	False	—
12.5/6/7	a	b d	True	True	Yes	True	True	Yes
13.1/2	b d	a c	True	True	No	True	False	—
13.3	b c	a c	True	True	No	True	False	—
13.4/5	a b d	b d	False	True	—	False	False	—

Index